Complex Valued Nonlinear Adaptive Filters

Complex Valued Nonlinear Adaptive Filters

Noncircularity, Widely Linear and Neural Models

Danilo P. Mandic

Imperial College London, UK

Vanessa Su Lee Goh

Shell EP, Europe

A John Wiley and Sons, Ltd, Publication

This edition first published 2009

Registered office
John Wiley & Sons Ltd, The Atrium, Southern Gate, Chichester, West Sussex, PO19 8SQ, United Kingdom

For details of our global editorial offices, for customer services and for information about how to apply for permission to reuse the copyright material in this book please see our website at www.wiley.com.

Library of Congress Cataloging-in-Publication Data

Mandic, Danilo P.
Complex valued nonlinear adaptive filters : noncircularity, widely linear, and neural models / by Danilo P. Mandic, Vanessa Su Lee Goh, Shell EP, Europe.
p. cm.
Includes bibliographical references and index.
ISBN 978-0-470-06635-5 (cloth)
1. Functions of complex variables. 2. Adaptive filters–Mathematical models. 3. Filters (Mathematics) 4. Nonlinear theories. 5. Neural networks (Computer science) I. Goh, Vanessa Su Lee. II. Holland, Shell. III. Title.
TA347.C64.M36 2009
621.382'2–dc22 2009001965

A catalogue record for this book is available from the British Library.

ISBN: 978-0-470-06635-5 (H/B)

Typeset in 10/12 pt Times by Thomson Digital, Noida, India

Printed in Great Britain by CPI Antony Rowe, Chippenham, Wiltshire

The real voyage of discovery consists
not in seeking new landscapes
but in having new eyes
Marcel Proust

Contents

Preface

This book was written in response to the growing demand for a text that provides a unified treatment of complex valued adaptive filters, both linear and nonlinear, and methods for the processing of both complex circular and complex noncircular signals. We believe that this is the first attempt to bring together established adaptive filtering algorithms in $\mathbb{C}$ and the recent developments in the statistics of complex variable under the umbrella of powerful mathematical frameworks of $\mathbb{CR}$ (Wirtinger) calculus and augmented complex statistics. Combining the results from the authors' original research and current established methods, this books serves as a rigorous account of existing and novel complex signal processing methods, and provides next generation solutions for adaptive filtering of the generality of complex valued signals. The introductory chapters can be used as a text for a course on adaptive filtering. It is our hope that people as excited as we are by the possibilities opened by the more advanced work in this book will further develop these ideas into new and useful applications.

The title reflects our ambition to write a book which addresses several major problems in modern complex adaptive filtering. Real world data are non-Gaussian, nonstationary and generated by nonlinear systems with possibly long impulse responses. For the processing of such signals we therefore need *nonlinear* architectures to deal with nonlinearity and non-Gaussianity, *feedback* to deal with long responses, and *adaptive* mode of operation to deal with the nonstationary nature of the data. These have all been brought together in this book, hence the title "*Complex Valued Nonlinear Adaptive Filters*". The subtitle reflects some more intricate aspects of the processing of complex random variables, and that the class of nonlinear filters addressed in this work can be viewed as temporal neural networks. This material can also be used to supplement courses on neural networks, as the algorithms developed can be used to train neural networks for pattern processing and classification.

Complex valued signals play a pivotal role in communications, array signal processing, power, environmental, and biomedical signal processing and related fields. These signals are either complex by design, such as symbols used in data communications (e.g. quadrature phase shift keying), or are made complex by convenience of representation. The latter class includes analytic signals and signals coming from many important modern applications in magnetic source imaging, interferometric radar, direction of arrival estimation and smart antennas, mathematical biosciences, mobile communications, optics and seismics. Existing books do not take into account the effects on performance of a unique property of complex statistics – complex noncircularity, and employ several convenient mathematical shortcuts in the treatment of complex random variables.

Adaptive filters based on widely linear models introduced in this work are derived rigorously, and are suited for the processing of a much wider class of *complex noncircular signals* (directional processes, vector fields), and offer a number of theoretical performance gains.

Perhaps the first time we became involved in practical applications of complex adaptive filtering was when trying to perform short term wind forecasting by treating wind speed and direction, which are routinely processed separately, as a unique complex valued quantity. Our results outperformed the standard approaches. This opened a can of worms, as it became apparent that the standard techniques were not adequate, and that mathematical foundations and practical tools for the applications of complex valued adaptive filters to the generality of complex signals are scattered throughout the literature. For instance, an often confusing aspect of complex adaptive filtering is that the cost (objective) function to be minimised is a real function (measure of error power) of complex variables, and is not analytic. Thus, standard complex differentiability (Cauchy-Riemann conditions) does not apply, and we need to resort to pseudoderivatives. We identified the need for a rigorous, concise, and unified treatment of the statistics of complex variables, methods for dealing with nonlinearity and noncircularity, and enhanced solutions for adaptive signal processing in $\mathbb{C}$, and were encouraged by our series editor Simon Haykin and the staff from Wiley Chichester to produce this text.

The first two chapters give the introduction to the field and illustrate the benefits of the processing in the complex domain. Chapter 1 provides a personal view of the history of complex numbers. They are truly fascinating and, unlike other number systems which were introduced as solutions to practical problems, they arose as a product of intellectual exercise. Complex numbers were formalised in the mid-19th century by Gauss and Euler in order to provide solutions for the fundamental theorem of algebra; within 50 years (and without the Internet) they became a linchpin of electromagnetic field and relativity theory. Chapter 2 offers theoretical and practical justification for converting many apparently real valued signal processing problems into the complex domain, where they can benefit from the convenience of representation and the power and beauty of complex calculus. It illustrates the duality between the processing in $\mathbb{R}^2$ and $\mathbb{C}$, and the benefits of complex valued processing – unlike $\mathbb{R}^2$ the field of complex numbers forms a division algebra and provides a rigorous mathematics framework for the treatment of phase, nonlinearity and coupling between signal components.

The foundations of standard complex adaptive filtering are established in Chapters 3–7. Chapter 3 provides an overview of adaptive filtering architectures, and introduces the background for their state space representations and links with polynomial filters and neural networks. Chapter 4 deals with the choice of complex nonlinear activation function and addresses the trade off between their boundedness and analyticity. The only continuously differentiable function in $\mathbb{C}$ that satisfies the Cauchy-Riemann conditions is a constant; to preserve boundedness some ad hoc approaches (also called split-complex) employ real valued nonlinearities on the real and imaginary parts. Our main interest is in complex functions of complex variables (also called fully complex) which are not bounded on the whole complex plane, but are complex differentiable and provide solutions which are generic extensions of the corresponding solutions in $\mathbb{R}$. Chapter 5 addresses the duality between gradient calculation in $\mathbb{R}^2$ and $\mathbb{C}$ and introduces the so called $\mathbb{CR}$ calculus which is suitable for general functions of complex variables, both holomorphic and non-holomorphic. This provides a unified framework for computing the Jacobians, Hessians, and gradients of cost functions, and serves as a basis for the derivation of learning algorithms throughout this book. Chapters 6 and 7 introduce standard complex valued adaptive filters, both linear and nonlinear; they are supported by rigorous proofs of convergence, and can be used to teach a course on adaptive filtering. The complex least mean square (CLMS) in Chapter 6 is derived step by step, whereas the learning algorithms for feedback structures in Chapter 7 are derived in a compact way, based on $\mathbb{CR}$

calculus. Furthermore, learning algorithms for both linear and nonlinear feedback architectures are introduced, starting from linear IIR filters to temporal recurrent neural networks.

Chapters 8–11 address several practical aspects of adaptive filtering, such as adaptive step-sizes, dynamical range extension, and a posteriori mode of operation. Chapter 8 provides a thorough review of adaptive step size algorithms and introduces the general normalised gradient descent (GNGD) algorithm for enhanced stability. Chapter 9 gives solutions for dynamical range extension of nonlinear neural adaptive filters, whereas Chapter 10 explains a posteriori algorithms and analyses them in the framework of fixed point theory. Chapter 11 rounds up the first part of the book and introduces fractional delay filters together with links between complex nonlinear functions and number theory.

Chapters 12–15 introduce linear and nonlinear adaptive filters based on widely linear models, which are suited to deal with complex noncircularity, thus providing theoretical and practical adaptive filtering solutions for the generality of complex signals. Chapter 12 provides a comprehensive overview of the latest results (2008) in the statistics of complex random signals, with a particular emphasis on complex noncircularity. It is shown that the standard complex Gaussian model is inadequate and the concepts of noise, stationarity, multicorrelation, and multispectra are re-introduced based on the augmented statistics. This has served as a basis for the development of the class of 'augmented' adaptive filtering algorithms, starting from the complex least square (ACLMS) algorithm through to augmented learning algorithms for IIR filters, recurrent neural networks, and augmented Kalman filters. Chapter 13 introduces the augmented least mean square algorithm, a quantum step in the adaptive signal processing of complex noncircular signals. It is shown that this approach is as good as standard approaches for circular data, whereas it outperforms standard filters for noncircular data. Chapter 14 provides an insight into the duality between complex valued linear adaptive filters and dual channel real adaptive filters. A correspondence is established between the ACLMS and the dual channel real LMS algorithms. Chapter 15 extends widely linear modelling in $\mathbb{C}$ to feedback and nonlinear architectures. The derivations are based on $\mathbb{CR}$ calculus and are provided for both the gradient descent and state space (Kalman filtering) models.

Chapter 16 addresses collaborative adaptive filtering in $\mathbb{C}$. It is shown that by employing collaborative filtering architectures we can gain insight into the nature of a signal in hand, and a simple test for complex noncircularity is proposed. Chapter 17 introduces complex empirical mode decomposition (EMD), a data driven time-frequency technique. This technique, when used for preprocessing complex valued data, provides a framework for "data fusion via fission", with a number of applications, especially in biomedical engineering and neuroscience. Chapter 18 provides a rigorous statistical testing framework for the validity of complex representation.

The material is supported by a number of Appendices (some of them based on [190]), ranging from the theory of complex variable through to fixed point theory. We believe this makes the book self-sufficient for a reader who has basic knowledge of adaptive signal processing. Simulations were performed for both circular and noncircular data sources, from benchmark linear and nonlinear models to real world wind and radar signals. The applications are set in a prediction setting, as prediction is at the core of adaptive filtering. The complex valued wind signal is our most frequently used test signal, due to its intermittent, non-Gaussian and noncircular nature. Gill Instruments provided ultrasonic anemometers used for our wind recordings.

Acknowledgements

Vanessa and I would like to thank our series editor Simon Haykin for encouraging us to write a text on modern complex valued adaptive signal processing. In addition, my own work in this area was inspired by the success of my earlier monograph "*Recurrent Neural Networks for Prediction*", Wiley 2001, co-authored with Jonathon Chambers, where some earlier results were outlined. Over the last seven years these ideas have matured greatly, through working with my co-author Vanessa Su Lee Goh and a number of graduate students, to a point where it was possible to write this book. I have had great pleasure to work with Temujin Gautama, Maciej Pedzisz, Mo Chen, David Looney, Phebe Vayanos, Beth Jelfs, Clive Cheong Took, Yili Xia, Andrew Hanna, Christos Boukis, George Souretis, Naveed Ur Rehman, Tomasz Rutkowski, Toshihisa Tanaka, and Soroush Javidi (who has also designed the book cover), who have all been involved in the research that led to this book. Their dedication and excitement have helped to make this journey through the largely unchartered territory of modern complex valued signal processing so much more rewarding.

Peter Schreier has provided deep and insightful feedback on several chapters, especially when it comes to dealing with complex noncircularity. We have enjoyed the interaction with Tülay Adalı, who also proofread several key chapters. Ideas on the duality between real and complex filters matured through discussions with Susanna Still and Jacob Benesty. The collaboration with Scott Douglas influenced convergence proofs in Chapter 6. The results in Chapter 18 arose from collaboration with Marc Van Hulle and his team. Tony Constantinides, Igor Aizenberg, Aurelio Uncini, Tony Kuh, Preben Kidmose, Maria Petrou, Isao Yamada, and Olga Boric Lubecke provided valuable comments.

Additionally, I would like to thank Andrzej Cichocki for invigorating discussions and the timely reminder that the quantum developments of science are in the hands of young researchers. Consequently, we decided to hurry up with this book while I can still (just) qualify. The collaboration with Kazuyuki Aihara and Yoshito Hirata helped us to hone our ideas related to complex valued wind forecasting.

It is not possible to mention all the colleagues and friends who have helped towards this book. Members of the IEEE Signal Processing Society Technical Committee on Machine Learning for Signal Processing have provided support and stimulating discussions, in particular, David Miller, Dragan Obradovic, Jose Principe, and Jan Larsen. We wish to express our appreciation to the signal processing tradition and vibrant research atmosphere at Imperial College London, which have made delving into this area so rewarding.

We are deeply indebted to Henry Goldstein, who tamed our immense enthusiasm for the subject and focused it to the needs of our readers.

Finally, our love and gratitude goes to our families and friends for supporting us since the summer of 2006, when this work began.

Danilo P. Mandic
Vanessa Su Lee Goh

1

The Magic of Complex Numbers

The notion of complex number is intimately related to the *Fundamental Theorem of Algebra* and is therefore at the very foundation of mathematical analysis. The development of complex algebra, however, has been far from straightforward.[1]

The human idea of 'number' has evolved together with human society. The *natural* numbers ($1, 2, \ldots \in \mathbb{N}$) are straightforward to accept, and they have been used for *counting* in many cultures, irrespective of the actual base of the number system used. At a later stage, for *sharing*, people introduced fractions in order to answer a simple problem such as 'if we catch $\mathcal{U}$ fish, I will have two parts $\frac{2}{5}\mathcal{U}$ and you will have three parts $\frac{3}{5}\mathcal{U}$ of the whole catch'. The acceptance of negative numbers and zero has been motivated by the emergence of economy, for dealing with profit and loss. It is rather impressive that ancient civilisations were aware of the need for irrational numbers such as $\sqrt{2}$ in the case of the Babylonians [77] and π in the case of the ancient Greeks.[2]

The concept of a new 'number' often came from the need to solve a specific practical problem. For instance, in the above example of sharing $\mathcal{U}$ number of fish caught, we need to solve for $2\mathcal{U} = 5$ and hence to introduce fractions, whereas to solve $x^2 = 2$ (diagonal of a square) irrational numbers needed to be introduced. Complex numbers came from the necessity to solve equations such as $x^2 = -1$.

[1] A classic reference which provides a comprehensive account of the development of numbers is *Number: The Language of Science* by Tobias Dantzig [57].

[2] The Babylonians have actually left us the basics of Fixed Point Theory (see Appendix P), which in terms of modern mathematics was introduced by Stefan Banach in 1922. On a clay tablet (YBC 7289) from the Yale Babylonian Collection, the Mesopotamian scribes explain how to calculate the diagonal of a square with base 30. This was achieved using a fixed point iteration around the initial guess. The ancient Greeks used π in geometry, although the irrationality of π was only proved in the 1700s. More information on the history of mathematics can be found in [34] whereas P. Nahin's book is dedicated to the history of complex numbers [215].

Complex Valued Nonlinear Adaptive Filters: Noncircularity, Widely Linear and Neural Models
Danilo P. Mandic and Vanessa Su Lee Goh

1.1 History of Complex Numbers

Perhaps the earliest reference to square roots of negative numbers occurred in the work of Heron of Alexandria[3], around 60 AD, who encountered them while calculating volumes of geometric bodies. Some 200 years later, Diophantus (about 275 AD) posed a simple problem in geometry,

Find the sides of a right–angled triangle of perimeter 12 units and area 7 squared units.

which is illustrated in Figure 1.1. To solve this, let the side $|AB| = x$, and the height $|BC| = h$, to yield

$$area = \frac{1}{2} x \; h$$

$$perimeter = x + h + \sqrt{x^2 + h^2}$$

In order to solve for x we need to find the roots of

$$6x^2 - 43x + 84 = 0$$

however this equation does not have real roots.

A similar problem was posed by Cardan[4] in 1545. He attempted to find two numbers a and b such that

$$a + b = 10$$

$$a\,b = 40$$

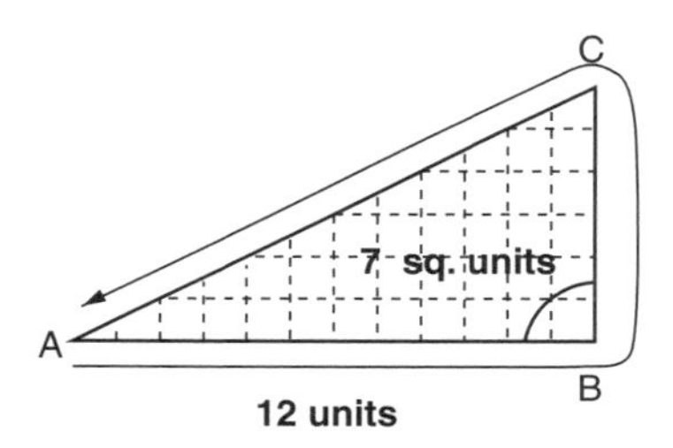

Figure 1.1 Problem posed by Diophantus (third century AD)

[3]Heron (or Hero) of Alexandria was a Greek mathematician and inventor. He is credited with finding a formula for the area of a triangle (as a function of the perimeter). He invented many gadgets operated by fluids; these include a fountain, fire engine and siphons. The aeolipile, his engine in which the recoil of steam revolves a ball or a wheel, is the forerunner of the steam engine (and the jet engine). In his method for approximating the square root of a number he effectively found a way round the complex number. It is fascinating to realise that complex numbers have been used, implicitly, long before their introduction in the 16th century.

[4]Girolamo or Hieronimo Cardano (1501–1576). His name in Latin was Hieronymus Cardanus and he is also known by the English version of his name Jerome Cardan. For more detail on Cardano's life, see [1].

These equations are satisfied for

$$a = 5 + \sqrt{-15} \quad \text{and} \quad b = 5 - \sqrt{-15} \tag{1.1}$$

which are clearly not real.

The need to introduce the complex number became rather urgent in the 16th century. Several mathematicians were working on what is today known as the *Fundamental Theorem of Algebra* (FTA) which states that

Every nth order polynomial with real[5] *coefficients has exactly n roots in* $\mathbb{C}$.

Earlier attempts to find the roots of an arbitrary polynomial include the work by Al-Khwarizmi (ca 800 AD), which only allowed for positive roots, hence being only a special case of FTA. In the 16th century Niccolo Tartaglia[6] and Girolamo Cardano (see Equation 1.1) considered closed formulas for the roots of third- and fourth-order polynomials. Girolamo Cardano first introduced complex numbers in his *Ars Magna* in 1545 as a tool for finding *real* roots of the 'depressed' cubic equation $x^3 + ax + b = 0$. He needed this result to provide algebraic solutions to the general cubic equation

$$ay^3 + by^2 + cy + d = 0$$

By substituting $y = x - \frac{1}{3}b$, the cubic equation is transformed into a depressed cubic (without the square term), given by

$$x^3 + \beta x + \gamma = 0$$

Scipione del Ferro of Bologna and Tartaglia showed that the depressed cubic can be solved as[7]

$$x = \sqrt[3]{-\frac{\gamma}{2} + \sqrt{\frac{\gamma^2}{4} + \frac{\beta^3}{27}}} + \sqrt[3]{-\frac{\gamma}{2} - \sqrt{\frac{\gamma^2}{4} + \frac{\beta^3}{27}}} \tag{1.2}$$

For certain problem settings (for instance $a = 1, b = 9, c = 24, d = 20$), and using the substitution $y = x - 3$, Tartaglia could show that, by symmetry, there exists $\sqrt{-1}$ which has mathematical meaning. For example, Tartaglia's formula for the roots of $x^3 - x = 0$ is given by

$$\frac{1}{\sqrt{3}}\left((\sqrt{-1})^{\frac{1}{3}} + \frac{1}{(\sqrt{-1})^{\frac{1}{3}}}\right)$$

[5]In fact, it states that every nth order polynomial with complex coefficients has n roots in $\mathbb{C}$, but for historical reasons we adopt the above variant.

[6]Real name Niccolo Fontana, who is known as Tartaglia (the stammerer) due to a speaking disorder.

[7]In modern notation this can be written as $x = (q + w)^{\frac{1}{3}} + (q - w)^{\frac{1}{3}}$.

Rafael Bombelli also analysed the roots of cubic polynomials by the 'depressed cubic' transformations and by applying the Ferro–Tartaglia formula (1.2). While solving for the roots of

$$x^3 - 15x - 4 = 0$$

he was able to show that

$$\left(2 + \sqrt{-1}\right) + \left(2 - \sqrt{-1}\right) = 4$$

Indeed $x = 4$ is a correct solution, however, in order to solve for the real roots, it was necessary to perform calculations in $\mathbb{C}$. In 1572, in his *Algebra*, Bombelli introduced the symbol $\sqrt{-1}$ and established rules for manipulating 'complex numbers'.

The term 'imaginary' number was coined by Descartes in the 1630s to reflect his observation that '*For every equation of degree n, we can imagine n roots which do not correspond to any real quantity*'. In 1629, Flemish mathematician[8] Albert Girard in his *L'Invention Nouvelle en l'Algèbre* asserts that there are n roots to an nth order polynomial, however this was accepted as self-evident, but with no guarantee that the actual solution has the form $a + \jmath b$, $a, b \in \mathbb{R}$.

It was only after their geometric representation (John Wallis[9] in 1685 in *De Algebra Tractatus* and Caspar Wessel[10] in 1797 in the *Proceedings of the Copenhagen Academy*) that the complex numbers were finally accepted. In 1673, while investigating geometric representations of the roots of polynomials, John Wallis realised that for a general quadratic polynomial of the form

$$x^2 + 2bx + c^2 = 0$$

for which the solution is

$$x = -b \pm \sqrt{b^2 - c^2} \tag{1.3}$$

a geometric interpretation was only possible for $b^2 - c^2 \geq 0$. Wallis visualised this solution as displacements from the point $-b$, as shown in Figure 1.2(a) [206]. He interpreted each solution as a vertex (A and B in Figure 1.2) of a right triangle with height c and side $\sqrt{b^2 - c^2}$. Whereas this geometric interpretation is clearly correct for $b^2 - c^2 \geq 0$, Wallis argued that for $b^2 - c^2 < 0$, since b is shorter than c, we will have the situation shown in Figure 1.2(b); this

[8]Albert Girard was born in France in 1595, but his family later moved to the Netherlands as religious refugees. He attended the University of Leiden where he studied music. Girard was the first to propose the fundamental theorem of algebra, and in 1626, in his first book on trigonometry, he introduced the abbreviations sin, cos, and tan. This book also contains the formula for the area of a spherical triangle.

[9]In his *Treatise on Algebra* Wallis accepts negative and complex roots. He also shows that equation $x^3 - 7x = 6$ has exactly three roots in $\mathbb{R}$.

[10]Within his work on geodesy Caspar Wessel (1745–1818) used complex numbers to represent directions in a plane as early as in 1787. His article from 1797 entitled 'On the Analytical Representation of Direction: An Attempt Applied Chiefly to Solving Plane and Spherical Polygons' (in Danish) is perhaps the first to contain a well-thought-out geometrical interpretation of complex numbers.

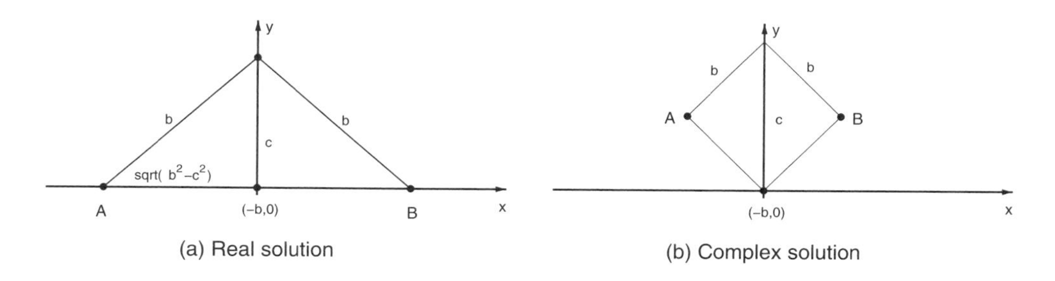

Figure 1.2 Geometric representation of the roots of a quadratic equation

way we can think of a complex number as a *point on the plane*.[11] In 1732 Leonhard Euler calculated the solutions to the equation

$$x^n - 1 = 0$$

in the form of

$$\cos\theta + \sqrt{-1}\sin\theta$$

and tried to visualise them as the vertices of a planar polygon. Further breakthroughs came with the work of Abraham de Moivre (1730) and again Euler (1748), who introduced the famous formulas

$$\begin{aligned} (\cos\theta + \jmath\sin\theta)^n &= \cos n\theta + \jmath\sin n\theta \\ \cos\theta + \jmath\sin\theta &= e^{\jmath\theta} \end{aligned}$$

Based on these results, in 1749 Euler attempted to prove FTA for real polynomials in *Recherches Sur Les Racines Imaginaires des Équations*. This was achieved based on a decomposition a monic polynomials and by using Cardano's technique from *Ars Magna* to remove the second largest degree term of a polynomial.

In 1806 the Swiss accountant and amateur mathematician Jean Robert Argand published a proof of the FTA which was based on an idea by d'Alembert from 1746. Argand's initial idea was published as *Essai Sur Une Manière de Représenter les Quantités Imaginaires Dans les Constructions Géométriques* [60, 305]. He simply interpreted $\jmath$ as a rotation by 90° and introduced the Argand plane (or Argand diagram) as a geometric representation of complex numbers. In Argand's diagram, $\pm\sqrt{-1}$ represents a unit line, perpendicular to the real axis. The notation and terminology we use today is pretty much the same. A complex number

$$z = x + \jmath y$$

[11]In his interpretation $-\sqrt{-1}$ is the same point as $\sqrt{-1}$, but nevertheless this was an important step towards the geometric representation of complex numbers.

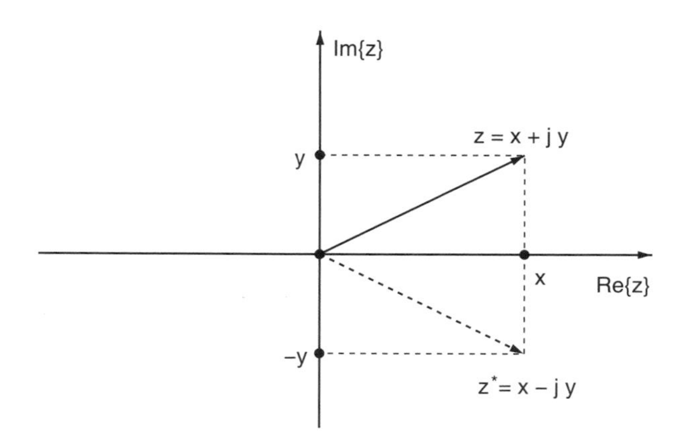

Figure 1.3 Argand's diagram for a complex number z and its conjugate z^*

is simply represented as a vector in the complex plane, as shown in Figure 1.3. Argand called $\sqrt{x^2 + y^2}$ the *modulus*, and Gauss introduced the term *complex number* and notation[12] $\imath = \sqrt{-1}$ (in signal processing we use $\jmath = \imath = \sqrt{-1}$). Karl Friedrich Gauss used complex numbers in his several proofs of the fundamental theorem of algebra, and in 1831 he not only associated the complex number $z = x + \jmath y$ with a point (x, y) on a plane, but also introduced the rules for the addition[13] and multiplication of such numbers. Much of the terminology used today comes from Gauss, Cauchy[14] who introduced the term 'conjugate', and Hankel who in 1867 introduced the term *direction coefficient* for $\cos\theta + \jmath \sin\theta$, whereas Weierstrass (1815–1897) introduced the term *absolute value* for the modulus.

Some analytical aspects of complex numbers were also developed by Georg Friedrich Bernhard Riemann (1826–1866), and those principles are nowadays the basics behind what is known as manifold signal processing.[15] To illustrate the potential of complex numbers in this context, consider the stereographic[16] projection [242] of the Riemann sphere, shown in Figure 1.4(a). In a way analogous to Cardano's 'depressed cubic', we can perform dimensionality reduction by embedding $\mathbb{C}$ in $\mathbb{R}^3$, and rewriting

$$Z = a + \jmath b, \quad (a, b, 0) \in \mathbb{R}^3$$

[12] There is a simple trap, that is, we cannot apply the identity of the type $\sqrt{ab} = \sqrt{a}\sqrt{b}$ to the 'imaginary' numbers, this would lead to the wrong conclusion $1 = \sqrt{(-1)(-1)} = \sqrt{-1}\sqrt{-1}$, however $\sqrt{-1}^2 = \sqrt{-1}\sqrt{-1} = -1$.

[13] So much so that, for instance, 3 remains a prime number whereas 5 does not, since it can be written as $(1 - 2\jmath)(1 + 2\jmath)$.

[14] Augustin Louis Cauchy (1789–1867) formulated many of the classic theorems in complex analysis.

[15] Examples include the Natural Gradient algorithm used in blind source separation [10, 49].

[16] The stereographic projection is a mapping that projects a sphere onto a plane. The mapping is smooth, bijective and conformal (preserves relationships between angles).

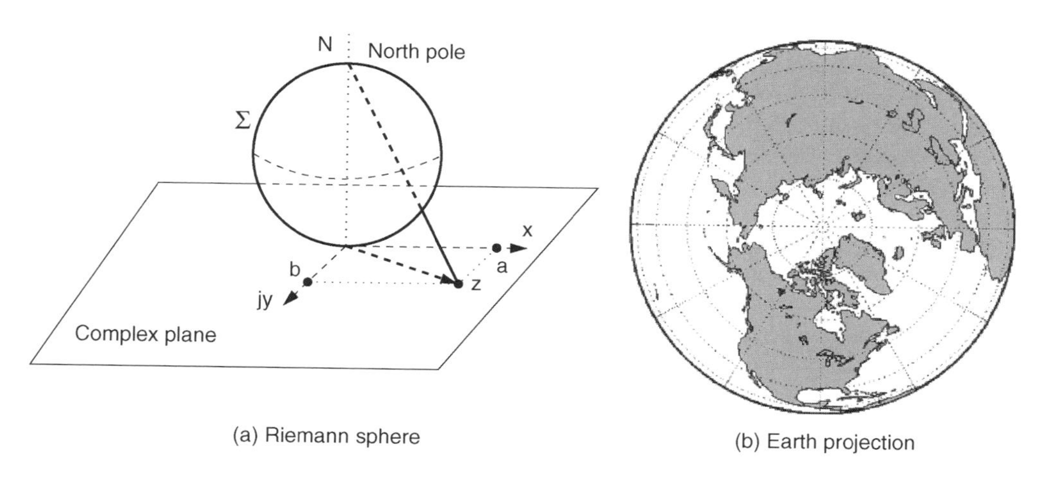

Figure 1.4 Stereographic projection and Riemann sphere: (a) the principle of the stereographic projection; (b) stereographic projection of the Earth (seen from the south pole S)

Consider a sphere Σ defined by

$$\Sigma = \left\{(x, y, u) \in \mathbb{R}^3 \ : \ x^2 + y^2 + (u - d)^2 = r^2\right\}, \quad d, r \in \mathbb{R}$$

There is a one-to-one correspondence between the points of $\mathbb{C}$ and the points of Σ, excluding N (the north pole of Σ), since the line from any point $z \in \mathbb{C}$ cuts $\Sigma \setminus \{N\}$ in precisely one point. If we include the point ∞, so as to have the *extended complex plane* $\mathbb{C} \cup \{\infty\}$, then the north pole N from sphere Σ is also included and we have a mapping of the Riemann sphere onto the extended complex plane. A stereographic projection of the Earth onto a plane tangential to the north pole N is shown in Figure 1.4(b).

1.1.1 Hypercomplex Numbers

Generalisations of complex numbers (generally termed 'hypercomplex numbers') include the work of Sir William Rowan Hamilton (1805–1865), who introduced the quaternions in 1843. A quaternion $\vec{q}$ is defined as [103]

$$\vec{q} = q_0 + q_1 \imath + q_2 \jmath + q_3 k \tag{1.4}$$

where the variables $\imath$, $\jmath$, k are all defined as $\sqrt{-1}$, but their multiplication is not commutative.[17]

Pivotal figures in the development of the theory of complex numbers are Hermann Günther Grassmann (1809–1877), who introduced multidimensional vector calculus, and James Cockle,

[17]That is: $\imath\jmath = -\jmath\imath = k$, $\jmath k = -k\jmath = \imath$, and $k\imath = -\imath k = \jmath$.

who in 1848 introduced split-complex numbers.[18] A split-complex number (also known as motors, dual numbers, hyperbolic numbers, tessarines, and Lorenz numbers) is defined as [51]

$$z = x + \jmath y, \qquad \jmath^2 = 1$$

In 1876, in order to model spins, William Kingdon Clifford introduced a system of hypercomplex numbers (Clifford algebra). This was achieved by conveniently combining the quaternion algebra and split-complex numbers. Both Hamilton and Clifford are credited with the introduction of *biquaternions*, that is, quaternions for which the coefficients are complex numbers. A comprehensive account of *hypercomplex* numbers can be found in [143]; in general a hypercomplex number system has at least one non-real axis and is closed under addition and multiplication. Other members of the family of hypercomplex numbers include McFarlane's hyperbolic quaternion, hyper-numbers, multicomplex numbers, and twistors (developed by Roger Penrose in 1967 [233]).

1.2 History of Mathematical Notation

It is also interesting to look at the development of 'symbols' and abbreviations in mathematics. For books copied by hand the choice of mathematical symbols was not an issue, whereas for printed books this choice was largely determined by the availability of fonts of the early printers. Thus, for instance, in the 9th century in Al-Khwarizmi's *Algebra* solutions were descriptive rather than in the form of equations, while in Cardano's *Ars Magna* in the 16th century the unknowns were denoted by single roman letters to facilitate the printing process.

It was arguably Descartes who first established some general rules for the use of mathematical symbols. He used lowercase italic letters at the beginning of the alphabet to denote unknown constants (a, b, c, d), whereas letters at the end of the alphabet were used for unknown variables (x, y, z, w). Using Descartes' recommendations, the expression for a quadratic equation becomes

$$ax^2 + bx + c = 0$$

which is exactly the way we use it in modern mathematics.

As already mentioned, the symbol for imaginary unit $\imath = \sqrt{-1}$ was introduced by Gauss, whereas boldface letters for vectors were first introduced by Oliver Heaviside [115]. More details on the history of mathematical notation can be found in the two–volume book *A History of Mathematical Notations* [39], written by Florian Cajori in 1929.

In the modern era, the introduction of mathematical symbols has been closely related with the developments in computing and programming languages.[19] The relationship between computers and typography is explored in *Digital Typography* by Donald E. Knuth [153], who also developed the TeX typesetting language.

---

[18] Notice the difference between the split-complex *numbers* and split-complex *activation functions* of neurons [152, 190]. The term split-complex number relates to an alternative hypercomplex *number* defined by $x + \jmath y$ where $\jmath^2 = 1$, whereas the term split-complex function refers to *functions* $g : \mathbb{C} \rightarrow \mathbb{C}$ for which the real and imaginary part of the 'net' function are processed separately by a real function of real argument f, to give $g(net) = f(\Re(net)) + \jmath f(\Im(net))$.

[19] Apart from the various new symbols used, e.g. in computing, one such symbol is © for 'copyright'.

1.3 Development of Complex Valued Adaptive Signal Processing

The distinguishing characteristics of complex valued nonlinear adaptive filtering are related to the character of complex nonlinearity, the associated learning algorithms, and some recent developments in complex statistics. It is also important to notice that the universal function approximation property of some complex nonlinearities does not guarantee fast and efficient learning.

Complex nonlinearities. In 1992, Georgiou and Koutsougeras [88] proposed a list of requirements that a complex valued activation function should satisfy in order to qualify for the nonlinearity at the neuron. The calculation of complex gradients and Hessians has been detailed in work by Van Den Bos [30]. In 1995 Arena *et al.* [18] proved the *universal approximation property*[20] of a Complex Multilayer Perceptron (CMLP), based on the split-complex approach. This also gave theoretical justification for the use of complex neural networks (NNs) in time series modelling tasks, and thus gave rise to temporal neural networks. The split-complex approach has been shown to yield reasonable performance in channel equalisation applications [27, 147, 166], and in applications where there is no strong coupling between the real and imaginary part within the complex signal. However, for the common case where the inphase (I) and quadrature (Q) components have the same variance and are uncorrelated, algorithms employing split-complex activation functions tend to yield poor performance.[21] In addition, split-complex based algorithms do not have a generic form of their real-valued counterparts, and hence their signal flow-graphs are fundamentally different [220]. In the classification context, early results on Boolean threshold functions and the notion of multiple-valued threshold function can be found in [7, 8].

The problems associated with the choice of complex nonlinearities suitable for nonlinear adaptive filtering in $\mathbb{C}$ have been addressed by Kim and Adali in 2003 [152]. They have identified a class of 'fully complex' activation functions (differentiable and bounded almost everywhere in $\mathbb{C}$ such as tanh), as a suitable choice, and have derived the fully complex back-propagation algorithm [150, 151], which is a generic extension of its real-valued counterpart. They also provide an insight into the character of singularities of fully complex nonlinearities, together with their universal function approximation properties. Uncini *et al.* have introduced a 2D splitting complex activation function [298], and have also applied complex neural networks in the context of blind equalisation [278] and complex blind source separation [259].

Learning algorithms. The first adaptive signal processing algorithm operating completely in $\mathbb{C}$ was the complex least mean square (CLMS), introduced in 1975 by Widrow, Mc Cool and Ball [307] as a natural extension of the real LMS. Work on complex nonlinear architectures, such as complex neural networks (NNs) started much later. Whereas the extension from real LMS to CLMS was fairly straightforward, the extensions of algorithms for nonlinear adaptive filtering from $\mathbb{R}$ into $\mathbb{C}$ have not been trivial. This is largely due to problems associated with the

[20]This is the famous 13th problem of Hilbert, which has been the basis for the development of adaptive models for universal function approximation [56, 125, 126, 155].

[21]Split-complex algorithms cannot calculate the true gradient unless the real and imaginary weight updates are mutually independent. This proves useful, e.g. in communications applications where the data symbols are made orthogonal by design.

choice of complex nonlinear activation function.[22] One of the first results on complex valued NNs is the 1990 paper by Clarke [50]. Soon afterwards, the complex backpropagation (CBP) algorithm was introduced [25, 166]. This was achieved based on the so called split-complex[23] nonlinear activation function of a neuron [26], where the real and imaginary parts of the *net* input are processed separately by two real-valued nonlinear functions, and then combined together into a complex quantity. This approach produced bounded outputs at the expense of closed and generic formulas for complex gradients. Fully complex algorithms for nonlinear adaptive filters and recurrent neural networks (RNNs) were subsequently introduced by Goh and Mandic in 2004 [93, 98]. As for nonlinear sequential state estimation, an extended Kalman filter (EKF) algorithm for the training of complex valued neural networks was proposed in [129].

Augmented complex statistics. In the early 1990s, with the emergence of new applications in communications and elsewhere, the lack of general theory for complex-valued statistical signal processing was brought to light by several authors. It was also realised that the statistics in $\mathbb{C}$ are not an analytical continuation of the corresponding statistics in $\mathbb{R}$. Thus for instance, so called 'conjugate linear' (also known as widely linear [240]) filtering was introduced by Brown and Crane in 1969 [38], generalised complex Gaussian models were introduced by Van Den Bos in 1995 [31], whereas the notions of 'proper complex random process' (closely related[24] to the notion of 'circularity') and 'improper complex random process' were introduced by Neeser and Massey in 1993 [219]. Other important results on 'augmented complex statistics' include work by Schreier and Scharf [266, 268, 271], and Picinbono, Chevalier and Bondon [237–240]. This work has given rise to the application of augmented statistics in adaptive filtering, both supervised and blind. For supervised learning, EKF based training in the framework of complex-valued recurrent neural networks was introduced by Goh and Mandic in 2007 [95], whereas augmented learning algorithms in the stochastic gradient setting were proposed by the same authors in [96]. Algorithms for complex-valued blind separation problems in biomedicine were introduced by Calhoun and Adali [40–42], whereas Eriksson and Koivunen focused on communications applications [67, 252]. Notice that properties of complex signals are not only varying in terms of their statistical nature, but also in terms of their 'dual univariate', 'bivariate', or 'complex' nature. A statistical test for this purpose based on hypothesis testing was developed by Gautama, Mandic and Van Hulle [85], whereas a test for complex circularity was developed by Schreier, Scharf and Hanssen [270]. The recent book by Schreier and Scharf gives an overview of complex statistics [269].

Hypercomplex nonlinear adaptive filters. A comprehensive introduction to hypercomplex neural networks was provided by Arena, Fortuna, Muscato and Xibilia in 1998 [17], where special attention was given to quaternion MLPs. Extensions of complex neural networks include

[22] We need to make a choice between boundedness for differentiability, since by Liouville's theorem the only continuously differentiable function on $\mathbb{C}$ is a constant.

[23] The reader should not mistake split-complex numbers for split-complex nonlinearities.

[24] Terms *proper random process* and *circular random process* are often used interchangeably, although strictly speaking, 'properness' is a second-order concept, whereas 'circularity' is a property of the probability density function, and the two terms are not completely equivalent. For more detail see Chapter 12.

neural networks whose operation is based on the geometric (Clifford) algebra, proposed by Pearson [230]. The Clifford MLPs are based on a variant of the complex activation function from [88], where the standard product of two scalar variables is replaced by a special product of two multidimensional quantities [17, 18].

A comprehensive account of standard linear and nonlinear adaptive filtering algorithms in $\mathbb{C}$, which are based on the assumption of second-order circularity of complex processes, can be found in *Adaptive Filter Theory* by Simon Haykin [113]. Complex-valued NNs in the context of classification and pattern recognition have been addressed in an edited book and a monograph by Akira Hirose [119, 120], and in work by Naum Aizenberg [6, 7], Igor Aizenberg [4, 5] and Tohru Nitta [221].

The existing statistical signal processing algorithms are based on standard complex statistics, which is a direct extension of real statistics, and boils down to exactly the same expressions as those in $\mathbb{R}$, if we

- Remove complex conjugation whenever it occurs in the algorithm;
- Replace the Hermitian transpose operator with the ordinary transpose operator.

This, however, applies only to the rather limited class of circular complex signals, and such solutions when applied to general complex data are suboptimal.

This book provides a comprehensive account of so-called augmented complex statistics, and offers solutions to a general adaptive filtering problem in $\mathbb{C}$.

2

Why Signal Processing in the Complex Domain?

Applications of adaptive systems normally use the signal magnitude as the main source of information [227]. Real world processes with the 'intensity' and 'direction' components (radar, sonar, vector fields), however, require also the phase information to be considered. In the complex domain $\mathbb{C}$, this phase information is accounted for naturally, and this chapter illustrates the duality between the processing in $\mathbb{R}$ and $\mathbb{C}$ for several classes of real world processes. It is shown that the advantages of using complex valued solutions for real valued problems arise not only from the full utilisation of the phase information (e.g. time-delay converted into a phase shift), but also from the use of different algebra and statistics.

2.1 Some Examples of Complex Valued Signal Processing

Fourier analysis. Perhaps the most frequently used form of complex valued modelling of real valued data is the Fourier series, introduced in 1807 by Joseph Fourier, whereby a real function $f(t)$ is represented as[1]

$$f(t) = \sum_{n=-\infty}^{\infty} c_n e^{\jmath\omega_n t} \tag{2.1}$$

where coefficients $\{c_n\}$ are calculated as

$$c_n = \frac{1}{T}\int_{t_1}^{t_2} f(t)e^{-\jmath\omega_n t}dt$$

and $T = t_2 - t_1$ is the period of function $f(t)$.

[1]This is the more compact complex form of the original expressions. In his *Théorie Analytique de la Chaleur* (1822), Fourier showed how the conduction of heat in solid bodies may be analysed in terms of infinite mathematical series.

Complex Valued Nonlinear Adaptive Filters: Noncircularity, Widely Linear and Neural Models
Danilo P. Mandic and Vanessa Su Lee Goh

Phasor representation of harmonic signals. Another classical example comes from electronics, where complex numbers are used in signal analysis as a convenient description for periodically varying signals. For a sinusoidal signal $x(t) = |x|\cos(\omega t + \Phi)$, the *phasor*, or complex amplitude, is defined as

$$\mathbf{X} = |x|\, e^{\jmath\Phi} \qquad \leftrightarrow \qquad x(t) = \Re\{\mathbf{X}e^{\jmath\omega t}\} = |x|\,\cos(\omega t + \Phi) \tag{2.2}$$

The absolute value $|x|$ is interpreted as the amplitude and the argument $\Phi = \arg(x)$ as the phase of a sine wave of given frequency. Using the phasor representation of the voltage $v(t) = V\cos(\omega t + \Phi_v) \to \mathbf{V}e^{\jmath\Phi_v}$ and current $i(t) = I\cos(\omega t + \Phi_i) \to \mathbf{I}e^{\jmath\Phi_i}$, we can extend the concept of resistance to AC circuits, where the complex impedance

$$\mathbf{Z} = \frac{\mathbf{V}}{\mathbf{I}}e^{\jmath(\Phi_v - \Phi_i)} \quad \leftrightarrow \quad Z(\jmath\omega) = R + \jmath\omega L + \frac{1}{\jmath\omega C} \quad = \quad |Z(\jmath\omega)|\; e^{\jmath\Phi(\omega)}$$

describes not only the relative magnitudes of the voltage and current, but also their phase difference $\Phi(\omega) = \tan^{-1}(\omega L - \frac{1}{\omega C}/R)$, that is, the angle between the dissipative part of the impedance (real resistance R) and the frequency-dependent imaginary part of the impedance $(\omega L - 1/\omega C)$.

It may not be immediately obvious that it is complex numbers $Z \in \mathbb{C}$, rather than two-dimensional vectors $(R, \omega L - 1/\omega C) \in \mathbb{R}^2$, that are appropriate for this purpose. It is the *convenience of dealing with the phase information* and the computational power of complex algebra that makes it so, and such phase information is best visualised through the notion of *phasor*, a rotating vector in $\mathbb{C}$.

Complex step derivative approximation. Finite differencing formulas are commonly used for estimating the value of a derivative of a function, one such example is the first-order approximation

$$f'(x_0) = \frac{f(x_0 + h) - f(x_0)}{h} + \mathcal{O}(h) \tag{2.3}$$

where h is the finite difference interval and the truncation error is $\mathcal{O}(h)$. We can obtain a simpler and more accurate estimate of the first derivative using complex calculus [204, 280]. Using the first Cauchy–Riemmann equation (Chapter 5, Equation 5.6) and function $f(x + \jmath y) = u(x, y) + \jmath v(x, y)$, and noting that the real valued $f(x)$ is obtained for $y = 0$, we have (for more detail see Chapter 5)

$$\frac{\partial u(x, y)}{\partial x} = \frac{\partial v(x, y)}{\partial y} = \lim_{h \to 0} \frac{v(x + \jmath(y + h)) - v(x + \jmath y)}{h} \quad \Rightarrow \quad \frac{\partial f}{\partial x} \approx \frac{\Im\{f(x + \jmath h)\}}{h} \tag{2.4}$$

Unlike the real valued derivative approximation (Equation 2.3), the *complex step derivative approximation* (Equation 2.4) does not involve a difference operation and we can choose very small stepsizes h with no loss of accuracy due to subtractive cancellation.

To assess the error in the approximation (2.4), following the original derivation by Squire and Trapp [280], replace $f(x_0 + h)$ in Equation (2.3) with $f(x_0 + \jmath h)$ and apply a Taylor series expansion, to give

$$f(x_0 + \jmath h) = f(x_0) + \jmath h f'(x_0) - \frac{1}{2!}h^2 f''(x_0) - \frac{1}{3!}\jmath h^3 f^{(3)}(x_0) + \cdots \tag{2.5}$$

Take the imaginary parts on both sides of Equation (2.5) and divide by h to yield[2]

$$f'(x_0) = \frac{\Im\{f(x_0 + \jmath h)\}}{h} + \mathcal{O}(h^2) \tag{2.6}$$

that is, by using complex variables to estimate derivatives of real functions, the accuracy is increased by an order of magnitude.

Analytic signals. One convenient way to obtain the phase and instantaneous frequency information from a single channel recording x is by means of an *analytic* extension of a real valued signal. The basic idea behind analytic signals is that due to the symmetry of the spectrum, the negative frequency components of the Fourier transform of a real valued signal can be discarded without loss of information. For instance, a real valued cosine wave $x(t) = \cos(\omega t + \phi)$, can be converted into the complex domain by adding the phase shifted 'phase-quadrature' component $y(t) = \sin(\omega t + \phi)$ as an imaginary part, to give

$$z(t) = x(t) + \jmath y(t) = e^{\jmath(\omega t + \phi)} \tag{2.7}$$

The cosine and sine have spectral components in both the positive and negative frequency range, whereas their analytic counterpart $z(t) = e^{\jmath(\omega t + \phi)}$ has only one spectral component in the positive frequency range. Clearly, the *phase-quadrature* component can be generated from the *in-phase* component by a phase shift by $\pi/2$, and the original real valued signal $x(t)$ is simply the real part of the analytic signal $z(t)$.

Since an arbitrary signal can be represented as a weighted sum of orthogonal harmonic signals, the analytic transform also applies to any general signal, however, instead of a simple phase shift (Equation 2.7) we need to employ a filter called the Hilbert transform, to give

$$z = x + \jmath \mathcal{H}(x) = x + \jmath y \tag{2.8}$$

where $\mathcal{H}(\cdot)$ denotes the Hilbert transform [218].

The Hilbert transform performs a time domain filtering operation in the form of the convolution of signal $x(t)$ with the impulse response of the Hilbert filter $h(t) = 1/(\pi t)$, that is

$$y(t) = \mathcal{H}(x(t)) = h(t) * x(t) = \int_{-\infty}^{\infty} h(\tau)\, s(t - \tau)\, \mathrm{d}\tau$$

[2]For more detail on $\mathcal{O}$ notation see Appendix F.

Table 2.1 The XOR operation

x_1	x_2	$x_1 \oplus x_2$
1	1	1
1	−1	−1
−1	1	−1
−1	−1	1

hence, the transfer function of a Hilbert filter[3] is given by

$$H(\omega) = \begin{cases} e^{+\jmath\pi/2}, & \text{for } \omega < 0 \\ 0, & \text{for } \omega = 0 \\ e^{-\jmath\pi/2}, & \text{for } \omega > 0 \end{cases}$$

The operator $\mathcal{H}(\cdot)$ is an all–pass filter (see Section 11.4), given by

$$H(f) = -\jmath sgn(f), \quad |H(f)| = 1, \quad \angle H(f) = -\frac{\pi}{2} sgn(f)$$

which introduces a phase shift of $-\pi/2$ at positive frequencies and $+\pi/2$ at negative frequencies. The analytic signal can also be expressed in terms of the polar coordinates

$$z(t) = A(t)e^{\jmath\Phi(t)} \quad where \quad A(t) = |z(t)|, \quad \Phi(t) = \arg(z(t)) \tag{2.9}$$

where $A(t)$ is called the *amplitude envelope* and $\Phi(t)$ the *instantaneous phase* of the signal $z(t)$. An insight into the time–frequency characteristics is provided via so-called 'instantaneous frequency', that is, the first derivative of the phase of an analytic signal [128].

Classification and nonlinear separability. One example illustrating the usefulness of complex valued representation in neural network classification problems is the XOR problem. For binary variables $x_1, x_2 \in \{-1, 1\}$, the XOR operation $x_1 \oplus x_2$ performs the mapping shown in Table 2.1. This problem is nonlinearly separable in $\mathbb{R}$ only when using networks with more than one neuron. On the other hand, this is possible to achieve in $\mathbb{C}$ with a single neuron.[4] For instance, the complex nonlinearity at a neuron given by (see also Figure 2.1)

$$P(z) = \begin{cases} 1, & 0 \le \arg(z) < \pi/2 \quad or \quad \pi \le \arg(z) < 3\pi/2 \\ -1, & \pi/2 \le \arg(z) < \pi \quad or \quad 3\pi/2 \le \arg(z) < 2\pi \end{cases} \tag{2.10}$$

[3] In the discrete time we have

$$H(\omega) = \begin{cases} e^{+\jmath\pi/2}, & -\pi \le \omega < 0 \\ e^{-\jmath\pi/2}, & 0 \le \omega < \pi \end{cases}$$

whereas the impulse response takes values $h[k] = 0$ for k even, and $h[k] = 2/\pi k$ for k odd.

[4] Example provided by Igor Aizenberg.

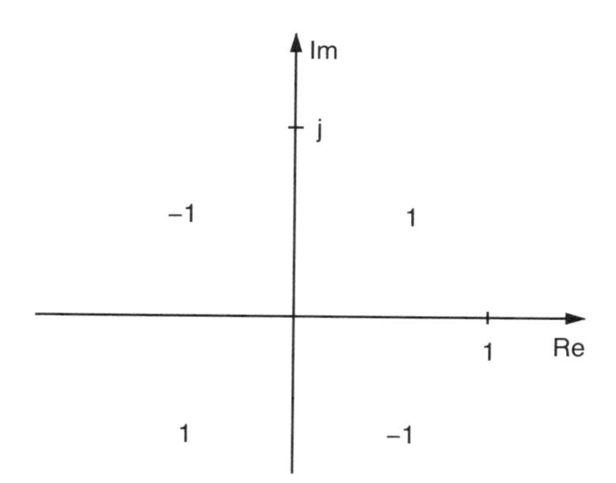

Figure 2.1 Nonlinear mapping at a complex neuron performed by function $P(z)$ given in (2.10). The quarter-planes in the $\mathbb{C}$ plane are mapped on the discrete set of numbers $\{-1, 1\}$

splits the complex plane into four parts. In order to transform the pairs (x_1, x_2) into $\{1 + \jmath, \ldots, -1 - \jmath\}$, we can linearly combine the variables x_1 and x_2 with the weighting coefficients $\mathbf{w} = [w_1, w_2] = \left[1, \jmath\right]$, to give

$$z = w_1 x_1 + w_2 x_2 \tag{2.11}$$

From Equation (2.11), depending on the combination of x_1 and x_2, variable z takes four discrete values $\{1 + \jmath, 1 - \jmath, -1 + \jmath, -1 - \jmath\}$, which when passed through $P(z)$ from Equation (2.10), give the correct XOR solution, as shown in Figure 2.1 and Table 2.2.

Modelling of three-dimensional problems in $\mathbb{C}$. Convenience of complex representations of some three-dimensional problems may be illustrated by the example of stereographic projections[5] [242] (see also Figure 1.4a), where there is a one-to-one correspondence between the points of $\mathbb{C} \setminus \infty$ and the points on sphere $\Sigma \in \mathbb{R}^3$ (excluding the north pole N); the line between N and any point $z \in \mathbb{C}$ cuts $\Sigma \setminus \{N\}$ in precisely one point. Figure 2.2 shows the mapping of the points on a circle on sphere Σ onto a curve in the complex plane $\mathbb{C}$. In general, the point

Table 2.2 The complex valued realisation of the XOR problem

x_1	x_2	z	$P(z)$
1	1	$1 + \jmath$	1
1	−1	$1 - \jmath$	−1
−1	1	$-1 + \jmath$	−1
−1	−1	$-1 - \jmath$	1

[5] Angle-preserving (conformal) projections are preferred for 'navigation' applications, such as in the Smith chart, which is of considerable use in transmission line theory. Other applications of stereographic projections include those in crystallography and in photography (fisheye lenses).

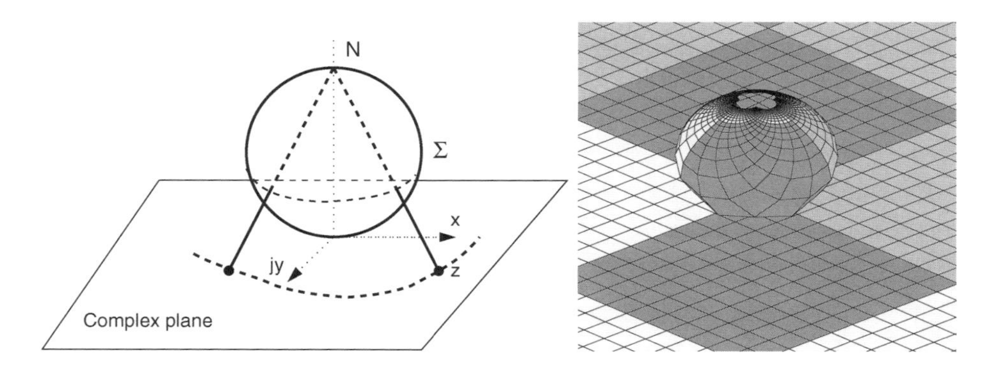

Figure 2.2 Stereographic projections and Riemann sphere. Left: stereographic projection of a circle; right: geometric interpretation

(X, Y, Z) on the Riemann sphere corresponds to the point $(x, y) \in \mathbb{C}$, such that [206]

$$X = \frac{x}{x^2 + y^2 + 1}, \quad Y = \frac{y}{x^2 + y^2 + 1}, \quad Z = \frac{x^2 + y^2}{x^2 + y^2 + 1} \tag{2.12}$$

This type of projection is in fact a Möbius transformation (see Chapter 11), and has found numerous applications, such as those in quantum mechanics (photon polarisation states), and in relativity theory, where the Riemann sphere is used to model the celestial sphere.[6] One much more obvious application is in cartography, where the charts of 3D surfaces (such as the Earth) are produced as 2D projections, similarly to the situation in Figure 2.2. The distance between two points on a curved space such as the sphere is called a *geodesic* (the shortest path between two points). This term obviously comes from geodesy, as the shortest route between two points on the surface of the Earth.[7]

2.1.1 Duality Between Signal Representations in $\mathbb{R}$ and $\mathbb{C}$

Figure 2.3(a) summarizes the duality between the processing in $\mathbb{R}$ and $\mathbb{C}$ and can be explained as follows:

- The nature of purely *real* and *complex* signals is obvious: *real* signals are magnitude-only whereas *complex* signals comprise both magnitude and phase components;
- *Phase only* signals are real signals formed from the phase of a complex signal or from the phase of an analytic signal (Equation 2.8);
- *Dual univariate* signals are the real and imaginary component of a complex signal that are processed separately as real valued quantities.

[6]Ptolemy (ca AD 125) was first to plot the positions of heavenly bodies on the 'celestial sphere', his method is called *stereographic projection*.

[7]It is important to notice that the shortest route on a sphere is a segment of the great circle; when mapped onto the plane, in general, this is not a straight line (see Figure 2.2).

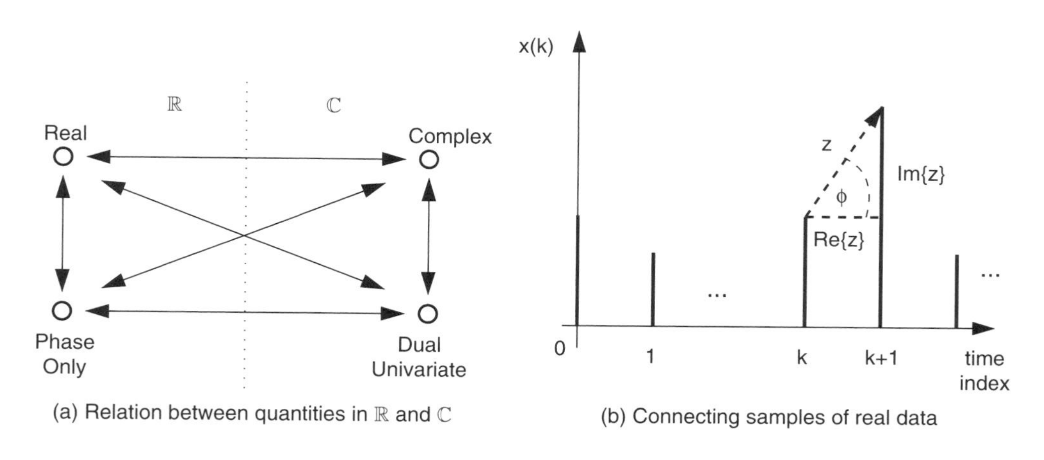

Figure 2.3 Duality between signal processing in $\mathbb{R}$ and $\mathbb{C}$

In order to transform a real signal (say coming from an array of sensors [200]) into its complex valued counterpart, it is often convenient to make use of the delay (or phase) associated with the time of arrival of the real valued signal (or vector field) at sensors (see also Figure 2.8a and Figure 2.11). Another interesting scenario is the 'complex to phase only' transformation for phase-only signal processing; this is very practical in cases where the magnitude of the signal has little or no variation [227, 285].

In the 'mirroring' approach, the extension from $\mathbb{R}$ to $\mathbb{C}$ is performed by producing a complex signal z from two real valued processes x and y, as

$$z = x + \jmath y$$

This approach is convenient for the detection of synchronisation within multichannel recordings, as shown in Section 2.4.1. Alternatively, for every time instant k, we may connect $x(k)$ and $x(k+1)$, to obtain a complex vector $z(k)$, as shown in Figure 2.3(b).

2.2 Modelling in ℂ is Not Only Convenient But Also Natural

The examples above illustrate that in practical applications we encounter complex processes in two general cases:

- Real–life quantities that are naturally described by complex numbers;
- Real–life quantities which although real, are best understood through complex analysis.

In other words, in engineering and computing, complex quantities are produced either by *design* or by *convenience of representation*, as shown in Figure 2.4.

- Signals made complex by design include symbols used in data communications; for instance in the quadrature phase shift keying (QPSK) constellation, the symbols are located on the unit circle in the $\mathcal{Z}$ plane, so that they can be transmitted with the same energy (Figure 2.4a);

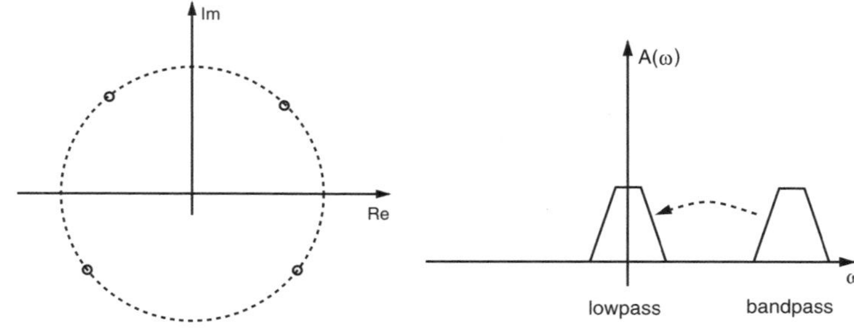

Figure 2.4 Examples of complex signals: (a) by design; (b) by convenience of representation

- By convenience of representation, for instance, a general real valued bandpass[8] signal $x(t)$ has a complex valued lowpass equivalent[9] $z(t)$, that is

$$x(t) = A(t)\cos(\omega_c t + \Phi(t)) \quad \leftrightarrow \quad z(t) = A(t)e^{\jmath\Phi(t)} \quad \Leftrightarrow \quad x(t) = \Re\{z(t)e^{\jmath\omega_c t}\} \tag{2.13}$$

where $A(t)$ is the amplitude, $\Phi(t)$ the phase, and ω_c the modulation (carrier) frequency (see Figure 2.4b and also Figure 2.8b). Alternatively, an equivalent lowpass equivalent can be obtained by Hilbert transform, as shown in Equation (2.8).

2.3 Why Complex Modelling of Real Valued Processes?

Complex valued representations may not have direct physical relevance (only their real parts do), but they can provide a general and mathematically more tractable framework for the analysis of several important classes of real processes. Two aspects of this duality between real and complex valued processes are particularly important:

- The *importance of phase* (time delay) information (communications, array signal processing, beamforming);
- The advantages arising from the simultaneous modelling of the '*intensity*' and '*direction*' component of vector field processes in $\mathbb{C}$ (radar, sonar, vector fields, wind modelling).

2.3.1 Phase Information in Imaging

While the phase information in 1D signals is subtly hidden, in 2D signals such as images, the role of the phase of a signal is more obvious[10] [227]. To illustrate this, consider a scenario

[8]A signal whose bandwidth is much smaller than its centre frequency, for instance an AM signal.

[9]Notice similarity with phasors (2.2), except that lowpass equivalent signals are functions of time, whereas phasors are not. Lowpass equivalent signals can therefore represent general bandpass signals, whereas phasors represent only sinusoidal signals.

[10]The importance of the phase spectrum is intimately related with the nonlinearity within a signal [83, 144]. The 'linear' properties of a signal are the mean and variance (or equivalently the covariance and power spectrum), whereas the 'nonlinear' signal properties are related to higher-order statistical moments and phase spectra.

Figure 2.5 Surrogate images. *Top:* original images I_1 and I_2; *bottom:* images $\hat{I}_1$ and $\hat{I}_2$ generated by exchanging the amplitude and phase spectra of the original images

similar to that in the surrogate data generation method in nonlinear time series analysis.[11] The top panel in Figure 2.5 shows two greyscale images denoted by I_1 (wheel) and I_2 (child); the 2D Fourier transform is applied and the phase spectra of I_1 and I_2 are swapped to give

$$S_1(x, y) = |\mathcal{F}(I_1)|\, e^{\jmath\angle\mathcal{F}(I_2)}$$
$$S_2(x, y) = |\mathcal{F}(I_2)|\, e^{\jmath\angle\mathcal{F}(I_1)} \tag{2.14}$$

Thus, spectrum S_1 has the magnitude spectrum of I_1 and phase spectrum of I_2, whereas spectrum S_2 has the magnitude spectrum of I_2 and phase spectrum of I_1. The bottom panel of Figure 2.5 shows the situation after the inverse Fourier transform is applied to S_1 and S_2 to obtain respectively $\hat{I}_1$ and $\hat{I}_2$. Observe that image $\hat{I}_1$, for which the magnitude spectrum is that

[11]There, to produce a 'surrogate' signal which has the same statistics as the original, the phase spectrum of the original signal is randomised and the surrogate is obtained from the inverse Fourier transform of the correct amplitude spectrum and randomised phase spectrum of the original signal [83].

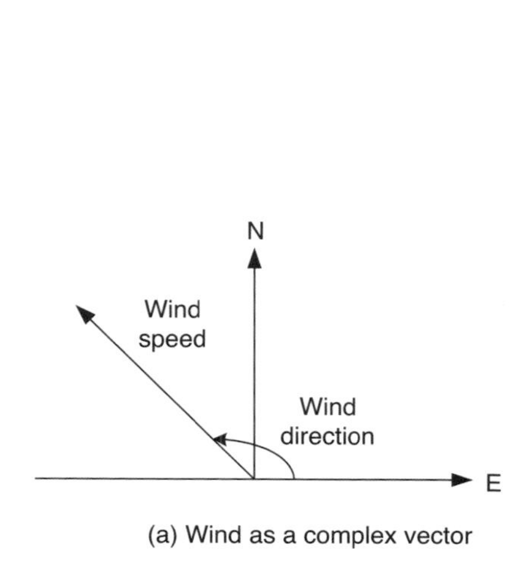

(a) Wind as a complex vector

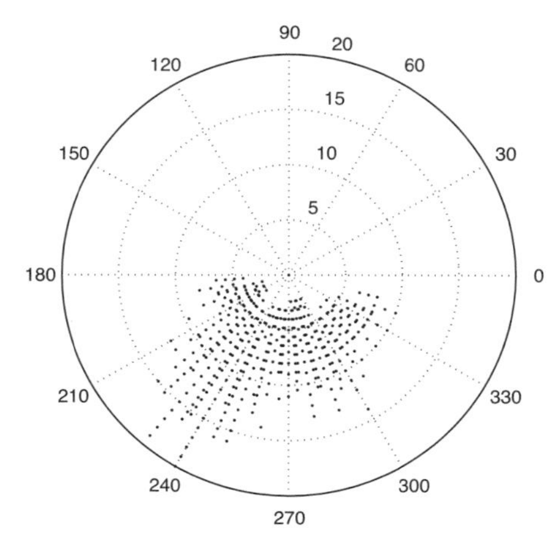

(b) Wind lattice (rose) – distribution of wind speeds as a function of wind directions

Figure 2.6 Wind recordings: a) complex–valued representation; b) wind lattice

of the wheel and phase spectrum is that of the child, is recognised as a child (and vice versa for $\hat{I}_2$); this clearly shows that for the human visual system the information is predominantly encoded in the phase.

In some situations, it is therefore convenient to consider images based on their complex valued representation [14, 15]. One way to achieve this would be to consider a greyscale or 'intensity' image

$$I = \begin{bmatrix} a_{11} & a_{12} & \cdots & a_{1m} \\ a_{21} & a_{22} & \cdots & a_{2m} \\ \vdots & \vdots & \ddots & \vdots \\ a_{n1} & a_{n2} & \cdots & a_{nm} \end{bmatrix} \tag{2.15}$$

where for an image with n intensity levels, elements a_{ij} take values from $\{0, 1, \ldots, n-1\}$. Alternatively, these elements can be modelled as nth (complex) roots on the unit circle in the $\mathcal{Z}$-plane, thus giving a complex valued, phase described, representation of the image. A review of the benefits of complex valued modelling of real valued processes can be found in [195].

2.3.2 Modelling of Directional Processes

Consider a class of processes with 'intensity' and 'direction' components, such as wind, radar, sonar, or sensor array measurements. Figure 2.6 represents a wind measurement[12] as a vector

[12]The wind signal used was obtained from readings from the Iowa (USA) Department of Transport http://mesonet.agron.iastate.edu/request/awos/1min.php database.

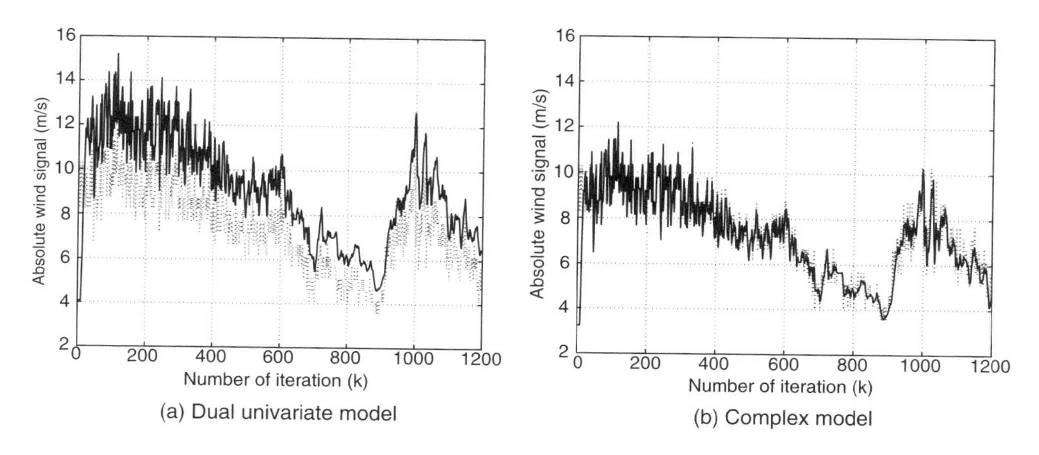

Figure 2.7 Modelling of wind profile: dual univariate vs complex model. Original signal (dashed line) and one step ahead prediction (solid line)

$\mathbf{v}(t) = v(t)e^{\jmath d(t)}$ of its speed $v(t)$ and direction $d(t)$ components, in the $N - E$ coordinate system, together with the distribution of wind speeds over various directions. Despite the clear interdependence between the wind speed and direction (from the wind lattice in Figure 2.6b, the distribution of significant wind speeds is in the range of 190° to 340°), in most practical applications, these are treated as independent real quantities, that is, as *dual univariate* time series, hence introducing error in both the models of wind dynamics and associated forecasts. In the polar representation, the wind speed v corresponds to the modulus, and the direction d to the angle of a complex vector $\mathbf{v}$. In the Cartesian representation, we may exploit the natural coupling between the real and imaginary part in order simultaneously to track the changes in the dynamics of both the speed and direction component of wind.

Figure 2.7 illustrates the performance gain obtained by using the complex wind model, when applied to wind forecasting. The dual univariate approach (see Figure 2.3a) is based on the modelling of speed and direction as two independent real valued processes; the outputs of those univariate models are then combined into a single complex quantity at the output. Clearly, the dual univariate approach (Figure 2.7a) was not able to track the wind dynamics, whereas a complex model Figure 2.7b), which simultaneously modelled the wind speed and direction, exhibited excellent performance.[13]

2.4 Exploiting the Phase Information

An important class of problems which benefits from complex representation is in array signal processing [200]. One such application is in beamforming, where the phase and the amplitude of the received array signals are normally modelled as complex quantities [292]. Figure 2.8(a)

[13]The simulations were based on two separate real FIR filters trained with LMS (for the dual univariate case) and a complex FIR filter trained with complex LMS (for the complex representation).

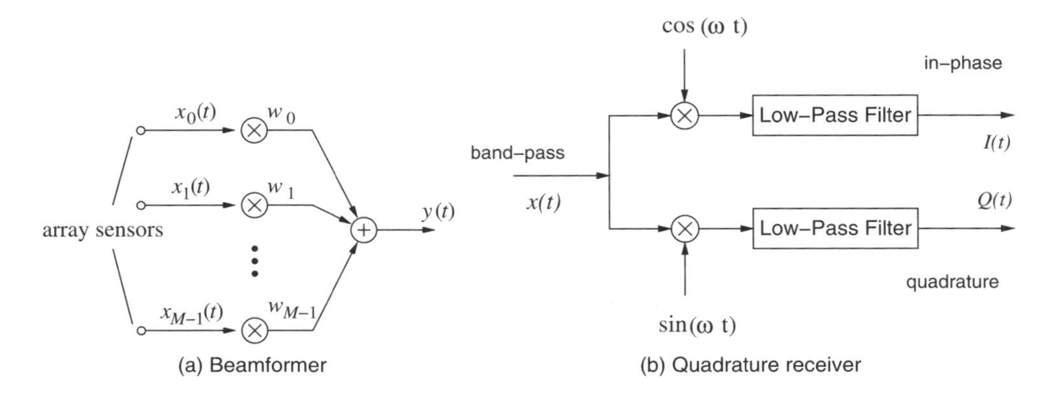

Figure 2.8 Beamformer and quadrature receiver. (a) Beamformer based on a linear array with M sensors; For simplicity, a single filter coefficient is associated with each of the received array signals. (b) General structure of the quadrature receiver

shows a simple beamformer for narrowband signals, for which the output $y(t)$ is given by

$$y(t) = \sum_{m=0}^{M-1} x_m(t)\, w_m \tag{2.16}$$

where $x_m(t)$ and w_m, $m = 0, 1, \ldots, M-1$ are respectively the mth received array signal and the corresponding filter coefficient. The received array signals are made complex valued with the help of a quadrature receiver (in order to separate them and convert into baseband, see Equation 2.13), as shown in Figure 2.8(b), where ω is the centre frequency of the received bandpass signal $x(t)$ [291]. The in-phase output $I(t)$ becomes the real part of the complex lowpass equivalent signal, whereas the quadrature component $Q(t)$ is the imaginary part.

2.4.1 Synchronisation of Real Valued Processes

Another recent application where complex domain processing of real data has significant potential is the design of Brain Computer Interface (BCI). In the design of brain prosthetics, a major problem in the processing information from a microarray of electrodes which are implanted into the cortex of the brain is the modelling of neuronal spiking activity. Whereas this is very difficult to solve in $\mathbb{R}$ (due only partly to high levels of noise), the synchronisation of spike events is straightforward to model in $\mathbb{C}$. An approach for converting multichannel real valued sequences of spiking neuronal recordings (point processes coming from implanted microarrays into the brain cortex) into their complex–valued counterparts is elaborated in [296]. The main underlying idea is to code the spike events of interest as complex processes, where the phase encodes the interspike interval.

For illustration, consider two artificially generated real valued point (spiky) processes x_1 and x_2 [130], depicted in Figure 2.9(a), which are contaminated by independent realisations of white Gaussian noise and are shifted by ten samples. We can convert the spike synchronisation

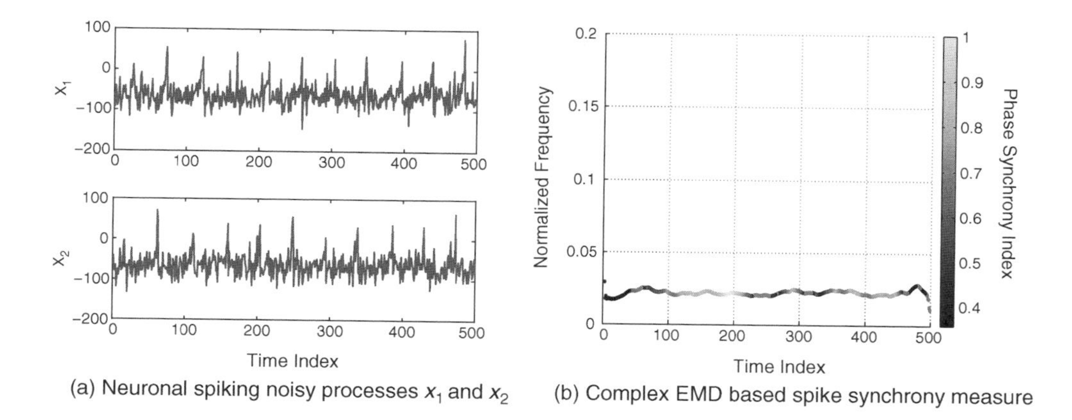

(a) Neuronal spiking noisy processes x_1 and x_2 (b) Complex EMD based spike synchrony measure

Figure 2.9 Synchronisation of neural point processes

problem into one of phase matching by first constructing the complex signal

$$z = x_1 + \jmath x_2 \tag{2.17}$$

The noise introduces spurious spikes at multiple time intervals, which affects the temporal structures of the original signals. In order to detect the synchronised spike events, a complex extension of Empirical Mode Decomposition (EMD) was employed [9, 196]. The spike synchrony detected is shown in Figure 2.9(b), where the time, frequency, and the spike synchrony (calculated using the phase coherence value [286]) between the signals x_1 and x_2 are respectively represented by the x, y, and z (colour-coded) axes. For more detail on empirical mode decomposition see Chapter 17.

2.4.2 Adaptive Filtering by Incorporating Phase Information

In several applications of adaptive filtering, such as those in data communications (MPSK, QPSK) the information is encoded in the phase and the amplitude is kept constant. However, because signal propagation causes distortion, when performing adaptive filtering (for instance for adaptive equalisation), both the magnitude and phase of the received symbols should be considered. For instance, the Least Mean Phase Least Mean Square (LMP–LMS) algorithm [285] deals simultaneously with the magnitude and phase, and is given by

$$J(\mathbf{w}) = k_1 \underbrace{E\left|d - \mathbf{x}^{\mathrm{T}}\mathbf{w}\right|^2}_{LMS} + k_2 \underbrace{E\left|\angle d - \angle \mathbf{x}^{\mathrm{T}}\mathbf{w}\right|^m}_{LMP} \tag{2.18}$$

where $\mathbf{x}$ is the input signal, d is the teaching signal, $\mathbf{w}$ are the filter coefficients, $m \in \{1, 2\}$, and k_1 and k_2 are mixing coefficients. Cost function (2.18) simultaneously minimises a measure of both the phase and magnitude error, resulting in the weight update

$$\mathbf{w}(k+1) = \mathbf{w}(k) + \mu_1 e(k)\mathbf{x}^*(k) + \mu_2 \angle e(k) \frac{\jmath \mathbf{x}^*(k)}{y^*(k)}, \tag{2.19}$$

where y denotes the output of the filter, and μ_1 and μ_2 are positive learning rates.

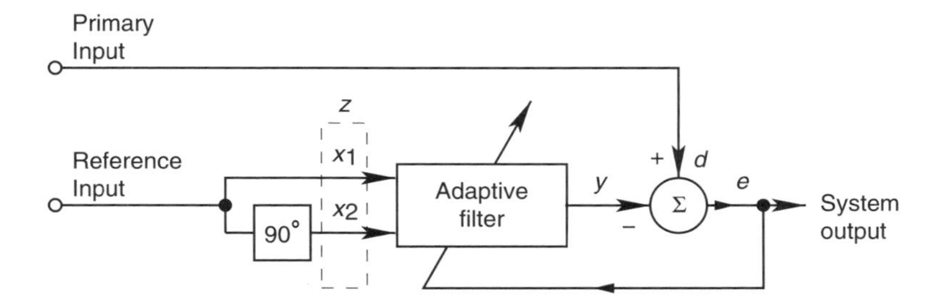

Figure 2.10 Adaptive noise cancelling adaptive filtering configuration

Other real valued adaptive filtering problems that can be converted into the complex domain include the standard noise cancellation adaptive filtering configuration, shown in Figure 2.10. The primary input is the useful signal corrupted by additive noise, whereas the reference input is any noise correlated with the noise within the primary input. In [308] it was shown that the scheme from Figure 2.10 outperforms the standard noise cancellation scheme in Electroencephalogram (EEG) applications. By means of the phase shift of $\pi/2$ within the filter, a complex reference input $z = x_1 + \jmath x_2$ (see Figure 2.3) can be produced and noise cancellation may be performed by the complex LMS algorithm, given by [307]

$$\mathbf{w}(k+1) = \mathbf{w}(k) + \mu e(k)\mathbf{x}^*(k) \tag{2.20}$$

2.5 Other Applications of Complex Domain Processing of Real Valued Signals

Complex valued representations have been instrumental to the advances in diverse fields of electronics, physics and biomedicine. Some of the applications which have benefited greatly from this approach include

- *Magnetic Source Imaging (MRI, fMRI, and MEG)*. A magnetic resonance imaging (MRI) signal is acquired as a quadrature signal, by means of two orthogonally placed detectors and is then Fourier transformed for further processing [253]. Standard approaches for the enhancement of MRI images consider only the magnitude spectra of such images. The phase information, however, can be obtained as a function of the difference in magnetic susceptibility between a blood vessel and the surrounding tissue as well as from the orientation of the blood vessel with respect to the static magnetic field. Recent results by Calhoun and Adali [40–42] illustrate the benefits of incorporating both the magnitude and phase information into the processing of functional MRI (fMRI) data. Magnetoencephalography (MEG) is a noninvasive neurophysiological technique that measures the magnetic fields generated by neuronal activity of the brain. By Fourier transforming the MEG data, they can be conveniently analysed in $\mathbb{C}$ [276]. In addition, the electric and magnetic field at a neuron are orthogonal and obey the Maxwell equations, this facilitates the combined analysis of the electroencephalogram (EEG) and MEG [315].
- *Interferometric radar*. Electromagnetic wave imaging technology such as Synthetic Aperture Radar (SAR) has a wide range of applications since the relatively long

wavelength used enables reduced absorption and scattering by clouds and smoke. Most such radar systems use electromagnetic waves whose coherence is high so that they can obtain both the phase (altitude) and amplitude information (airplane/satellite surface observation radars). The high coherence, however, brings problems of interference and noise, such as speckles. The restoration and clustering of images of the reflected waves can be conveniently cast as a complex valued problem, as shown in work by Akira Hirose [120, 121, 282].

- *Direction of arrival (DoA) estimation and smart antennas.* These are major problems in array signal processing [319], which are traditionally solved by maximum likelihood or linear prediction methods. It has been shown, however, that DoA estimation is a complex valued optimisation problem [137, 138]. Thus for instance, in the research monograph by A. Manikas [200], the role of differential geometry as an analytical tool in array processing and array communications is highlighted and this theoretical framework is extended to complex spaces.
- *Mathematical biosciences.* In functional genomics, problems such as temporal classification of confocal images of expression patterns of genes can be solved by neural networks with multivalued neurons (MVN). MVNs are processing elements with complex valued weights and high functionality which have proved to be efficient in image recognition problems [4]. The goal of temporal classification is to obtain groups of embryos which are indistinguishable with respect to the temporal development of expression patterns.
- *Transform domain signal processing.* In adaptive filtering and elsewhere, it is often beneficial to process data in a *transform domain*, for instance, by means of the Discrete Fourier Transform (DFT) or Discrete Cosine Transform (DCT) [113]. This way, the real valued input is effectively pre-whitened in the complex domain; this then speeds up the convergence of the adaptive filtering algorithms employed [149]. This is particularly convenient in *separation of real valued convolutive mixtures*, where the idea is to perform the Fourier transform of the observed mixtures, in order to convert real convolution into complex multiplication [231]. This way, the problem is transformed into source separation of complex instantaneous mixtures, for which there are many established algorithms [49].
- *Mobile communications and interference cancellation in broadcasting.* In digital data transmission, the working algorithms are almost always derived in $\mathbb{C}$. This way, both the data model and the channel model are complex functions of complex variables [113], and signal detection, channel estimation and equalisation are all performed in $\mathbb{C}$; one such algorithm for blind channel equalisation is the constant modulus algorithm (CMA) [89]. In interference cancellation applications in broadcasting, real valued signals do not satisfy some of the requirements,[14] whereas complex polyphase sequences (elements of which are roots of unity) are naturally suited for this purpose.
- *Homomorphic filtering.* Homomorphic filtering is particularly suited for real valued multiplicative models and for processes where the dynamical range of the signal is large [232, 318]. Homomorphic filters are based on a combination of a logarithmic input layer

[14]The reference signal should have a flat spectrum, the autocorrelation only at zero–lag, and the signal energy should be as large as possible [302].

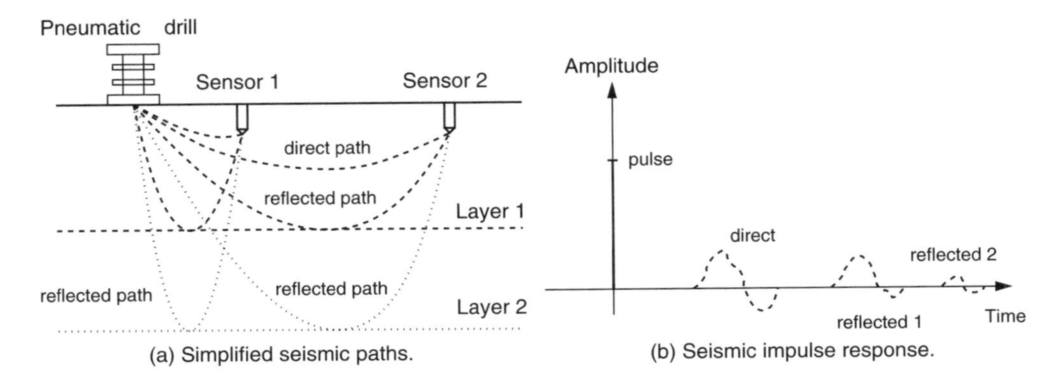

Figure 2.11 A simplified model of seismic wave propagation

and an exponential output layer, these are connected by adaptive weights. Since the logarithm in $\mathbb{R}$ does not exist for signal values $x \leq 0$, it is convenient to employ complex logarithms on analytic inputs $z = x + \jmath\mathcal{H}(x)$ to obtain $\log(z) = \log(|z|) + \jmath \arg(z)$, perform complex filtering, and then convert the outputs back to the real domain.

- *Optics and seismics.* Reflections and refractions introduce both the attenuation of the amplitude and phase shifts, as illustrated on a simplified diagram of the propagation of seismic waves through two layers of soil, shown in Figure 2.11. Figure 2.11 shows a pneumatic drill producing a pulse that approximates the Dirac function, the signals received at sensors $S1$ and $S2$ are altered both in terms of the filtering through the soil and phase change (soil impulse response). These are best modelled simultaneously in $\mathbb{C}$.
- *Fractals and complex iterated maps.* Fractals are self-similar infinitely repeating[15] mathematical objects, that is, the parts of the object are similar to the whole. This self-similar structure implies that fractals are scale-invariant, and we cannot distinguish a small part from the larger structure [176]. Approximate fractals are easily found in nature,[16] examples include clouds, snowflakes, crystals, mountain ranges, lightning, river networks, and systems of blood and pulmonary vessels. Applications of fractal modelling include the analysis of medical tissues [43], signal and image compression [75], seismology [257], virtual environments design, and music and art applications. For more detail see Appendix P.
- *Other applications.* Complex valued processing has found applications in other areas, including the analysis of cyclostationary processes [80, 131], semiconductor design [277], chaos engineering [3, 208], acoustic microscopy [36], stochastic mechanics [229], and nuclear physics [81, 202]. In addition, in applied fields complex analysis is used to compute certain real valued improper integrals, for instance by contour integration [206, 218].

[15]Complex extensions of Newton's iteration

$$x_{i+1} = x_i - \frac{f(x_i)}{f'(x_i)}$$

have attracted considerable interest, but are not straightforward. For instance, Newton's method fails for the function $f(z) = z^2 + 1$ if the initial guess is a real number (see Appendix P).

[16]These objects display self-similar structure over an extended, but finite, scale range.

2.6 Additional Benefits of Complex Domain Processing

In addition to the examples of complex domain processing of real valued signals introduced so far, signal processing in $\mathbb{C}$ possesses several distinctive features which are not present in $\mathbb{R}$. These include:

- *More powerful statistics*. Although it is often assumed that the statistics in $\mathbb{C}$ is a straightforward extension[17] of the statistics in $\mathbb{R}$, some recent results in complex statistics have exploited complex circularity and introduced the notions of proper ($E\{\mathbf{x}\mathbf{x}^{\mathrm{T}}\} = 0$) and improper ($E\{\mathbf{x}\mathbf{x}^{\mathrm{T}}\} \neq 0$) complex random variables [219]. This gives us more degrees of freedom and hence greater potential for improved performance and resolution, compared with the standard modelling in $\mathbb{C}$. To derive so called *augmented learning algorithms*, consider a complex vector $\mathbf{x}$ and its conjugate to produce the augmented vector $\mathbf{x}^{\mathrm{a}} = [\mathbf{x}; \mathbf{x}^*]$. To illustrate the benefits of using the augmented complex statistics, Table 2.3 shows the improved quantitative performance of an augmented over a standard learning algorithm[18] for the task of one step ahead adaptive prediction of a stable complex $AR(4)$ process and complex valued radar data.[19] For more detail see Chapter 12.
- *Complex nonlinearity*. Apart from the *dual univariate* approach (Figure 2.3a) which applies mainly to processes where the real and imaginary part are predominantly independent and linear in nature), ways to introduce nonlinearity in $\mathbb{C}$ include the so called *split-complex* and *fully complex* approach [152]. Within the split-complex approach, a pair of real valued nonlinear activation functions is employed to separately process the real and imaginary components of the net input $net(k) = \mathbf{x}^{\mathrm{T}}(k)\mathbf{w}(k)$. Such a split-complex activation function is given by

$$f(net) = f_R(\Re(net)) + \jmath f_I(\Im(net)) \tag{2.21}$$

  where $f_R = f_I$ are real functions of real variable. Fully complex functions are standard complex functions of complex variables, such as the complex tanh. Owing to the intricate properties of complex nonlinearities in $\mathbb{C}$ (only a constant is continuously differentiable), this gives us the opportunity to trade between boundedness and differentiability of complex nonlinear activation functions when designing learning algorithms. For more detail see Chapter 4.

Table 2.3 Comparison of prediction gains R_p [dB] for the various classes of signals

R_p [dB]	Standard algorithm	Augmented algorithm
Linear $AR(4)$ process	3.22	4.10
Radar (low)	11.40	13.57
Radar (high)	4.56	5.41

[17] Normally, simply replace $(\cdot)^{\mathrm{T}}$ by the Hermitian $(\cdot)^{\mathrm{H}}$ in the corresponding statistical expressions.
[18] We used the complex real time recurrent learning (CRTRL) [93] as a standard complex algorithm and the augmented CRTRL (ACRTRL) [96] as the corresponding algorithm based on augmented complex statistics.
[19] Radar (high) is referred to as 'high sea state data' and radar (low) is referred to as 'low sea state data'. The data used in simulations are publicly available from http://soma.ece.mcmaster.ca/ipix.

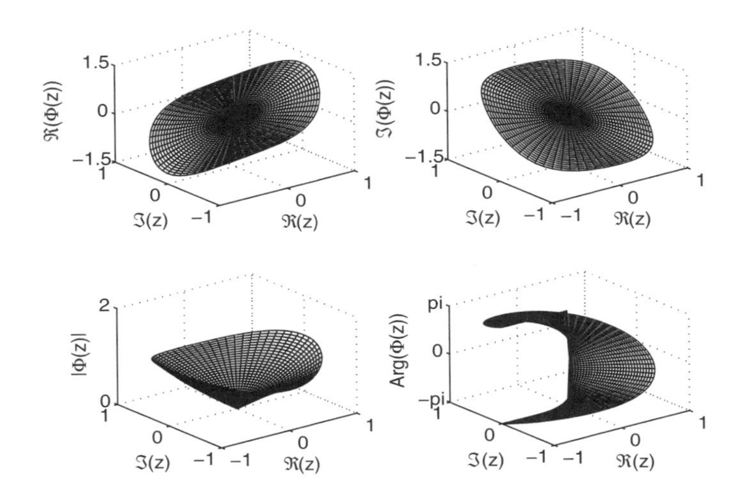

Figure 2.12 Visualisation of complex function $z \to \Phi(z) = \sinh(z)$. *Top panel:* the real and imaginary part; *Bottom panel:* the magnitude and phase functions

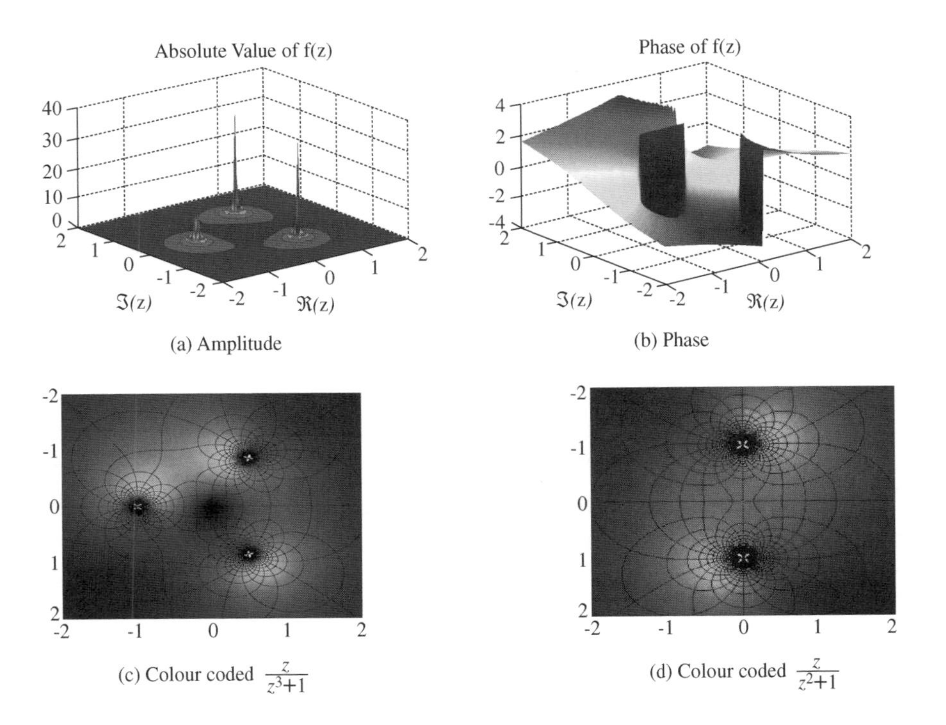

(c) Colour coded $\frac{z}{z^3+1}$

(d) Colour coded $\frac{z}{z^2+1}$

Figure 2.13 Panels (a), (b), and (c): Visualisation of the complex function $f(z) = z/(1 + z^3)$ by colour coding. Panel (d): Function $f(z) = 1/(z^2 + 1)$

- *Simultaneous modelling of heterogeneous variables and data fusion.* When dealing with processes with intensity and direction components, such as the wind field from Figure 2.6, complex domain modelling not only provides a simultaneous and compact representation of the wind speed and direction, but also it performs *sequential data fusion*. The wind speed and direction are of different natures (heterogeneous data) and by virtue of complex valued representation they are naturally *fused* into a single quantity [192].
- *Visualisation.* Unlike the real valued functions which are represented as two-dimensional graphs, complex functions are represented by four-dimensional graphs (two axes for the real and imaginary part of the argument and two axes for the real and imaginary part of the evaluated function). To visualise complex functions we therefore either consider the amplitude and phase (both 3D graphs with a complex argument) separately, or we perform colour coding of a three-dimensional graph to suggest the fourth dimension. Figure 2.12 shows the $\sinh(z)$ function plotted in terms if its real and imaginary part (top panel) and as a modulus–phase plot (bottom panel). A colour-coded visualisation of complex functions $f(z) = z/(z^2 + 1)$ and $f(z) = 1/(z^2 + 1)$ is illustrated in Figure 2.13. The phase of the complex number is represented by the hue and the absolute value by the lightness of the colour.
- *Compact and natural mathematical representation.* When dealing with complex numbers, we think of $a + \jmath b$ as an entire thing on its own, and all the standard rules of algebra are satisfied. For instance, a complex number can be represented in matrix form as

$$a + \jmath b \rightarrow \begin{bmatrix} a & -b \\ b & a \end{bmatrix} = a \begin{bmatrix} 1 & 0 \\ 0 & 1 \end{bmatrix} + b \begin{bmatrix} 0 & -1 \\ 1 & 0 \end{bmatrix} \tag{2.22}$$

  A comprehensive account of matrix representation of complex numbers and the representation of complex mappings by Möbius transformations is provided in Chapter 11, whereas Appendix A illustrates some differences between basic operations in $\mathbb{R}$ and $\mathbb{C}$.

It is fitting to end this Chapter with the quote from Richard Penrose's *The Road to Reality: A Complete Guide to the Laws of the Universe* [233].

> *We shall find that complex numbers, as much as reals, and perhaps even more, find a unity with nature that is truly remarkable. It is as though Nature herself is as impressed by the scope and consistency of the complex–number system as we are ourselves, and has entrusted to these numbers the precise operations of her world at its minutest scales.*

3

Adaptive Filtering Architectures

The architecture of a digital filter underpins its capacity to represent the dynamic properties of an input signal and hence its ability to estimate some future value. Linear filtering structures are very well established in digital signal processing and are classified either as finite impulse response (FIR) or infinite impulse response (IIR) digital filters [225]. FIR filters are generally realised without feedback, whereas IIR filters[1] utilise feedback to limit the number of parameters necessary for their realisation. There is an intimate link between the architecture of an adaptive filter and a stochastic model it represents:

- In statistical signal modelling, FIR filters are also known as moving average (MA) filters and IIR filters are named autoregressive (AR) or autoregressive moving average (ARMA) filters;
- Nonlinear filter architectures can be readily formulated by including a nonlinear operation in the output stage of an FIR or an IIR filter. These represent examples of nonlinear moving average (NMA), nonlinear autoregressive (NAR), or nonlinear autoregressive moving average (NARMA) structures [28];
- The model of an artificial neuron comprises a linear FIR filter whose coefficients are termed synaptic weights (or simply weights), and has a zero-memory nonlinearity;
- Different neural network architectures are designed by the combination of multiple neurons with various interconnections. Feedforward neural networks have no feedback within their structure; recurrent neural networks, on the other hand, exploit feedback and hence have the ability to model rich nonlinear dynamics [190].

The foundations for linear estimators of statistically stationary signals are found in the work of Yule [320], Kolmogorov [154], and Wiener [309]. The later studies of Box and Jenkins [33] and Makhoul [175] are built upon these fundamentals. This chapter provides a brief overview of linear and nonlinear, both feedforward and feedback, adaptive filtering architectures, for which learning algorithms will be presented throughout this book.

[1]FIR filters can be represented by IIR filters, however, in practice it is not possible to represent an arbitrary IIR filter with an FIR filter.

Complex Valued Nonlinear Adaptive Filters: Noncircularity, Widely Linear and Neural Models
Danilo P. Mandic and Vanessa Su Lee Goh

3.1 Linear and Nonlinear Stochastic Models

A general linear stochastic model is described by a difference equation with constant coefficients, given by

$$y(k) = \sum_{i=1}^{p} a_i y(k-i) + \sum_{j=1}^{q} b_j n(k-j) + b_0 n(k) \tag{3.1}$$

where $y(k)$ is the output, $n(k)$ are samples of doubly white Gaussian noise with zero mean and unit variance, $a_i,\ i = 1, 2, \ldots, p$ are the AR (feedback) coefficients, and $b_j,\ j = 0, 1, \ldots, q$ are the MA (feedforward) coefficients. Such a model is termed autoregressive moving average ARMA(p,q), where p is the order of the autoregressive, or feedback, part of the structure, and q is the order of the moving average, or feedforward, substructure. Due to the feedback present within this filter, the impulse response, namely the values of $y(k)$, $k \geq 0$, when $n(k)$ is a discrete time impulse, is infinite in duration (IIR).

The general ARMA form of Equation (3.1) can be simplified by removing the feedback terms to yield

$$y(k) = \sum_{j=1}^{q} b_j n(k-j) + b_0 n(k) \tag{3.2}$$

Such a model is termed moving average MA(q), and has a finite impulse response which is identical to the parameters $b_j,\ j = 0, 1, \ldots, q$. In digital signal processing, a filter based on this model is therefore named a finite impulse response (FIR) filter.

Alternatively, Equation (3.1) can be simplified by removing the delayed inputs, to yield an autoregressive AR(p) model given by

$$y(k) = \sum_{i=1}^{p} a_i y(k-i) + n(k) \tag{3.3}$$

which also has infinite impulse response.

The most straightforward way to test stability is to exploit the $\mathcal{Z}$-domain representation of the transfer function of Equation (3.1), that is

$$H(z) = \frac{Y(z)}{E(z)} = \frac{b_0 + b_1 z^{-1} + \cdots + b_q z^{-q}}{1 - a_1 z^{-1} - \cdots - a_p z^{-p}} = \frac{N(z)}{D(z)} \tag{3.4}$$

To guarantee stability, the p roots of the denominator polynomial of $H(z)$, must lie within the unit circle in the z-plane, $|z| < 1$.

Nonlinear stochastic models are formally obtained by applying a nonlinear function to the corresponding linear stochastic models, that is

$$\begin{aligned}
\text{NMA(q)} \quad & y(k) = \Phi\big(b_1 n(k-1) + \cdots + b_q n(k-q)\big) + b_0 n(k) \\
\text{NAR(p)} \quad & y(k) = \Phi\big(a_1 y(k-1) + \cdots + a_p y(k-p)\big) + b_0 n(k) \\
\text{NARMA(p,q)} \quad & y(k) = \Phi\big(a_1 y(k-1) + \cdots + a_p y(k-p) + b_1 n(k-1) \\
& \qquad + \cdots + b_q n(k-q)\big) + b_0 n(k)
\end{aligned} \tag{3.5}$$

where the nonlinearity $\Phi(\cdot)$ typically has isolated singularities, as explained in Chapter 4.

3.2 Linear and Nonlinear Adaptive Filtering Architectures

For the modelling of statistically nonstationary signals, the coefficients of the ARMA and NARMA functional expressions can be made adaptive, thus representing feedforward and feedback adaptive filters. For convenience, we will consider these filters in the prediction setting.

The input–output relationship for a linear adaptive FIR filter is given by

$$y(k) = \sum_{m=1}^{M} w_m(k)x(k-m) = \mathbf{x}^T(k)\mathbf{w}(k) \tag{3.6}$$

where $\mathbf{w}(k) = [w_1(k), \ldots, w_M(k)]^T$ are filter coefficients and the tap input vector is given by $\mathbf{x}(k) = [x(k-1), \ldots, x(k-M)]^T$. The input–output relationship for an IIR adaptive filter, shown in Figure 3.1(a), is given by

$$y(k) = \sum_{n=1}^{N} a_n(k)y(k-n) + \sum_{m=1}^{M} b_m(k)x(k-m) = \mathbf{a}^{\mathrm{T}}(k)\mathbf{y}(k) + \mathbf{b}^{\mathrm{T}}(k)\mathbf{x}(k) \tag{3.7}$$

where $\mathbf{a}$ and $\mathbf{b}$ are coefficient vectors and $\mathbf{y}$ is a vector of delayed feedback. The standard FIR filter is obtained upon removing the feedback.

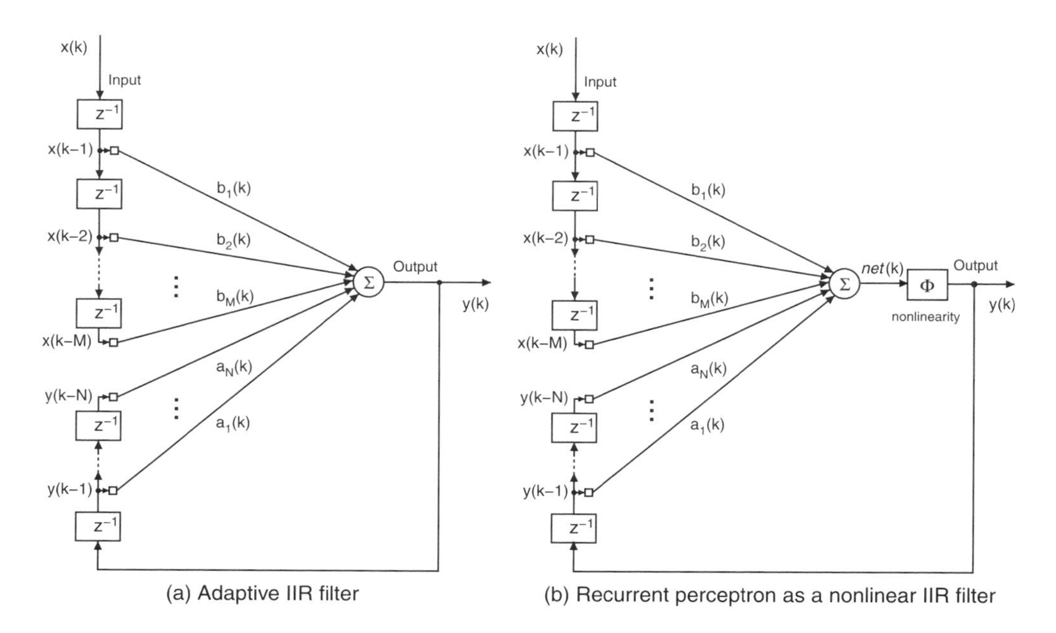

Figure 3.1 Linear and nonlinear adaptive infinite impulse response filters (in the prediction setting)

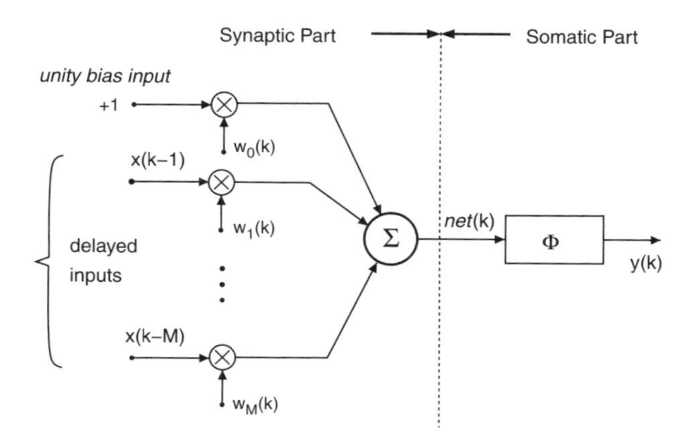

Figure 3.2 Model of an artificial neuron

The input–output expressions for a nonlinear FIR filter (dynamical perceptron) and a nonlinear IIR filter (recurrent perceptron) are respectively given by

$$y(k) = \Phi\left(\sum_{m=1}^{M} w_m(k)x(k-m)\right)$$

$$y(k) = \Phi\left(\sum_{n=1}^{N} a_n(k)y(k-n) + \sum_{m=1}^{M} b_m(k)x(k-m)\right) \tag{3.8}$$

Figure 3.1(b) shows a block diagram of such a NARMA(M,N) recurrent perceptron. Depending on the application, an external input set to unity, called the bias, may also be included. A nonlinear FIR filter is obtained upon removing the feedback.

The basic building block of larger nonlinear structures – neural networks – is an artificial neuron, shown in Figure 3.2, which is functionally similar to the nonlinear FIR adaptive filter described above. The net input $net(k)$ is also called the activation, and the nonlinearity Φ the nonlinear activation function. The bias input is set to unity and reflects a biological motivation for this architecture.

3.2.1 Feedforward Neural Networks

A multilayer feedforward neural network is shown in Figure 3.3. It consists of a number of interconnected neurons, for which the bias input can be removed from the neuron architecture (Figure 3.2) and included externally to the whole network. For successful operation, the inputs to the multilayer feedforward neural network must capture sufficient information about the time evolution of the underlying discrete time random signal. The simplest situation is to have time-delayed versions of the input signal, i.e. $x(k-i)$, $i = 1, 2, \ldots, M$, commonly termed a

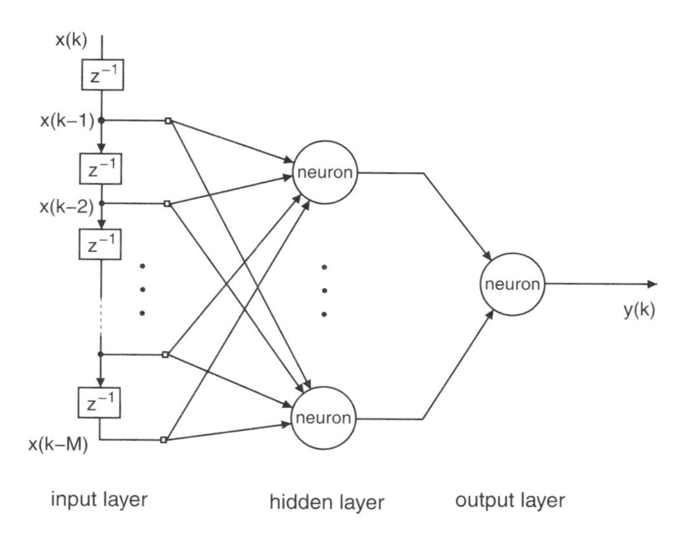

Figure 3.3 Multilayer feedforward neural network

tapped delay line,[2] which provides the network with short-term memory. For the prediction application only a single output is assumed.

For other applications, such as classification, where the input is not temporal (a tap delay line), we typically have more than one output neuron. Notice the outputs of each layer are connected only to the inputs of the adjacent layer. The nonlinearity inherent in the network is due to the overall action of all the activation functions of the neurons within the structure.

3.2.2 Recurrent Neural Networks

Two ways to include recurrent connections in neural networks are activation feedback and output feedback,[3] as shown respectively in Figures 3.4(a) and 3.4(b). The blocks labelled 'linear dynamical system' comprise delays and multipliers, hence providing linear combination of their input signals.

The output of the activation feedback scheme can be expressed as

$$\begin{aligned} net(k) &= \sum_{i=1}^{M} w_{x,i}(k)x(k-i) + \sum_{j=1}^{N} w_{\mathrm{n},j}(k)net(k-j) \\ y(k) &= \Phi\big(net(k)\big) \end{aligned} \tag{3.9}$$

where $w_{x,i}$ and $w_{\mathrm{n},j}$ correspond respectively to the weights associated with x and net.

[2]The optimal length of the tap delay line is closely related to delay space embedding [214].

[3]These schemes are closely related to the state space representations of neural networks. A comprehensive account of canonical forms and state space representation of general neural networks is given in [64].

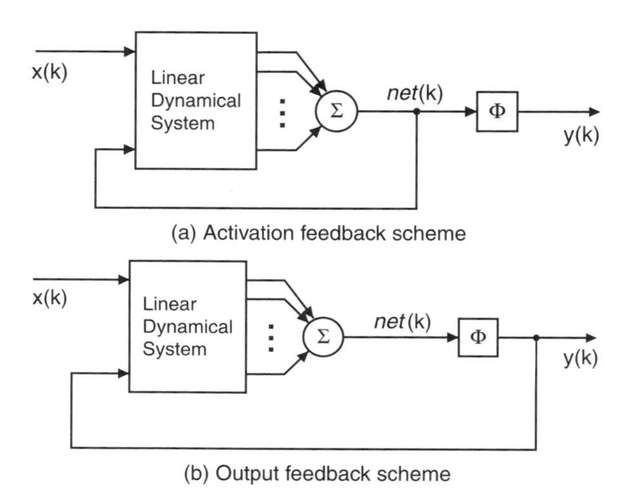

Figure 3.4 Recurrent neural network architectures

The input–output function of the output feedback scheme can be expressed as

$$net(k) = \sum_{i=1}^{M} w_{x,i}(k)x(k-i) + \sum_{j=1}^{N} w_{y,j}(k)y(k-j)$$
$$y(k) = \Phi\big(net(k)\big) \tag{3.10}$$

where $w_{y,j}$ correspond to the weights associated with the delayed outputs.

The provision of feedback introduces longer-term memory to recurrent neural networks – it is the feedback, together with the nonlinearity within the network, that makes RNNs so powerful for the modelling of nonlinear dynamical systems [214]. Figure 3.5 shows an RNN with one output neuron; in classification applications we may have as many output neurons as the number of classes.

An example of an RNN for which the feedback is also provided in hidden layers is the Elman RNN, whereas the Jordan RNN has both local and global feedback.

3.2.3 Neural Networks and Polynomial Filters

Nonlinear system identification has been traditionally based upon the Kolmogorov approximation theorem which states that a neural network with a hidden layer can approximate an arbitrary function (nonlinear system). The problem, however, is that inner functions in Kolmogorov's formula, although continuous, are non-smooth.

Another convenient form of nonlinear systems is the bilinear (truncated Volterra) system, described by

$$y(k) = \sum_{j=1}^{N} c_i y(k-j) + \sum_{i=0}^{N-1}\sum_{j=1}^{N} b_{i,j} y(k-j)x(k-i) + \sum_{i=0}^{N-1} a_i x(k-i) \tag{3.11}$$

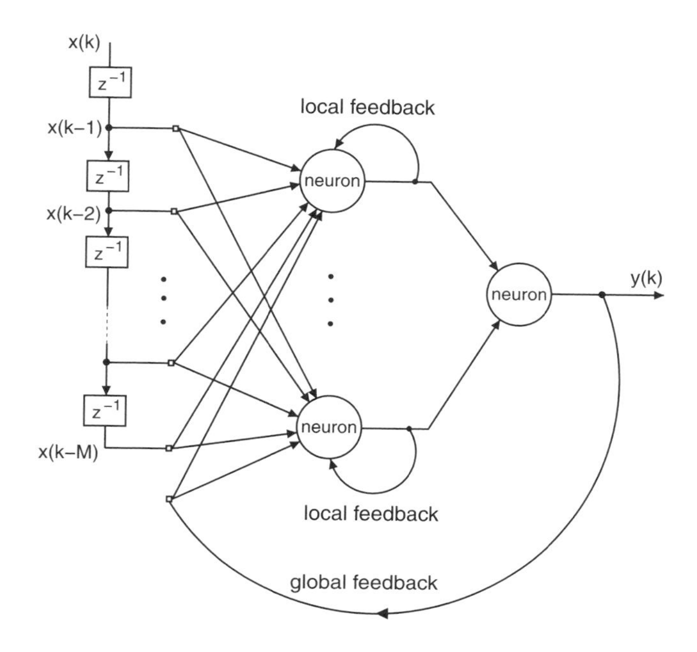

Figure 3.5 A recurrent neural network with local and global feedback

This is a powerful nonlinear model which can approximate arbitrarily well a large class of nonlinear systems. An example of a recurrent neural network that realises the bilinear model given by

$$y(k) = c_1 y(k-1) + b_{0,1}x(k)y(k-1) + b_{1,1}x(k-1)y(k-1) + a_0 x(k) + a_1 x(k-1) \tag{3.12}$$

is depicted in Figure 3.6. As seen from Figure 3.6, multiplicative synapses have to be introduced in order to represent the bilinear model.

3.3 State Space Representation and Canonical Forms

It is often convenient to consider the analysis of adaptive filters in terms of state space, as it is always possible to rewrite a nonlinear input–output model in a state space form, whereas for any given state space model an I–O model may not exist. This is of fundamental importance when only a limited number of data samples is available. Figure 3.7 shows a canonical form of an RNN, where state vector is denoted by $\mathbf{s}(k) = [s_1(k), s_2(k), \ldots, s_N(k)]^{\mathrm{T}}$, and a vector of M external inputs is given by $\mathbf{x}(k) = [x(k-1), x(k-2), \ldots, x(k-M)]^{\mathrm{T}}$.

For a recurrent neural network, the state evolution and network output equations are given by

$$\begin{aligned} \mathbf{s}(k+1) &= \boldsymbol{\Theta}\big(\mathbf{s}(k), \mathbf{x}(k)\big) \\ \hat{y}(k) &= \boldsymbol{\Psi}\big(\mathbf{s}(k), \mathbf{x}(k)\big) \end{aligned} \tag{3.13}$$

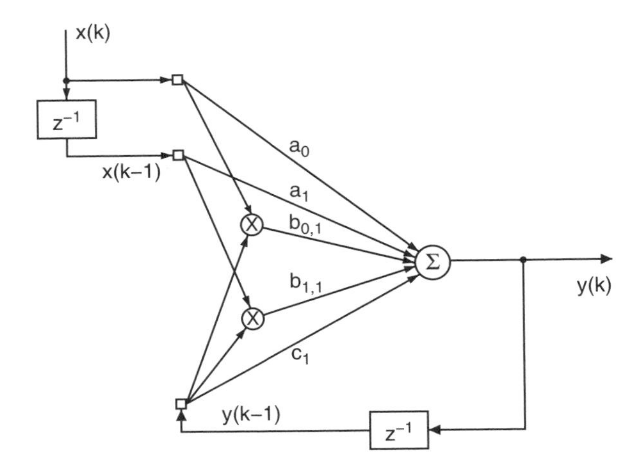

Figure 3.6 Recurrent neural network representation of the bilinear model

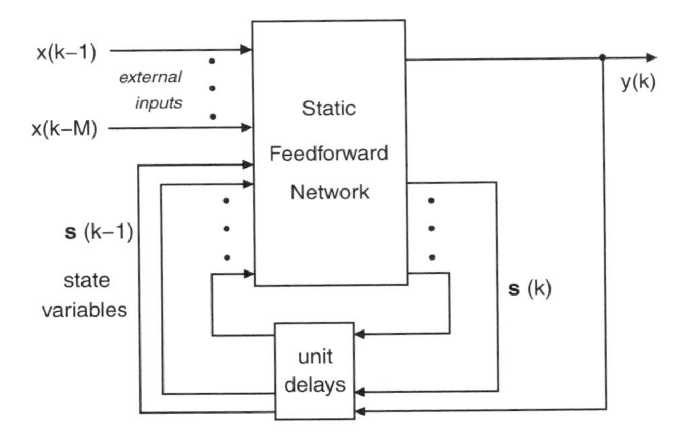

Figure 3.7 Canonical form of a recurrent neural network

where $\mathbf{\Theta}$ and $\mathbf{\Psi}$ are vector valued nonlinear mappings. A particular choice of N minimal state variables is not unique, therefore several canonical forms may exist [235].

To summarise:

- Linear and nonlinear adaptive filtering architectures have been introduced as realisations of standard stochastic ARMA and NARMA models;
- A model of an artificial neuron has been introduced and it has been shown how multiple neurons can be connected to form a neural network;

- Recurrent neural networks (RNN), that is, networks with feedback, have been demonstrated to be able to model complex nonlinear dynamics and several of their architectures have been introduced;
- Finally, it has been shown that the analysis of adaptive filters can be cast into a state space form.

4

Complex Nonlinear Activation Functions

Central to the development of complex nonlinear adaptive filters is their ability to perform universal function approximation. To select a nonlinear function in $\mathbb{C}$ to approximate another function to a desired degree of accuracy, and at the same time to make it suitable for nonlinear adaptive filtering applications, it would appear advantageous to consider continuous, bounded, and analytic functions. However, by Liouville's theorem (see Appendix B), the only function which is analytic everywhere in $\mathbb{C}$ is a constant. Hence it is not possible to use directly an analytic continuation of real valued functions for this purpose, and we need to restrict ourselves to a subset of the complex plane, where we can choose between boundedness and differentiability. This chapter deals with these issues and provides a comprehensive account of nonlinear activation functions used in complex valued nonlinear adaptive filters and neural networks.

4.1 Properties of Complex Functions

We are particularly interested in complex functions that have a derivative in a neighbourhood of a point $z_0 \in \mathbb{C}$, as this would ensure smoothness of the approximation. A complex derivative[1] is defined in Equation (5.3) in Chapter 5. The notion of an analytic function requires some attention, as shown below [206].

- **Analyticity at a point.** The function f is said to be analytic at $z_0 \in \mathbb{C}$ if it has a derivative at each point in some neighbourhood of z_0;
- **Analyticity on a set.** If f is analytic at each point in a set $S \subset \mathbb{C}$, then we say that f is analytic on S;

[1] Although differentiability implies continuity, the reverse is not true.

Complex Valued Nonlinear Adaptive Filters: Noncircularity, Widely Linear and Neural Models
Danilo P. Mandic and Vanessa Su Lee Goh

- **Analyticity on $\mathbb{C}$.** If f is analytic in the finite complex plane (everywhere except $z = \infty$), then f is said to be *entire*. An entire function can be represented by a Taylor series which has an infinite radius of convergence – some examples of entire functions are e^z and trigonometric functions.

Analytic functions are also called *regular functions* or *holomorphic functions*. It is shown in Chapter 5 that for a function $f(z) = u(x, y) + \jmath v(x, y)$ to be analytic on a region $S \subset \mathbb{C}$, the Cauchy–Riemann equations need to be satisfied, that is

$$\frac{\partial u}{\partial x} = \frac{\partial v}{\partial y}, \qquad \frac{\partial u}{\partial y} = -\frac{\partial v}{\partial x} \tag{4.1}$$

The analysis of analytic functions is very convenient through their Taylor and Laurent series representation.

Taylor series representation. For a function $f(z)$ which in analytic at $z = \alpha$, the series

$$f(z) = f(\alpha) + f'(\alpha)(z-\alpha) + \frac{f''(\alpha)}{2!}(z-\alpha)^2 + \cdots = \sum_{n=1}^{\infty} \frac{f^{(n)}(\alpha)}{n!}(z-\alpha)^{n-1} \tag{4.2}$$

is called the Taylor series for function f at point α. When $\alpha = 0$, the series (4.2) is called the Maclaurin series for function f [206]. Taylor series representations for several important functions expanded around $\alpha = 0$ are given below.

$$\begin{aligned}
e^z &= 1 + z + \frac{z^2}{2!} + \frac{z^3}{3!} + \cdots + \frac{z^n}{n!} + \cdots \\
\sin z &= z - \frac{z^3}{3!} + \frac{z^5}{5!} - \cdots + (-1)^{n-1}\frac{z^{2n-1}}{(2n-1)!} + \cdots \\
\cos z &= 1 - \frac{z^2}{2!} + \frac{z^4}{4!} + \cdots - (-1)^{n-1}\frac{z^{2n-2}}{(2n-2)!} + \cdots \\
ln(1+z) &= z - \frac{z^2}{2} + \frac{z^3}{3} - \cdots + (-1)^{n-1}\frac{z^n}{n!} + \cdots
\end{aligned} \tag{4.3}$$

Expansions for e^z, $\sin z$ and $\cos z$ are defined on $|z| < \infty$, whereas the expansion for $ln(1+z)$ is defined on $|z| < 1$.

Laurent series representation. Functions which are analytic in an open annulus[2] $A(R_1, R_2, \alpha) = \{z : R_1 < |z - \alpha| < R_2\}$, illustrated in Figure 4.1, have the Laurent series representation

$$f(z) = \sum_{n=-\infty}^{\infty} c_n (z-\alpha)^n = \sum_{n=-\infty}^{-1} c_n (z-\alpha)^n + \sum_{n=0}^{\infty} c_n (z-\alpha)^n \tag{4.4}$$

[2] For instance, function $f(z) = \frac{e^z}{z}$ is not analytic for $z = 0$, but it is analytic for $|z| > 0$, and its Taylor series representation is $f(z) = 1/z + 1 + z/2! + z^2/3! + \cdots$.

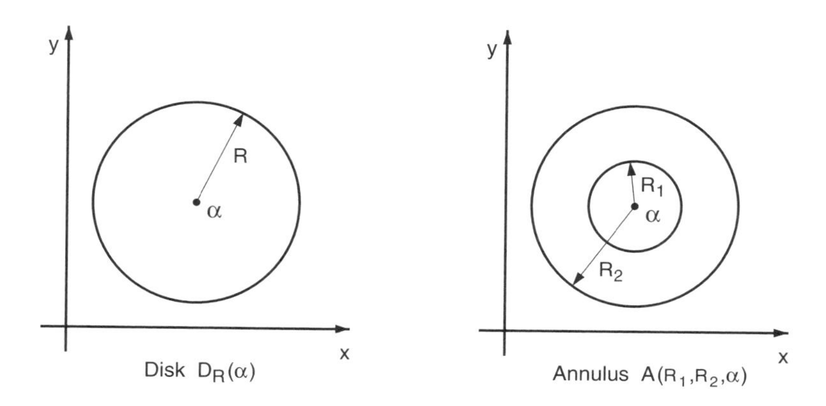

Figure 4.1 Left: disk $D_R(\alpha)$ with center α and radius R. Right: annulus $A(R_1, R_2, \alpha)$ with center α and radii R_1 and R_2

where the coefficients of the *principal part* c_{-n} and the coefficients of the *analytic part* c_n are given by

$$c_{-n} = \frac{1}{2\pi \jmath} \oint_{C_1} \frac{f(z)}{(z-\alpha)^{-n+1}} \mathrm{d}z \qquad c_n = \frac{1}{2\pi \jmath} \oint_{C_2} \frac{f(z)}{(z-\alpha)^{n+1}} \mathrm{d}z \tag{4.5}$$

and C_1 and C_2 are respectively concentric circles (contours) with center at α, radii R_1 and R_2, and a counterclockwise orientation.

4.1.1 Singularities of Complex Functions

Points at which a complex function is not analytic are called *singular points*, for instance, the function $f(z) = 2/(1-z)$ is not analytic at $z = 1$. There are several types of singularities of complex function, which can be roughly grouped into:

- **Isolated singularities.** Point $\alpha \in \mathbb{C}$ is called an isolated singularity if the function $f : \mathbb{C} \rightarrow \mathbb{C}$ is not analytic at α, but there exists a real number $R > 0$ such that f is analytic anywhere in the punctured disk[3] $D_R(\alpha) \setminus \{\alpha\} = D_R^*(\alpha)$.
- **Branch points.** A branch of a multiple valued function $f(z)$ is any single valued function that is continuous and analytic in some domain. For instance, branches of the function $f(z) = \sqrt{z} = \sqrt{re^{\jmath\theta}}$, in a region defined by $r > 0, \ -\pi < \theta \leq \pi$, are $f_1(z) = \sqrt{r}e^{\jmath\theta/2}$ and $f_2(z) = -\sqrt{r}e^{\jmath\theta/2}$; point $z = 0$ is common to both the branch cuts and is called a *branch point*.
- **Singularities at infinity.** If the nature of singularity of $f(z)$ at $z_0 = \infty$ is the same as that for $f(w) = f(1/z)$ at $w_0 = 1/z_0$, this type of singularity is called *singularity at infinity*.

[3]Functions with isolated singularities have a Laurent series representation, since the punctured disk $D_R^*(\alpha)$ is equivalent to the annulus $A(0, R, \alpha)$ (see Figure 4.1).

Table 4.1 Isolated singularities

Removable singularity. If $f(z)$ is not defined at $z = \alpha$, but $\lim_{z\to\alpha} f(z)$ exists, then $z = \alpha$ is called a removable singularity and $f(\alpha)$ is defined as $f(\alpha) = \lim_{z\to\alpha} f(z)$. For example, consider the function $f(z) = \sin z/z$ and its Taylor series expansion

$$f(z) = \frac{\sin z}{z} = \frac{1}{z}\left(z - \frac{z^3}{3!} + \frac{z^5}{5!} - \frac{z^7}{7!} + \cdots\right) = 1 - \frac{z^2}{3!} + \frac{z^4}{5!} - \frac{z^6}{7!} + \cdots$$

This function has a singularity at $z = 0$, which can be 'removed' by defining $f(0) = 1$, as shown in its Taylor series expansion. For removable singularities, the coefficients of the principal part of the Laurent series $c_n = 0, \ n = -1, -2, \ldots$ all vanish

Pole. If the principal part of the Laurent series expansion of function $f(z)$ has a finite number n of nonzero terms $c_{-1}, \ldots, c_{-n}$, that is

$$f(z) = \sum_{k=-n}^{\infty} c_k (z - \alpha)^k$$

then we say that $f(z)$ has a pole of order n at $z = \alpha$. For example, function

$$f(z) = \frac{sinz}{z^5} = \frac{1}{z^4} - \frac{1}{3!\,z^2} + \frac{1}{5!} - \frac{z^2}{7!} + \cdots$$

has a pole of order $n = 4$ at $z = 0$

Essential singularity. Any singularity which is not a removable singularity or a pole is an essential singularity. In this case the principal part of the Laurent expansion has infinitely many terms. For example, function

$$f(z) = e^{\frac{1}{z}} = 1 + \frac{1}{z} + \frac{1}{2!\,z^2} + \frac{1}{3!\,z^3} + \frac{1}{4!\,z^4} + \cdots = \sum_{n=0}^{\infty} \frac{1}{n!\,z^n}$$

has an essential singularity at $z = 0$

Classification of isolated singularities. Of particular interest in nonlinear adaptive filtering are isolated singularities, which can be classified into poles, removable singularities, and essential singularities. As the point $\alpha \in \mathbb{C}$ is an isolated singular point if the disk $D_R(\alpha)$ encloses no other singular point, we can distinguish between the three classes of isolated singularities based on the properties of their Laurent series representation on an annulus $A(0, R, \alpha)$, as shown in Table 4.1.

4.2 Universal Function Approximation

Learning an input–output relationship from examples using a neural network can be treated as the problem of approximating an unknown function from a set of data points [241]. The universal approximation ability of temporal neural networks makes them suitable for nonlinear adaptive filtering applications [190].

4.2.1 *Universal Approximation in* $\mathbb{R}$

The problem of approximation of a function of n variables by a linear combination of functions of $k < n$ variables is known as the 13th problem of Hilbert [172]. The existence of the solution is provided by Kolmogorov's theorem, which in the neural network community was first recognised by Hecht-Nielsen [116] and Lippmann [168]. The first constructive proof of universal function approximation ability of neural networks was given by Cybenko [56]. Based on the analysis of the denseness of nonlinear functions in the space where the original function resides, this result shows that if σ is a continuous discriminatory function,[4] then finite sums of the form ($w_n, b_n, \quad n = 1, \ldots, N$ are coefficients)

$$g(\mathbf{x}) = \sum_{n=1}^{N} w_n \sigma(\mathbf{a}_n^{\mathrm{T}}\mathbf{x} + b_n) \tag{4.6}$$

are dense[5] in the space of continuous functions defined on $[0, 1]^N$. Since every bounded and measurable sigmoidal function is discriminatory, a multilayer neural network can be used to learn an arbitrary function [56].

Funahashi [79] showed that an arbitrary continuous function can be approximated to any desired accuracy by a three-layer multilayer perceptron (MLP) with bounded and monotonically increasing activation functions within hidden units. Kuan and Hornik [159] extended the class of nonlinear activation functions allowed, and proved that a neural network can approximate simultaneously both a function and its derivative. Kurkova [161] addressed universal function approximation by a superposition of nonlinear functions within the constraints of neural networks, whereas Leshno *et al.* [165] relaxed the conditions for the activation function to a 'locally bounded piecewise continuous' (not a polynomial).

Based on the above results, for a neural network with N neurons, its activation functions $\sigma_n(x_n)$, $n = 1, \ldots, N$ should be sigmoid functions for which the desired properties are given in Table 4.2. For more detail on the properties of nonlinear activation functions in $\mathbb{R}$ see Appendix D.

---

[4]Function $\sigma(\cdot)$ is discriminatory if for a Borel measure μ on $[0, 1]^N$, $\int_{[0,1]^N} \sigma(\mathbf{a}^{\mathrm{T}}\mathbf{x} + b) d\mu(\mathbf{x}) = 0, \quad \forall \mathbf{a} \in \mathbb{R}^N$, $\forall b \in \mathbb{R}$ implies that $\mu = 0$. The functions Cybenko considered had limits

$$\sigma(t) = \begin{cases} 1, & t \rightarrow \infty \\ 0, & t \rightarrow -\infty \end{cases}$$

This justifies the use of the sigmoid (S - shaped) functions, such as $\sigma(x) = 1/1 + e^{-\beta x}$, in neural networks for universal function approximation.

[5]The denseness ensures that for any continuous function f defined on $[0, 1]^N$ and any $\varepsilon > 0$, there is a $g(\mathbf{x})$ defined in Equation (4.6), for which $|g(\mathbf{x}) - f(\mathbf{x})| < \varepsilon$ for all $\mathbf{x} \in [0, 1]^N$.

Table 4.2 Properties of real valued nonlinear activation functions

a) $\sigma_n(x_n)$ is a continuously differentiable function, to ensure the existence of the gradient
b) $\frac{d\sigma_n(x_n)}{dx_n} > 0$ for all $x_n \in \mathbb{R}$, to enable gradient descent based learning
c) $\sigma_n(\mathbb{R}) = (a_n, b_n),\ a_n, b_n \in \mathbb{R}, a_n \neq b_n$, to ensure that the function is bounded
d) $\sigma_n'(x_n) \to 0$ as $x_n \to \pm\infty$, to reduce the effect of artifacts in adaptive learning
e) A sigmoidal function σ_n should have only one inflection point, preferably at $x_n = 0$, ensuring that $\sigma_n'(x_n)$ has a global maximum, thus providing the existence of the solution for NN based learning [107]
f) σ_n should be monotonically nondecreasing and uniformly Lipschitz, that is, there exists a constant $L > 0$ such that $\| \sigma_n(x_1) - \sigma_n(x_2) \| \leq L \| x_1 - x_2 \|, \forall x_1, x_2 \in \mathbb{R}$, or in other words

$$\frac{\sigma_n(x_1) - \sigma_n(x_2)}{x_1 - x_2} \leq L, \quad \forall x_1, x_2 \in \mathbb{R}, \ x_1 \neq x_2$$

This property facilitates the use of NNs as iterative maps (see Appendix P).

4.3 Nonlinear Activation Functions for Complex Neural Networks

To illustrate the link between universal function approximation in $\mathbb{R}$ and $\mathbb{C}$, consider again the approximation

$$f(x) = \sum_{n=1}^{N} c_n \sigma(x - a_n) \tag{4.7}$$

Obviously, different choices of function σ will give different approximations; an extensive analysis of this problem is given in [312]. Upon swapping the variables $x \to 1/x$, the approximation formula (4.7) becomes (see Appendix E)

$$f(x) = z \sum_{n=1}^{N} \frac{c_n}{z + \alpha_n} \tag{4.8}$$

which is a partial fractional expansion with poles α_n and residuals c_n. Notice, however, that both α_n and c_n are allowed to be complex.

As the function approximation formula (4.8) has singularities (also supported by Liouville's theorem), complex valued nonlinear activation functions of a neuron *cannot* be obtained directly by analytic continuation of real valued sigmoids. To this end, a density[6] theorem [18] provides theoretical justification for the use of complex MLPs with non–analytic activation functions for universal function approximation.

For a complex nonlinear function to be suitable for an activation function of a complex neuron, it should be analytic in the finite complex plane, except for a limited number of singularities, for instance, a complex logistic function $\Phi(z) = 1/1 + e^{-z} = u + \jmath v$ has singularities at $0 \pm \jmath(2n+1)\pi,\ n \in \mathbb{Z}$ (see also Figure E.1 in Appendix E).

[6]The denseness conditions can be considered a special case of the Stone–Weierstrass theorem [55].

Table 4.3 Desired properties of complex valued nonlinear activation functions

i)	Φ is nonlinear in x and y
ii)	$\Phi(z)$ is bounded $\Rightarrow u$ and v are bounded
iii)	Partial derivatives u_x, u_y, v_x, v_y exist and are bounded (see Chapter 5)
iv)	To guarantee continuous learning, $u_x v_y \neq v_x u_y$, except when $u_x = 0, v_x = 0, u_y \neq 0, v_y \neq 0$ and $u_y = 0, v_y = 0, u_x \neq 0, v_x \neq 0$

Following on the similar analysis for real activation functions, it has been shown that for a function

$$\Phi(z) = u(x, y) + \jmath v(x, y)$$

to be a candidate for a complex valued nonlinear activation function [88, 152], it should satisfy the conditions given in Table 4.3.

As no function satisfies all the conditions in Table 4.3 on $\mathbb{C}$, in practical applications we need to choose between boundedness and differentiability [152].

4.3.1 Split-complex Approach

A simple way to ensure that the output of a complex neuron

$$y(k) = \Phi\big(net(k)\big) = \Phi\big(net_r(k) + \jmath net_i(k)\big) = \sigma_r\big(net_r(k)\big) + \jmath\sigma_i\big(net_i(k)\big) \qquad (4.9)$$

is bounded is to apply a real valued nonlinear activation function separately to the real and imaginary part of the complex net input *net*(k). Depending on the basic form of a complex number, the two possibilities for split complex nonlinearity are:

- Real–imaginary split-complex nonlinearities (RISC). To deal with signals which exhibit symmetry around the real and imaginary axes, it is convenient to define split-complex nonlinearities in terms of their real and imaginary parts [118, 119];
- Amplitude–phase split-complex nonlinearities (APSC). A complex nonlinearity expressed by its magnitude and phase is symmetric about the origin, this makes it suitable for processes which exhibit rotational symmetry [120, 121].

Split-complex nonlinear functions, however, are not analytic – they do not satisfy the Cauchy–Riemann equations and hence their use in nonlinear adaptive filtering is application specific.

Real–imaginary split-complex approach. In this approach, a split-complex nonlinearity comprises two identical real valued nonlinear activation functions [27, 166], that is

$$\Phi(z) = \sigma_r(u(x, y)) + \jmath\sigma_i(v(x, y)) = \sigma(u) + \jmath\sigma(v), \qquad \sigma(u), \sigma(v) : \mathbb{R} \rightarrow \mathbb{R} \qquad (4.10)$$

whereby the real and imaginary parts of the net input are processed separately. Function σ can be any suitable real valued nonlinear activation function, as outlined in Section 4.2.1 and

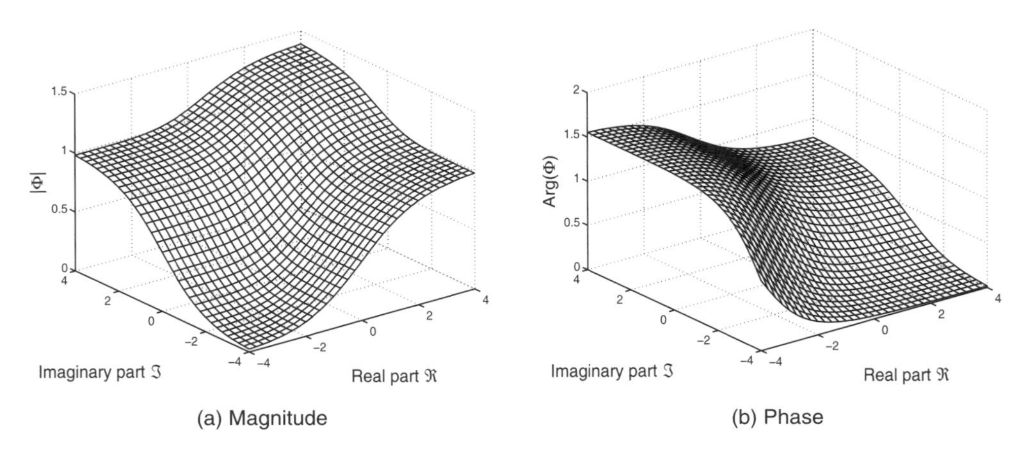

Figure 4.2 Real–imaginary split-complex logistic function $\Phi(z) = \sigma(\Re) + \jmath\sigma(\Im) = \frac{1}{1+e^{-\beta\Re}} + \jmath\frac{1}{1+e^{-\beta\Im}}$

Appendix D. Figure 4.2 shows the magnitude and phase functions for the split-complex logistic function – notice the symmetry of the magnitude function in Figure 4.2(a).

Amplitude–phase split-complex approach. Examples for this class of functions are the activation function introduced by Georgiou and Koutsougeras [88], given by

$$\Phi(z) = \frac{z}{c + |z|/r} \tag{4.11}$$

and the activation function proposed by Hirose [118], given by

$$\Phi(z) = \tanh(|z|/m)e^{\jmath \arg(z)} \tag{4.12}$$

where c, r, and m are real positive constants. Figure 4.3(a) illustrates the rotational symmetry of the magnitude function of Equation (4.11), which maps the complex plane onto an open

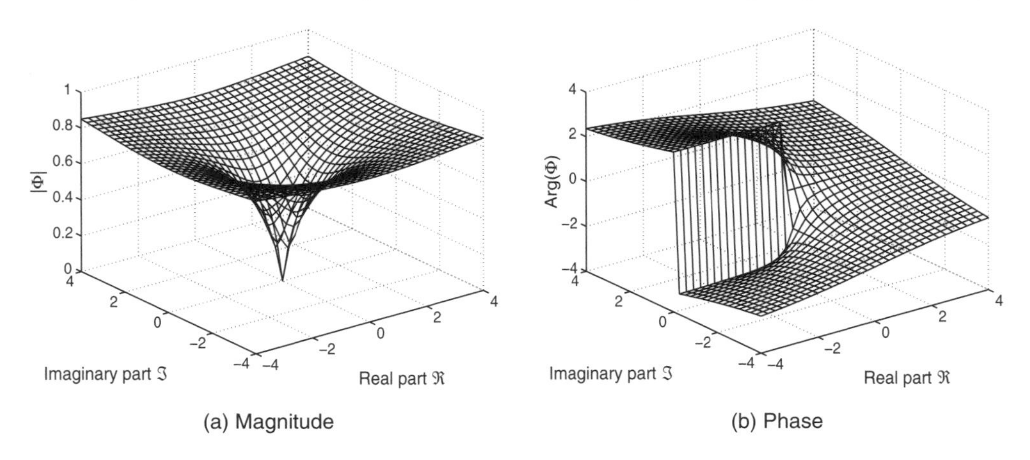

Figure 4.3 Amplitude–phase split-complex function (Equation 4.11)

disk $D_R(0) = \{z : |z| < R\}$. Due to their radial mapping, in the APSC approach the processing is limited to magnitude variations. Thus, these functions are not capable of learning the phase relationship between the input and desired signal, and are not suitable for restoring phase distortion, for example.

Dual–univariate approach. The dual–univariate approach splits the complex input into its real and imaginary part and processes them separately using real valued adaptive filters, both linear and nonlinear [156, 213] (see also Figure 14.2a). Since during the processing there is no mixing between the real and imaginary parts of the input, this approach achieves reasonable performance only when the real and imaginary parts of the complex input signal are independent.

4.3.2 Fully Complex Nonlinear Activation Functions

This is the most general case, where the nonlinear activation function Φ is a complex function of a complex variable, whose value is calculated using complex algebra. To illustrate fully complex nonlinearities, consider a net input to a neuron $net(k) = net_r(k) + \jmath net_i(k)$, given by

$$\begin{aligned} net(k) &= \sum_{n=1}^{N} w_n(k)x(k-n+1) = \sum_{n=1}^{N} \big(w_n^r(k) + \jmath w_n^i(k)\big)\big(x_r(k-n+1) + \jmath x_i(k-n+1)\big) \\ &= \big(\underbrace{\mathbf{x}_r^{\mathrm{T}}(k)\mathbf{w}_r(k) - \mathbf{x}_i^{\mathrm{T}}(k)\mathbf{w}_i(k)}_{net_r(k)}\big) + \jmath\big(\underbrace{\mathbf{x}_r^{\mathrm{T}}(k)\mathbf{w}_i(k) + \mathbf{x}_i^{\mathrm{T}}(k)\mathbf{w}_r(k)}_{net_i(k)}\big) \end{aligned} \tag{4.13}$$

The net input is passed through a general fully complex nonlinear activation function Φ to produce the output

$$y(k) = \Phi(net(k)) = \phi(net_r(k), net_i(k)) + \jmath\psi\big(net_r(k), net_i(k)\big) \tag{4.14}$$

where $\phi(net_r(k), net_i(k))$ and $\psi(net_r(k), net_r(k))$ are two–dimensional.

To enable gradient based learning, function $\Phi(v(k))$ must be holomorphic, that is, it must satisfy the Cauchy–Riemann equations [203]

$$\frac{\partial\phi\big(net_r(k), net_i(k)\big)}{\partial net_r(k)} = \frac{\partial\psi\big(net_r(k), net_i(k)\big)}{\partial net_i(k)} \quad \text{and}$$
$$\frac{\partial\phi\big(net_r(k), net_i(k)\big)}{\partial net_i(k)} = -\frac{\partial\psi\big(net_r(k), net_i(k)\big)}{\partial net_r(k)}$$

Although, at first, this requirement seems rather stringent, the choice of fully complex activation functions is very rich, as for them the condition (iv) from Table 4.3 is irrelevant. For instance,

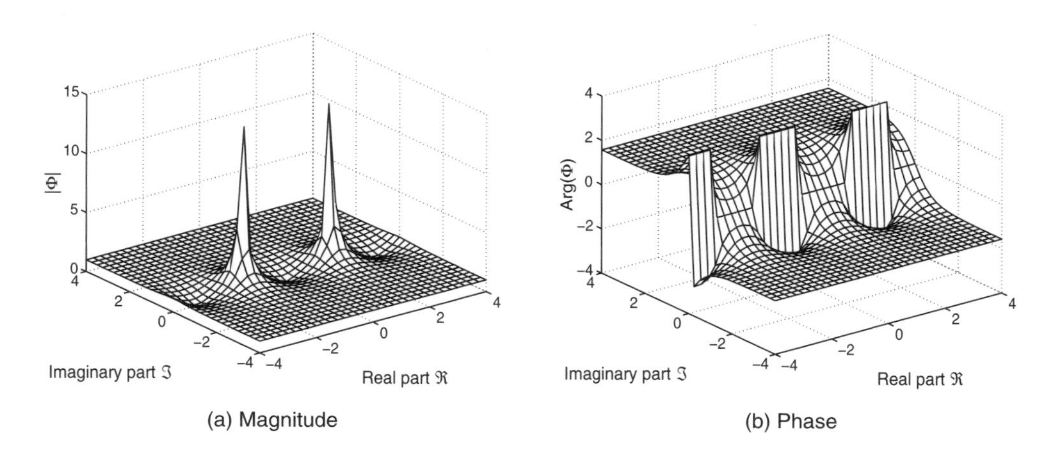

Figure 4.4 Magnitude and phase for the tan(z) activation function

as proposed in [152], we can employ the class of Elementary Transcendental Functions (ETF), which includes:

- Circular functions (Figure 4.4 illustrates function tan(z) and sin(z) is shown in Figure E.2 in Appendix E)

$$\Phi(z) = \tan(z), \quad \Phi'(z) = \sec^2(z) \tag{4.15}$$

$$\Phi(z) = \sin(z), \quad \Phi'(z) = \cos(z) \tag{4.16}$$

- Inverse circular functions (function arcsin(z) is shown in Figure E.3 in Appendix E)

$$\Phi(z) = \arctan(z), \quad \Phi'(z) = (1 + z^2)^{-1} \tag{4.17}$$

$$\Phi(z) = \arcsin(z), \quad \Phi'(z) = (1 - z^2)^{-1/2} \tag{4.18}$$

- Hyperbolic functions (function sinh(z) is shown in Figure E.4 in Appendix E)

$$\Phi(z) = \tanh(z), \quad \Phi'(z) = \mathrm{sech}^2(z) \tag{4.19}$$

$$\Phi(z) = \sinh(z), \quad \Phi'(z) = \cosh(z) \tag{4.20}$$

- Inverse hyperbolic functions (function *arctanh*(z) is shown in Figure 4.5)

$$\Phi(z) = \mathrm{arctanh}(z), \quad \Phi'(z) = (1 - z^2)^{-1} \tag{4.21}$$

$$\Phi(z) = \mathrm{arcsinh}(z), \quad \Phi'(z) = (1 + z^2)^{-1/2} \tag{4.22}$$

Table E.1 in Appendix E comprises elementary transcendental functions and their associated type of singularity.

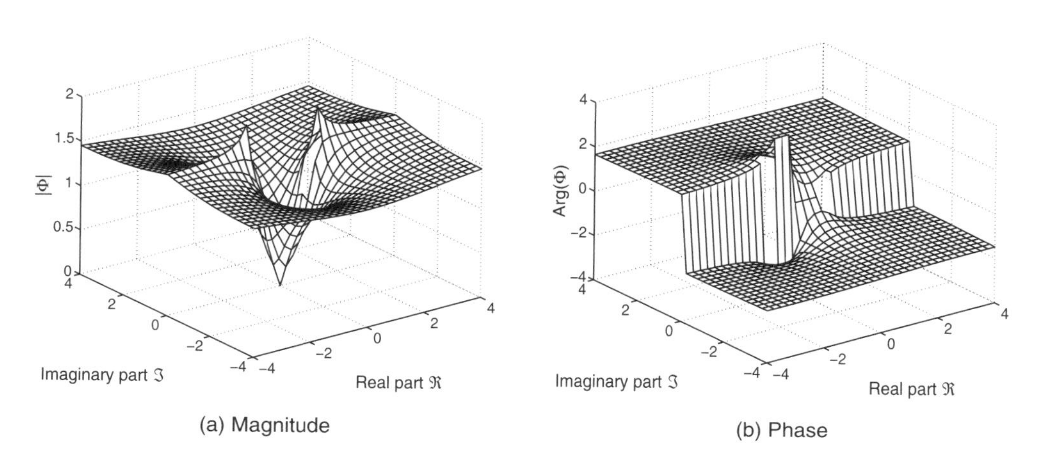

Figure 4.5 The inverse hyperbolic function $arctanh(z)$

4.4 Generalised Splitting Activation Functions (GSAF)

This class of activation functions, proposed in [295], are a hybrid between the split-complex and fully complex approach, in the sense that they have a form of a split-complex activation function, and yet satisfy the Cauchy–Riemann conditions. Figure 4.6 illustrates the generalised splitting approach–the flow of information between the real and imaginary channel is achieved using bidimensional splines to form functions $u(x, y)$ and $v(x, y)$ (for instance the Catmull–Rom cubic spline [45]). The use of splines ensures that the Cauchy–Riemann conditions are satisfied, whereas the boundedness is preserved by the split-complex nature of this approach.

4.4.1 The Clifford Neuron

The Clifford neuron is a special case of a standard neuron, whereby all the operations are performed based on Clifford algebra. The standard scalar product in the calculation of the net function $net(k) = \mathbf{w}^{\mathrm{T}}(k)\mathbf{x}(k)$ is replaced by the Clifford (or geometric) product, given by

$$\mathbf{w}^{\mathrm{T}}\mathbf{x} + \theta \quad \Rightarrow \quad \mathbf{w} \cdot \mathbf{x} + \mathbf{w} \wedge \mathbf{x} + \theta \tag{4.23}$$

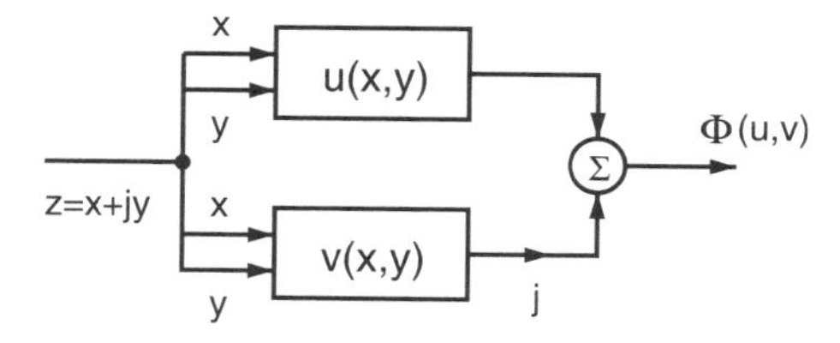

Figure 4.6 Generalised splitting activation function: $u(x, y)$ and $v(x, y)$ are built using adaptive bidimendionnal splines [295]

where the symbol $\wedge$ denotes the outer vector product.Clifford algebras form a basis for hypercomplex neural networks [230], for more detail see Appendix C.

4.5 Summary: Choice of the Complex Activation Function

Early extensions of temporal neural networks to the complex domain used analytic continuation of standard activation functions in $\mathbb{R}$, such as the complex logistic function [166] and complex tanh function [50], whereas neural networks for classification were based on split-complex activation functions [120] and multivalued neurons [5].

For universal function approximation and nonlinear adaptive filtering in $\mathbb{C}$, it has been shown that an ideal nonlinear activation function of a neuron should be both bounded and differentiable. This is, however, not possible as Liouville's theorem [255] states:

> *The only bounded differentiable (analytic) functions defined over the entire field of complex numbers are constants.*

Ways to partially circumvent these contrasting requirements include:

- Using complex nonlinear functions which are bounded almost everywhere in $\mathbb{C}$, allowing for isolated singularities (the fully complex approach).
- Admitting only functions which are bounded in the entire complex plane (the split-complex approach). These functions do not obey the Cauchy–Riemann conditions and both the real valued functions σ_r and σ_i in Equation (4.9) can be chosen individually.
- Separating the signal flow for the real and the imaginary parts of the input, and processing them either as independent data streams (dual univariate approach), or combining them using differentiable bivariate real functions, thus providing differentiability (generalised splitting activation functions).

In nonlinear adaptive filtering applications, differentiability is a prerequisite to designing learning algorithms. If all the variables within such filters are standardised so as not to approach singular points, the class of fully complex functions are the preferred choice as the Cauchy–Riemann equations facilitate gradient based learning. In addition, the algorithms derived based on the fully complex approach are generic extensions of the corresponding algorithms for real valued adaptive filters.

Chapter 11 presents a Möbius transformation framework for the analysis of mappings (nesting, invertibility) performed by fully complex nonlinearities, whereas Chapter 6 presents gradient based learning algorithms for a class of finite impulse response nonlinear adaptive filters with fully complex nonlinear activation functions (dynamical perceptron).

5

Elements of $\mathbb{CR}$ Calculus

The design of adaptive learning algorithms is based on the minimisation of a suitable objective (cost) function, typically a function of the output error of an adaptive filter. This optimisation problem is well understood for real valued adaptive filters where, for instance, the steepest descent approach is based on the iteration

$$\mathbf{w}(k+1) = \mathbf{w}(k) - \mu \nabla_{\mathbf{w}} J(k)$$

where $\mathbf{w}(k)$ is the vector of filter coefficients, μ is a parameter (learning rate) and $J = \frac{1}{2}e^2(k)$ is the cost function, a quadratic function of the output error of the filter.

Although a formalism similar to that used for real valued adaptive filters can also be used for complex valued adaptive filters, notice that in this case the cost function $J(k) = \frac{1}{2}|e(k)|^2 = \frac{1}{2}e(k)e^*(k) = \frac{1}{2}(e_r^2 + e_i^2)$ is a real valued function of complex variable, which gives rise to several important issues:

- Standard complex differentiability is based on the Cauchy–Riemann equations and imposes a stringent structure on complex holomorphic functions;
- Cost functions are real functions of complex variable, that is $J : \mathbb{C} \mapsto \mathbb{R}$, and so they are not differentiable in the complex sense, the Cauchy–Riemann equations do not apply, and we need to develop alternative, more general and relaxed, ways of calculating their gradients;
- It is also desired that these generalised gradients are equivalent to standard complex gradients when applied to holomorphic (analytic) functions.

This chapter provides an overview of complex differentiability for both real valued and complex valued functions of complex variables. The concepts of complex continuity, differentiability and Cauchy–Riemann conditions are first introduced; this is followed by more general $\mathbb{R}$-derivatives. The duality between gradient calculation in $\mathbb{R}$ and $\mathbb{C}$ is then addressed and the so called $\mathbb{CR}$ calculus is introduced for general functions of complex variables. It is shown that the $\mathbb{CR}$ calculus provides a unified framework for computing the Jacobians, Hessians, and gradients of cost functions. This chapter will serve as a basis for the derivation of learning algorithms throughout this book.

Complex Valued Nonlinear Adaptive Filters: Noncircularity, Widely Linear and Neural Models
Danilo P. Mandic and Vanessa Su Lee Goh

5.1 Continuous Complex Functions

Complex analyticity and singularities of complex functions have been introduced in Section 4.1 and Section 4.1.1. To address complex continuity, consider complex functions $f : D \mapsto \mathbb{C}$, $D \subset \mathbb{C}$

$$f(z) = f(x, y) = u(x, y) + \jmath v(x, y) = \big(u(x, y), v(x, y)\big) \tag{5.1}$$

Limits of complex functions are defined similarly to those in $\mathbb{R}$, $\lim_{z \to z_0} f(z) = \varsigma$ that is, for any $\epsilon > 0$ there exists $\delta > 0$ such that $|f(z) - \varsigma| < \epsilon$ when $0 < |z - z_0| < \delta$.

To define a continuous complex function we need to show that at any $z_0 \in D$

$$\lim_{z \to z_0} f(z) = f(z_0) \tag{5.2}$$

5.2 The Cauchy–Riemann Equations

For a complex function $f(z) = u + \jmath v$, to be differentiable at z, limit (5.3) must converge to a unique complex number no matter how $\Delta z \to 0$. In other words, for $f(z)$ to be analytic, the limit

$$f'(z) = \lim_{\Delta x \to 0, \Delta y \to 0} \frac{\big[u(x + \Delta x, y + \Delta y) + \jmath v(x + \Delta x, y + \Delta y)\big] - \big[u(x, y) + \jmath v(x, y)\big]}{\Delta x + \jmath \Delta y} \tag{5.3}$$

must exist regardless of how Δz approaches zero.

It is convenient to consider the two following cases [206, 218]

Case 1: $\Delta y = 0$ and $\Delta x \to 0$, which yields

$$\begin{aligned} f'(z) &= \lim_{\Delta x \to 0} \frac{\big[u(x + \Delta x, y) + \jmath v(x + \Delta x, y)\big] - \big[u(x, y) + \jmath v(x, y)\big]}{\Delta x} \\ &= \lim_{\Delta x \to 0} \frac{u(x + \Delta x, y) - u(x, y)}{\Delta x} + \jmath \frac{v(x + \Delta x, y) - v(x, y)}{\Delta x} \\ &= \frac{\partial u(x, y)}{\partial x} + \jmath \frac{\partial v(x, y)}{\partial x} \end{aligned} \tag{5.4}$$

Case 2: $\Delta x = 0$ and $\Delta y \to 0$, which yields

$$\begin{aligned} f'(z) &= \lim_{\Delta y \to 0} \frac{\big[u(x, y + \Delta y) + \jmath v(x, y + \Delta y)\big] - \big[u(x, y) + \jmath v(x, y)\big]}{\jmath \Delta y} \\ &= \lim_{\Delta y \to 0} \frac{u(x, y + \Delta y) - u(x, y)}{\jmath \Delta y} + \frac{v(x, y + \Delta y) - v(x, y)}{\Delta y} \\ &= \frac{\partial v(x, y)}{\partial y} - \jmath \frac{\partial u(x, y)}{\partial y} \end{aligned} \tag{5.5}$$

For continuity (Section 5.1), the limits from Case 1 and Case 2 must be identical, which yields

$$\frac{\partial u(x, y)}{\partial x} = \frac{\partial v(x, y)}{\partial y}, \qquad \frac{\partial v(x, y)}{\partial x} = -\frac{\partial u(x, y)}{\partial y} \tag{5.6}$$

that is, the expressions for the Cauchy-Riemann equations. Therefore, for a function $f(z) : \mathbb{C} \mapsto \mathbb{C}$ to be holomorphic (analytic in z), the partial derivatives $\partial u(x, y)/\partial x$, $\partial u(x, y)/\partial y$, $\partial v(x, y)/\partial x$, and $\partial v(x, y)/\partial y$, must not only exist – they must also satisfy the Cauchy–Riemann (C–R) conditions. This imposes a great amount of structure on holomorphic functions, which may prove rather stringent in practical applications. To avoid this, a more relaxed definition of a derivative is introduced in Section 5.3, based on the duality between the spaces $\mathbb{C}$ and $\mathbb{R}^2$. A comprehensive account of complex vector and matrix differentiation is given in Appendix A.

The Jacobian matrix of a complex function $f(z) = u + \jmath v$, where $z = x + \jmath y$, is given by

$$\mathbf{J} = \begin{bmatrix} \frac{\partial u}{\partial x} & \frac{\partial u}{\partial y} \\ \frac{\partial v}{\partial x} & \frac{\partial v}{\partial y} \end{bmatrix} \tag{5.7}$$

If z and $f'(z)$ were vectors in $\mathbb{R}^2$, say $z = [x, y]$, $dz = [dx, dy]$, and $df(z) = [du, dv]$, they would have to satisfy

$$df(z) = f'(z)\, dz = dz\, f'(z)$$

As the multiplication in the complex domain is commutative, and a 2×2 dimensional Jacobian matrix $\mathbf{J} = f'(z)$ cannot premultiply a row vector dz, in general function $f'(z)$ cannot lie in the same space as z and $f(z)$, and hence the Jacobian matrix cannot be an arbitrary matrix. We have already shown (see also Chapters 12 and 11) that special 2×2 matrices, such as

$$\begin{bmatrix} 1 & 1 \\ -1 & 1 \end{bmatrix} \quad \text{and} \quad \begin{bmatrix} x & y \\ -y & x \end{bmatrix}$$

are algebraically *isomorphic* with complex variables. These matrices commute only with the matrices of their own kind, and hence the Jacobian matrix in Equation (5.7) has to be of this kind too, thus conforming with the Cauchy–Riemann equations.

5.3 Generalised Derivatives of Functions of Complex Variable

In practical applications, we often need to perform optimisation on complex functions which are not directly analytic in $\mathbb{C}$[1]. Frequently encountered complex functions of complex variable which are not analytic include those which depend on the complex conjugate, and those which use absolute values of complex numbers. For instance, the complex conjugate $f(z) = z^*$ is

[1] For instance, in the Wiener filtering problem, the aim is to find the set of coefficients which minimise the total error power, a real function of complex variable. This function clearly has a minimum, but is not differentiable in $\mathbb{C}$. Similar problems arise in power engineering [100].

not analytic, and the Cauchy–Riemann conditions are not satisfied, as can be seen from its Jacobian

$$\mathbf{J} = \begin{bmatrix} 1 & 0 \\ 0 & -1 \end{bmatrix} \tag{5.8}$$

This is also the case with the class of functions which depend on both $z = x + \jmath y$ and $z^* = x - \jmath y$, for instance

$$J(z, z^*) = zz^* = x^2 + y^2 \quad \Rightarrow \quad \mathbf{J} = \begin{bmatrix} 2x & 2y \\ 0 & 0 \end{bmatrix} \quad \Leftrightarrow \quad \frac{\partial u}{\partial x} \neq \frac{\partial v}{\partial y} \quad \frac{\partial v}{\partial x} \neq -\frac{\partial u}{\partial y} \tag{5.9}$$

As a consequence, any polynomial in both z and z^*, or any polynomial depending on z^* alone, is not analytic. **Therefore, our usual cost function $J(k) = \frac{1}{2}e(k)e^*(k) = \frac{1}{2}[e_r^2 + e_i^2]$, a real function of a complex variable, is not analytic or differentiable in the complex sense, and does not satisfy the Cauchy–Riemann conditions.**

The Cauchy–Riemann conditions therefore impose a very stringent structure on functions of complex variables, and several attempts have been made to introduce more convenient derivatives. As functions of complex variable

$$f(z) \quad \leftrightarrow \quad g(x, y) \quad \text{(real valued bivariate)}$$

can also be viewed as functions of its real and imaginary components, it is natural to ask whether the rules of real gradient calculation may be somehow applied. This way, we may be able to replace the stringent conditions of the standard complex derivative ($\mathbb{C}$-derivative) of a holomorphic function $f : \mathbb{C} \to \mathbb{C}$ with the more relaxed conditions of a real derivative ($\mathbb{R}$-derivative) of bivariate function $g(x, y) : \mathbb{R}^2 \to \mathbb{R}^2$. For convenience, we would like the $\mathbb{R}$-derivative to be equivalent to the $\mathbb{C}$-derivative when applied to holomorphic functions.

Based on our earlier examples of nonanalytic functions $f(z) = z^*$ and $f(z) = |z|^2 = zz^*$, observe that:-

- A function $f(z)$ can be non-holomorphic in the complex variable $z = x + \jmath y$, but still be analytic in real variables x and y, as for instance, $f(z) = z^*$ and $f(z) = zz^* = x^2 + y^2$;
- Both $f(z) = z^*$ and $f(z) = zz^*$ are holomorphic in z for $z^* = const$, and are also holomorphic in z^* when $z = const$.

The main idea behind $\mathbb{CR}$ calculus (also known as Wirtinger calculus[2] [313]) and Brandwood's result [35], is to introduce so-called *conjugate coordinates*, a concept applicable to any complex valued or real valued function of a complex variable, whereby a complex function is formally expressed as a function of both z and z^*, that is[3]

$$f(z) = f(z, z^*) = \Re\{f\} + \jmath\Im\{f\} = u(x, y) + \jmath v(x, y) = g(x, y) \tag{5.10}$$

[2] Wirtinger's result has been used largely by the German speaking DSP community [78].

[3] For an excellent overview we refer to the lecture material 'The Complex Gradient Operator and the $\mathbb{CR}$ Calculus', (ECE275CG-F05v1.3d) by Kenneth Kreutz–Delgado.

Notice that $g(x, y)$ is a real bivariate function associated with the complex univariate function $f(z)$, and that the total differential of the function $g(x, y)$ can be expressed as

$$dg(x, y) = \frac{\partial g(x, y)}{\partial x} dx + \frac{\partial g(x, y)}{\partial y} dy$$

We then have

$$dg(x, y) = \frac{\partial u(x, y)}{\partial x} dx + \jmath \frac{\partial v(x, y)}{\partial x} dx + \frac{\partial u(x, y)}{\partial y} dy + \jmath \frac{\partial v(x, y)}{\partial y} dy$$

and the variable swap

$$dz = dx + \jmath dy \qquad dz^* = dx - \jmath dy$$

$$dx = \frac{1}{2}\left[dz + dz^*\right] \qquad dy = \frac{1}{2\jmath}\left[dz - dz^*\right]$$

yields

$$dg(x, y) = \frac{1}{2}\left[\frac{\partial u(x, y)}{\partial x} + \frac{\partial v(x, y)}{\partial y} + \jmath\left(\frac{\partial v(x, y)}{\partial x} - \frac{\partial u(x, y)}{\partial y}\right)\right] dz + \frac{1}{2}\left[\frac{\partial u(x, y)}{\partial x} - \frac{\partial v(x, y)}{\partial y} + \jmath\left(\frac{\partial v(x, y)}{\partial x} + \frac{\partial u(x, y)}{\partial y}\right)\right] dz^*$$

The differential $dg(x, y)$ now becomes

$$dg(x, y) = \frac{1}{2}\left[\frac{\partial g(x, y)}{\partial x} - \jmath \frac{\partial g(x, y)}{\partial y}\right] dz + \frac{1}{2}\left[\frac{\partial g(x, y)}{\partial x} + \jmath \frac{\partial g(x, y)}{\partial y}\right] dz^*$$

and hence the differential of the complex function $f(z)$ can be written as

$$df(z) = df(z, z^*) = \frac{\partial f(z)}{\partial z} dz + \frac{\partial f(z)}{\partial z^*} dz^*$$

5.3.1 $\mathbb{CR}$ Calculus

Using the above formalism [158], for a function $f(z) = f(z, z^*) = g(x, y)$, where f can be either complex valued or real valued, we can formally[4] introduce the $\mathbb{R}$-derivatives:-

- The $\mathbb{R}$-derivative of a real function of a complex variable $f = f(z, z^*)$ is given by

$$\frac{\partial f}{\partial z}\Big|_{z^*=const} = \frac{1}{2}\left(\frac{\partial f}{\partial x} - \jmath \frac{\partial f}{\partial y}\right) \tag{5.11}$$

  where the partial derivatives $\partial f/\partial x$ and $\partial f/\partial y$ are **true (non–formal)** partial derivatives of the function $f(z) = f(z, z^*) = g(x, y)$;

[4] As z is not independent of z^* this is only a formalism; a similar formalism has been used to introduce the augmented complex statistics in chapter 12.

- The conjugate ℝ-derivative (ℝ*-derivative) of a function $f(z) = f(z, z^*)$ is given by

$$\frac{\partial f}{\partial z^*}\bigg|_{z=const} = \frac{1}{2}\left(\frac{\partial f}{\partial x} + \jmath\frac{\partial f}{\partial y}\right) \tag{5.12}$$

  - For these generalised derivatives, it is assumed that z and z^* are mutually independent, that is

$$\frac{\partial z}{\partial z} = \frac{\partial z^*}{\partial z^*} = 1 \qquad \frac{\partial z}{\partial z^*} = \frac{\partial z^*}{\partial z} = 0$$

  Then, the ℝ-derivatives can be straightforwardly calculated by replacing

$$x = (z + z^*)/2 \qquad y = -\jmath(z - z^*)/2$$

  and using the chain rule, as shown above;

- The ℝ-derivatives can be expressed in terms of $g(x, y) = f(z, z^*)$ as

$$\frac{\partial f(z, z^*)}{\partial z} = \frac{1}{2}\left(\frac{\partial g(x, y)}{\partial x} - \jmath\frac{\partial g(x, y)}{\partial y}\right) \qquad \frac{\partial f(z, z^*)}{\partial z^*} = \frac{1}{2}\left(\frac{\partial g(x, y)}{\partial x} + \jmath\frac{\partial g(x, y)}{\partial y}\right)$$

  In other words, the analyticity of $f(z) = f(z, z^*)$ with respect to both z and z^* independently is equivalent to the ℝ-differentiability of $g(x, y)$;

- As a consequence, in terms of ℝ-derivatives, function $f(z)$ has two stationary points, at $\partial f(z, z^*)/\partial z = 0$ and $\partial f(z, z^*)/\partial z^* = 0$.

Thus, although real functions of complex variable are not differentiable in ℂ (the ℂ-derivative does not exist), they are generally differentiable in both x and y and their ℝ-derivatives do exist.

5.3.2 Link between ℝ- and ℂ-derivatives

When considering holomorphic complex functions of complex variables in light of the ℂℝ-derivatives [158], we can observe that:-

- If a function $f = f(z, z^*) = g(x, y) = u(x, y) + \jmath v(x, y)$ is holomorphic, then the Cauchy–Riemann conditions are satisfied, that is, $\partial u(x, y)/\partial x = \partial v(x, y)/\partial y$ and $\partial v(x, y)/\partial x = -\partial u(x, y)/\partial y$, and

$$\begin{aligned}
\mathbb{R}-\text{derivative} \quad \frac{1}{2}\left[\frac{\partial f}{\partial x} - \jmath\frac{\partial f}{\partial y}\right] &= \frac{1}{2}\left[\frac{\partial u}{\partial x} + \jmath\frac{\partial v}{\partial x} - \jmath\frac{\partial u}{\partial y} + \frac{\partial v}{\partial y}\right] \\
&= \frac{1}{2}\left[2\frac{\partial u}{\partial x} + 2\jmath\frac{\partial v}{\partial x}\right] = f'(z) \\
\mathbb{R}^*-\text{derivative} \quad \frac{1}{2}\left[\frac{\partial f}{\partial x} + \jmath\frac{\partial f}{\partial y}\right] &= \frac{1}{2}\left[\frac{\partial u}{\partial x} + \jmath\frac{\partial v}{\partial x} + \jmath\frac{\partial u}{\partial y} - \frac{\partial v}{\partial y}\right] = 0
\end{aligned}$$

  that is, for holomorphic functions the ℝ*-derivative vanishes and the ℝ-derivative is equivalent to the standard complex derivative $f'(z)$;

- In other words, if an ℝ-differentiable function $f(z, z^*)$ is independent of z^*, then the ℝ-derivative of $f(z)$ is equivalent to the standard ℂ-derivative;

- Since for a complex holomorphic function $f(z)$ the $\mathbb{R}^*$-derivative vanishes ($\partial f(z)/\partial z^* = 0$), we can state an *alternative, generalised, form of the Cauchy–Riemann conditions* as

$$\frac{\partial f(z)}{\partial z^*} = 0 \tag{5.13}$$

Thus, holomorphic functions are essentially those which can be written without z^* terms.

Therefore, the $\mathbb{R}$-derivatives are a natural generalisation of the standard complex derivative, and apply to both holomorphic and nonholomorphic functions. We say that complex functions are *real analytic* ($\mathbb{R}$-analytic) over $\mathbb{R}^2$ if they are both $\mathbb{R}$-differentiable and $\mathbb{R}^*$-differentiable [158, 250].

Based on the $\mathbb{CR}$-derivatives, rules of complex differentiation can be obtained by replacing $dz = dx + \jmath dy$ and $dz^* = dx - \jmath dy$. Several of these rules are listed below.

$$\begin{aligned}
\frac{\partial f^*(z)}{\partial z^*} &= \left(\frac{\partial f(z)}{\partial z}\right)^* \\
\frac{\partial f^*(z)}{\partial z} &= \left(\frac{\partial f(z)}{\partial z^*}\right)^* \\
df(z) &= \frac{\partial f(z)}{\partial z}dz + \frac{\partial f(z)}{\partial z^*}dz^* && \text{differential} \\
\frac{\partial f(g(z))}{\partial z} &= \frac{\partial f}{\partial g}\frac{\partial g}{\partial z} + \frac{\partial f}{\partial g^*}\frac{\partial g^*}{\partial z} && \text{chain rule}
\end{aligned} \tag{5.14}$$

In the particular case of cost functions, for instance, the nonholomorphic $f(z, z^*) : \mathbb{C} \times \mathbb{C} \mapsto \mathbb{R}$ given by $f(z) = zz^* = x^2 + y^2 = g(x, y)$, the $\mathbb{R}^*$-derivative

$$\frac{1}{2}\left[\frac{\partial g(x,y)}{\partial x} + \jmath\frac{\partial g(x,y)}{\partial y}\right] = x + \jmath y = z = \frac{\partial f(z,z^*)}{\partial z^*}_{|z=const} = \left(\frac{\partial f^*(z,z^*)}{\partial z}_{|z^*=const}\right)^* \tag{5.15}$$

that is, for a real function of complex variable we have

$$\left(\frac{\partial f}{\partial z}\right)^* = \frac{\partial f}{\partial z^*} \tag{5.16}$$

It is important to highlight again that for general holomorphic complex functions of complex variable $f(z) : \mathbb{C} \mapsto \mathbb{C}$

$$\frac{\partial f(z)}{\partial z} = \underbrace{\frac{\partial f_r(z)}{\partial x} + \jmath\frac{\partial f_i(z)}{\partial x}}_{\mathbb{C}-derivative} \neq \frac{\partial f(z)}{\partial z^*}_{|z=const.} = \frac{1}{2}\underbrace{\left[\frac{\partial f(z)}{\partial x} + \jmath\frac{\partial f(z)}{\partial y}\right]}_{\mathbb{R}^*-derivative} = 0 \tag{5.17}$$

and that in this case the $\mathbb{R}^*$-derivative vanishes.

For instance, for a holomorphic, $\mathbb{C}$-differentiable, function $f(z) = z = x + \jmath y$, we have

$$\begin{array}{ll} \mathbb{C} \;-\text{derivative} & f'(z) = 1 \\ \mathbb{R}^* \;-\text{derivative} & \dfrac{1}{2}\left[\dfrac{\partial(x+\jmath y)}{\partial x} + \jmath\dfrac{\partial(x+\jmath y)}{\partial y}\right] = 0 \\ \mathbb{R} \;-\text{derivative} & \dfrac{1}{2}\left[\dfrac{\partial(x+\jmath y)}{\partial x} - \jmath\dfrac{\partial(x+\jmath y)}{\partial y}\right] = 1 \end{array}$$

5.4 $\mathbb{CR}$-derivatives of Cost Functions

In complex valued adaptive signal processing, it is common to perform optimisation based on scalar functions of complex variables. These are typically quadratic functions of the output error of an adaptive filter, for instance

$$J(k) = \frac{1}{2}|e(k)|^2 = \frac{1}{2}e(k)e^*(k) \tag{5.18}$$

that is, $J(e, e^*) : \mathbb{C} \mapsto \mathbb{R}$, where the error $e(k)$ is a function of the vector of filter parameters $\mathbf{w}(k) = \mathbf{w}^r(k) + \jmath\mathbf{w}^i(k)$, $e(k) = d(k) - \mathbf{x}^T(k)\mathbf{w}(k)$, and the symbols $d(k)$ and $\mathbf{x}(k)$ denote respectively the teaching signal and the tap input vector. We have shown in Equation (5.9) that such functions are not holomorphic, hence the Cauchy–Riemann conditions are not satisfied, and their $\mathbb{C}$-derivative is not defined. However, $J(k)$ is differentiable in both the $\mathbf{w}^r(k)$ and $\mathbf{w}^i(k)$, and to find its stationary points (extrema), both the $\mathbb{R}$- and $\mathbb{R}^*$-derivatives must vanish; this is the essence of Brandwood's result [35].

5.4.1 The Complex Gradient

We shall now provide a step by step derivation of the expressions for gradients of cost functions, as their understanding is critical to the derivation of adaptive filtering algorithms in $\mathbb{C}$. Our aim is to show that

$$\nabla_{\mathbf{w}} J(k) = \frac{\partial J(k)}{\partial \mathbf{w}(k)} = \frac{\partial J(k)}{\partial \mathbf{w}^r(k)} + \jmath\frac{\partial J(k)}{\partial \mathbf{w}^i(k)} \tag{5.19}$$

To this end, recall that the relationship between the complex number $z = x + \jmath y$ and a composite real variable $\omega = (a, b) \in \mathbb{R}^2$ is described by[5] (also shown in chapters 11 and 12)

$$\begin{bmatrix} z_k \\ z_k^* \end{bmatrix} = \mathbb{J}\begin{bmatrix} x_k \\ y_k \end{bmatrix} = \begin{bmatrix} 1 & \jmath \\ 1 & -\jmath \end{bmatrix}\begin{bmatrix} x_k \\ y_k \end{bmatrix}$$

[5]We have introduced the relationship between the real and complex gradients and Hessians based on the work by Van Den Bos [30]. As points where the gradient vanishes (stationary points) give us the location of the extrema of the cost function, and the Hessian helps to differentiate between the minimum and maximum, it is important to provide further insight into their calculation.

whereas the relationship between the composite real vector $\boldsymbol{\omega} = [x_1, y_1, \ldots, x_N, y_N]^T \in \mathbb{R}^{2N\times 1}$ and the "augmented" complex vector $\mathbf{v} = \left[z_1, z_1^*, \ldots, z_N, z_N^*\right]^T \in \mathbb{C}^{2N\times 1}$ is given by

$$\mathbf{v} = \mathbf{A}\boldsymbol{\omega} \tag{5.20}$$

where the matrix $\mathbf{A} = diag(\mathbb{J}, \ldots, \mathbb{J}) \in \mathbb{C}^{2N\times 2N}$ is block diagonal. Thus, we have

$$\boldsymbol{\omega} = \mathbf{A}^{-1}\mathbf{v} = \frac{1}{2}\mathbf{A}^H\mathbf{v} = \frac{1}{2}\mathbf{A}^T\mathbf{v}^* \tag{5.21}$$

To establish a relation between the $\mathbb{R}$- and $\mathbb{C}$-derivatives, consider a Taylor series expansion of a real function $f(\boldsymbol{\omega})$ around $\boldsymbol{\omega} = \mathbf{0}$, given by

$$f + \frac{\partial f}{\partial \boldsymbol{\omega}^T}\boldsymbol{\omega} + \frac{1}{2!}\boldsymbol{\omega}^T\frac{\partial^2 f}{\partial \boldsymbol{\omega}\partial \boldsymbol{\omega}^T}\boldsymbol{\omega} + \cdots \tag{5.22}$$

where the vector $\partial f/\partial \boldsymbol{\omega}$ is called the gradient. The connection between the real and complex gradient is then given by

$$\frac{\partial f}{\partial \boldsymbol{\omega}^T}\boldsymbol{\omega} = \frac{1}{2}\frac{\partial f}{\partial \boldsymbol{\omega}^T}\mathbf{A}^H\mathbf{v} \quad \Rightarrow \quad \frac{\partial f}{\partial \mathbf{v}} = \frac{1}{2}\mathbf{A}^*\frac{\partial f}{\partial \boldsymbol{\omega}} \quad \Rightarrow \quad \frac{\partial f}{\partial \boldsymbol{\omega}} = \mathbf{A}^T\frac{\partial f}{\partial \mathbf{v}} \tag{5.23}$$

that is, the real and complex gradient are related by a linear transformation. For illustration, consider the elements of the gradient vector $\partial f/\partial \mathbf{v}$, that is, $\partial f/\partial z$ and $\partial f/\partial z^*$. From Equations (5.11) and (5.12), we then immediately have

$$\begin{aligned} &\mathbb{R}-\text{derivative:} \qquad \frac{\partial f}{\partial z_n} = \frac{1}{2}\left(\frac{\partial f}{\partial x_n} - \jmath\frac{\partial f}{\partial y_n}\right) \\ &\mathbb{R}^*-\text{derivative:} \qquad \frac{\partial f}{\partial z_n^*} = \frac{1}{2}\left(\frac{\partial f}{\partial x_n} + \jmath\frac{\partial f}{\partial y_n}\right) \end{aligned} \tag{5.24}$$

For the particular case of cost functions, real functions of complex variable, we have

$$\left(\frac{\partial f}{\partial z_n}\right)^* = \left(\frac{\partial f}{\partial z_n^*}\right) \qquad \frac{\partial f}{\partial \mathbf{v}^*} = \left(\frac{\partial f}{\partial \mathbf{v}}\right)^*$$

and also from Equation (5.23)

$$\frac{\partial f}{\partial \boldsymbol{\omega}} = \mathbf{A}^H\frac{\partial f}{\partial \mathbf{v}^*} \tag{5.25}$$

These relationships are very useful in the derivation of the learning algorithms in the steepest descent setting, and especially when addressing their convergence.[6]

[6]Relationships between complex random variables and composite real random variables in terms of their respective probability density functions are given in chapter 12 and Appendix A.

5.4.2 The Complex Hessian

To establish the ℂℝ relationships in terms of second order derivatives, consider the quadratic term in the Taylor series expansion (TSE) in Equation (5.22), given by

$$\frac{1}{2!}\boldsymbol{\omega}^T \mathbf{H}\boldsymbol{\omega} \quad \text{where} \quad \mathbf{H} = \frac{\partial^2 f}{\partial \boldsymbol{\omega}\, \partial \boldsymbol{\omega}^T} \tag{5.26}$$

Using the substitution (5.21), this can be rewritten as

$$\frac{1}{2!}\mathbf{v}^H \mathbf{G}\mathbf{v} \quad \text{where} \quad \mathbf{G} = \frac{\partial^2 f}{\partial \mathbf{v}^*\, \partial \mathbf{v}^T} = \frac{1}{4}\mathbf{A}\mathbf{H}\mathbf{A}^H \tag{5.27}$$

thus, giving the relation between the real Hessian $\mathbf{H}$ and the complex Hessian $\mathbf{G}$.

As $\mathbf{A}^{-1} = \frac{1}{2}\mathbf{A}^H$ and therefore $\mathbf{I} = \mathbf{A}\mathbf{A}^{-1} = \frac{1}{2}\mathbf{A}\mathbf{A}^H$, the characteristic equation for the eigenvalues of the complex Hessian $\mathbf{G}$ is given by

$$\mathbf{G} - \lambda\mathbf{I} = \frac{1}{4}\mathbf{A}\big(\mathbf{H} - 2\lambda\mathbf{I}\big)\mathbf{A}^H \tag{5.28}$$

The roots of $\mathbf{G} - \lambda\mathbf{I}$ are the same as the roots of $\mathbf{H} - 2\lambda\mathbf{I}$, and therefore the eigenvalues of the complex Hessian λ_n^c and the eigenvalues of the real Hessian λ_n^r are related as

$$\lambda_n^r = 2\lambda_n^c, \qquad\qquad n = 1, \ldots, N \tag{5.29}$$

An important consequence is that the conditioning of the matrices $\mathbf{G}$ and $\mathbf{H}$ is the same – this is very useful when studying the relationship between complex valued algorithms and their counterparts in $\mathbb{R}^2$.

In the particular case of the Newton type of optimisation, the real and complex Newton steps are given by

$$\text{Real:} \quad \mathbf{H}\Delta\boldsymbol{\omega} = -\frac{\partial f}{\partial \boldsymbol{\omega}} \qquad\qquad \text{Complex:} \quad \mathbf{G}\Delta\mathbf{v} = -\frac{\partial f}{\partial \mathbf{v}^*} \tag{5.30}$$

where $\partial f/\partial\boldsymbol{\omega}$ and $\partial f/\partial\mathbf{v}^*$ denote respectively the real gradient and the conjugate of the complex gradient.

5.4.3 The Complex Jacobian and Complex Differential

Based on Equation (5.14), the differential of the complex valued vector function F can now be defined as

$$dF(\mathbf{z}, \mathbf{z}^*) = \frac{\partial F(\mathbf{z}, \mathbf{z}^*)}{\partial \mathbf{z}} d\mathbf{z} + \frac{\partial F(\mathbf{z}, \mathbf{z}^*)}{\partial \mathbf{z}^*} d\mathbf{z}^* \tag{5.31}$$

where $\mathbf{z} = [z_1, \ldots, z_N]^T$ and $\mathbf{z}^* = [z_1^*, \ldots, z_N^*]^T$, and

$$F(\mathbf{z}, \mathbf{z}^*) = [f_1(\mathbf{z}, \mathbf{z}^*), \ldots, f_N(\mathbf{z}, \mathbf{z}^*)]^T \tag{5.32}$$

whereas the complex Jacobians $\mathbf{J} = \partial F(\mathbf{z}, \mathbf{z}^*)/\partial \mathbf{z}$ and $\mathbf{J}^c = \partial F(\mathbf{z}, \mathbf{z}^*)/\partial \mathbf{z}^*$ are defined as

$$\mathbf{J} = \begin{bmatrix} \frac{\partial f_1}{\partial z_1} & \cdots & \frac{\partial f_1}{\partial z_N} \\ \vdots & \ddots & \vdots \\ \frac{\partial f_N}{\partial z_1} & \cdots & \frac{\partial f_N}{\partial z_N} \end{bmatrix} \quad \text{and} \quad \mathbf{J}^c = \begin{bmatrix} \frac{\partial f_1}{\partial z_1^*} & \cdots & \frac{\partial f_1}{\partial z_N^*} \\ \vdots & \ddots & \vdots \\ \frac{\partial f_N}{\partial z_1^*} & \cdots & \frac{\partial f_N}{\partial z_N^*} \end{bmatrix} \tag{5.33}$$

Notice that for holomorphic functions $\mathbf{J}^* \neq \mathbf{J}^c$, whereas for real functions of complex variable $\mathbf{J}^* = \mathbf{J}^c$. The complex differential of a real function of complex variable thus becomes

$$dF(\mathbf{z}, \mathbf{z}^*) = \mathbf{J}d\mathbf{z} + \mathbf{J}^*d\mathbf{z}^* = 2\Re\{\mathbf{J}d\mathbf{z}\} = 2\Re\{\mathbf{J}^*d\mathbf{z}^*\} \tag{5.34}$$

The chain rule from Equation (5.14) can also be extended to the vector case as

$$\frac{\partial F(\mathbf{g})}{\partial \mathbf{z}} = \frac{\partial F}{\partial \mathbf{g}}\frac{\partial \mathbf{g}}{\partial \mathbf{z}} + \frac{\partial F}{\partial \mathbf{g}^*}\frac{\partial \mathbf{g}^*}{\partial \mathbf{z}} \quad \text{and} \quad \frac{\partial F(\mathbf{g})}{\partial \mathbf{z}^*} = \frac{\partial F}{\partial \mathbf{g}}\frac{\partial \mathbf{g}}{\partial \mathbf{z}^*} + \frac{\partial F}{\partial \mathbf{g}^*}\frac{\partial \mathbf{g}^*}{\partial \mathbf{z}^*} \tag{5.35}$$

5.4.4 Gradient of a Cost Function

As $\mathbb{C}$-derivatives are not defined for real functions of complex variable, generalised gradient operators can be defined based on the $\mathbb{R}$- and $\mathbb{R}^*$-derivatives in Equations (5.11–5.12) with respect to vectors $\mathbf{z} = [z_1, \ldots, z_N]^T$ and $\mathbf{z}^* = [z_1^*, \ldots, z_N^*]^T$, that is

$$\begin{aligned} \mathbb{R} - \text{derivative:} \quad & \frac{\partial}{\partial \mathbf{z}} = \frac{1}{2}\left[\frac{\partial}{\partial \mathbf{x}} - \jmath\frac{\partial}{\partial \mathbf{y}}\right] \\ \mathbb{R}^* - \text{derivative:} \quad & \frac{\partial}{\partial \mathbf{z}^*} = \frac{1}{2}\left[\frac{\partial}{\partial \mathbf{x}} + \jmath\frac{\partial}{\partial \mathbf{y}}\right] \end{aligned} \tag{5.36}$$

where $\mathbf{z} = \mathbf{x} + \jmath\mathbf{y}$ and $\mathbf{z}^* = \mathbf{x} - \jmath\mathbf{y}$ and the elements of the gradient vectors $\partial/\partial\mathbf{z}$ and $\partial/\partial\mathbf{z}^*$ are given by

$$\frac{\partial}{\partial z_n} = \frac{1}{2}\left[\frac{\partial}{\partial x_n} - \jmath\frac{\partial}{\partial y_n}\right] \quad \text{and} \quad \frac{\partial}{\partial z_n^*} = \frac{1}{2}\left[\frac{\partial}{\partial x_n} + \jmath\frac{\partial}{\partial y_n}\right] \tag{5.37}$$

Thus, in expressing the gradient of a scalar function with respect to a complex vector, the derivatives are applied component–wise, for instance, the gradient of $J(e, e^*)$ with respect to the complex weight vector $\mathbf{w} = [w_1, \ldots, w_N]^T$ is given by

$$\nabla_{\mathbf{w}} J(e, e^*) = \frac{\partial J(e, e^*)}{\partial \mathbf{w}} = \left[\frac{\partial J(e, e^*)}{\partial w_1}, \ldots, \frac{\partial J(e, e^*)}{\partial w_N}\right]^T \tag{5.38}$$

Therefore, to optimise a real function J with respect to a complex valued parameter vector $\mathbf{w}$, two conditions must be satisfied

$$\frac{\partial J(e, e^*)}{\partial \mathbf{w}} = \mathbf{0} \quad \text{and} \quad \frac{\partial J(e, e^*)}{\partial \mathbf{w}^*} = \mathbf{0} \tag{5.39}$$

To determine the extrema of real valued cost functions, apply the rules of complex differentiation from Equation (5.31) to the first term of the Taylor series expansion in Equation (5.22), to yield

$$\Delta J(e, e^*) = \left[\frac{\partial J}{\partial \mathbf{w}}\right]^T \Delta \mathbf{w} + \left[\frac{\partial J}{\partial \mathbf{w}^*}\right]^T \Delta \mathbf{w}^* = 2\Re\left\{\left[\frac{\partial J}{\partial \mathbf{w}}\right]^H \Delta \mathbf{w}^*\right\} = 2\Re\left\{\left[\frac{\partial J}{\partial \mathbf{w}^*}\right]^T \Delta \mathbf{w}^*\right\}$$

Observe that, although the set of stationary points of the gradient has two solutions $\partial J/\partial \mathbf{w}$ and $\partial J/\partial \mathbf{w}^*$, since $df(z) = df(z, z^*) = dg(x, y)$ and

$$dg(x, y) = \frac{\partial g(x, y)}{\partial x} dx + \frac{\partial g(x, y)}{\partial y} dy$$

the differential df vanishes only if the $\mathbb{R}$-derivative is zero, and *the maximum change of the cost function $J(e, e^*)$ is in the direction of the conjugate gradient.* It is therefore natural to express the gradient of the cost function with respect to the filter parameter vector as

$$\nabla_{\mathbf{w}} J = 2\frac{\partial J}{\partial \mathbf{w}^*} = \frac{\partial J}{\partial \mathbf{w}^r} + \jmath \frac{\partial J}{\partial \mathbf{w}^i} \tag{5.40}$$

as this reflects the direction of the maximum change of the gradient.

The stochastic gradient based coefficient update of an adaptive filter can now be expressed as

$$\mathbf{w}(k+1) = \mathbf{w}(k) - \mu \nabla_{\mathbf{w}} J(k)$$

where

$$\mathbf{w}^r(k+1) = \mathbf{w}^r(k) - \mu \frac{\partial J(k)}{\partial \mathbf{w}^r(k)} \quad \text{and} \quad \mathbf{w}^i(k+1) = \mathbf{w}^i(k) - \mu \frac{\partial J(k)}{\partial \mathbf{w}^i(k)} \tag{5.41}$$

For the particular case of a complex linear FIR filter, described in chapter 3, the cost function $J = \frac{1}{2}e(k)e^*(k)$ needs to be minimised with respect to the filter coefficient vector $\mathbf{w}(k)$, where $e(k) = d(k) - \mathbf{x}^T(k)\mathbf{w}(k)$, and $\mathbf{x}(k)$ is the input vector. The stochastic gradient algorithm for the weight update (complex LMS) thus can be expressed as

$$\mathbf{w}(k+1) = \mathbf{w}(k) - \mu \frac{\partial \frac{1}{2} e(k)e^*(k)}{\partial \mathbf{w}^*} = \mathbf{w}(k) + \mu e(k)\mathbf{x}^*(k) \tag{5.42}$$

Derivatives and differentials with respect to complex vectors and matrices are given in Table A.1–A.3 in Appendix A.

Summary: This chapter has introduced elements of $\mathbb{CR}$ calculus for general functions of complex variable. Particular topics include:-

- Complex continuity, differentiability, and the derivation of Cauchy–Riemann equations. It has been shown that the Cauchy–Riemann equations provide a very elegant tool to calculate derivatives of complex holomorphic functions, however, they also impose a great amount of structure on such functions;
- It has been shown that typical cost functions used in adaptive filtering are real functions of complex variables, and as such they do not obey the Cauchy–Riemann conditions. To

deal with this problem, the ℝ-derivatives, that is, generalised derivatives which apply to both complex functions of complex variables and real functions of complex variables have been introduced;

- Relationships between the complex (ℂ-derivatives) and ℝ-derivatives have been established, and a generalised Cauchy–Riemann condition has been introduced;
- An insight into the calculation of complex Jacobians and Hessians has been provided and their relation with the corresponding Jacobians and Hessians in $\mathbb{R}^2$ has been established;
- The ℂℝ calculus has been applied to compute gradients of real functions of complex variable, and the complex steepest descent method has been introduced within this framework.

6

Complex Valued Adaptive Filters

This Chapter presents algorithms for the training of linear and nonlinear feedforward complex adaptive filters. Stochastic gradient learning algorithms are introduced for:

- linear transversal adaptive filters
- fully complex feedforward adaptive filters
- split-complex feedforward adaptive filters
- dual univariate adaptive filters

and are supported by convergence studies and simulations.

As adaptive filtering in the complex domain gives us more degrees of freedom than processing in the real domain, there are several equivalent formulations for the operation of such filters. For instance, the operation of the Complex Least Mean Square (CLMS) algorithm for transversal adaptive filters (adaptive linear combiner) in its original form [307] is given by

$$y = \mathbf{x}^{\mathrm{T}}(k)\mathbf{w}(k) = \mathbf{w}^{\mathrm{T}}(k)\mathbf{x}(k) \quad \rightarrow \quad \Delta\mathbf{w}(k) = \mu e(k)\mathbf{x}^{*}(k) \tag{6.1}$$

and two other frequently used forms are

$$y = \mathbf{w}^{\mathrm{H}}(k)\mathbf{x}(k) = \mathbf{x}^{\mathrm{T}}(k)\mathbf{w}^{*}(k) \quad \rightarrow \quad \Delta\mathbf{w}(k) = \mu e^{*}(k)\mathbf{x}(k) \tag{6.2}$$

and

$$y = \mathbf{x}^{\mathrm{H}}(k)\mathbf{w}(k) = \mathbf{w}^{\mathrm{T}}(k)\mathbf{x}^{*}(k) \quad \rightarrow \quad \Delta\mathbf{w}(k) = \mu e(k)\mathbf{x}(k) \tag{6.3}$$

The formulations (Equations 6.1–6.3) produce identical results and can be transformed from one into another by deterministic mappings. In the convergence analysis, the most convenient form will be used.

Complex Valued Nonlinear Adaptive Filters: Noncircularity, Widely Linear and Neural Models
Danilo P. Mandic and Vanessa Su Lee Goh

6.1 Adaptive Filtering Configurations

Figure 6.1 shows the block diagram of an adaptive filter as a closed loop system consisting of the filter architecture, which can be linear, nonlinear, feedforward or feedback, and the control algorithm. At every time instant k, the coefficients of the filter $\mathbf{w}(k)$ are adjusted based on the output from the control algorithm, thus providing a closed loop adaptation. The optimisation criterion within the control algorithm is a function of the instantaneous output error $e(k) = d(k) - y(k)$, where $d(k)$ is the desired response, $y(k)$ is the filter output, and $x(k)$ is the input signal.

The simplest adaptive filter is a feedforward linear combiner shown in Figure 6.2. The output of this adaptive finite impulse response (FIR) filter of length N is given by

$$y(k) = \sum_{n=1}^{N} w_n(k)x(k-n+1) = \mathbf{x}^{\mathrm{T}}(k)\mathbf{w}(k) \tag{6.4}$$

where $\mathbf{w}(k) = [w_1(k), \ldots, w_N(k)]^{\mathrm{T}}$, $\mathbf{x}(k) = [x(k), \ldots, x(k-N+1)]^{\mathrm{T}}$, and symbol $(\cdot)^{\mathrm{T}}$ denotes the vector transpose operator.

We distinguish between four basic adaptive filtering configurations:

- **System identification configuration, Figure 6.3(a).** In order to model the time varying parameters of an unknown system, the adaptive filter is connected in parallel to the system whose parameters are to be estimated. The unknown system and the adaptive filter share the same input $x(k)$, whereas the output of the unknown system serves as a desired response for the adaptive filter. This configuration is used typically in echo cancellation in acoustics and communications.
- **Noise cancellation configuration, Figure 6.3(b).** In this configuration, the noisy input $s(k) + N_0(k)$ serves as a 'desired' response, whereas the 'reference input' is an external noise source $N_1(k)$ which is correlated with the noise $N_0(k)$. The output of the filter $y(k)$

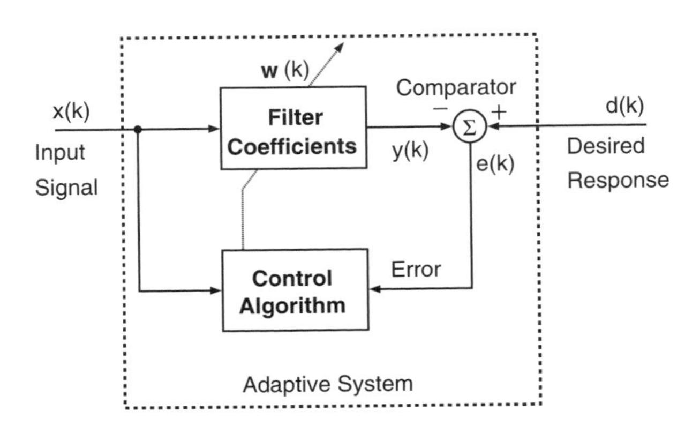

Figure 6.1 Block diagram of an adaptive filter as a closed loop system

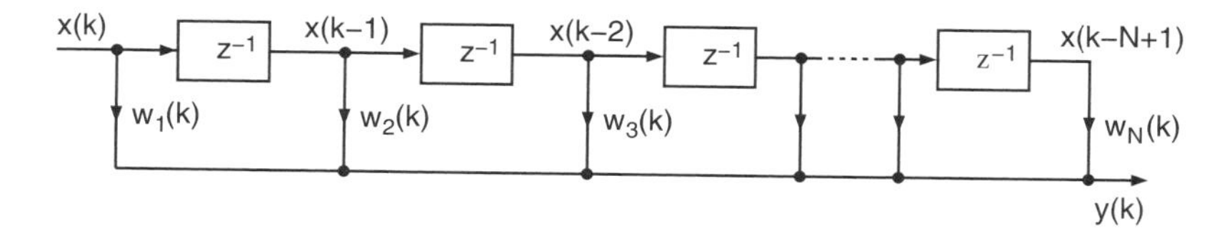

Figure 6.2 Linear adaptive filter

provides an estimate of the noise $\hat{N}_0(k)$, which is then subtracted from the primary input $s(k) + N_0(k)$ to produce a denoised signal $\hat{s}(k)$. This scheme is used in numerous noise cancellation applications in acoustics and biomedicine.

- **Inverse system modelling configuration, Figure 6.4(a).** The goal of inverse system modelling is to produce an estimate of the inverse of the transfer function of an unknown system. To achieve this, the adaptive filter is connected in series with the unknown system. The desired signal is the input signal delayed, due to signal propagation and the operation in discrete time. Typical applications of this scheme are in channel equalisation in digital communications and in control of industrial plants.
- **Adaptive prediction configuration, Figure 6.4(b).** In this configuration, the goal is to predict the value of the input signal M steps ahead, that is, to produce an estimate $\hat{x}(k + M)$, where M is the prediction horizon. The desired response (teaching signal) is the M steps ahead advanced version of the input signal. The range of applications of adaptive prediction are numerous, from quantitative finance to vehicle navigation systems.

Prediction is at the core of adaptive filtering and most of the simulations in this book will be conducted in the one step ahead adaptive prediction setting. In this case, the output of the linear adaptive filter from Figure 6.2 can be written as

$$y(k) = \sum_{n=1}^{N} w_n(k)x(k-n) = \mathbf{x}^{\mathrm{T}}(k)\mathbf{w}(k) \tag{6.5}$$

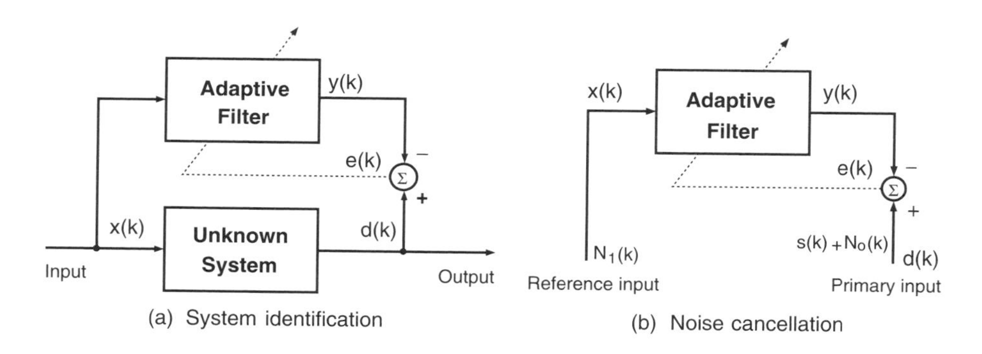

Figure 6.3 Adaptive filtering configurations

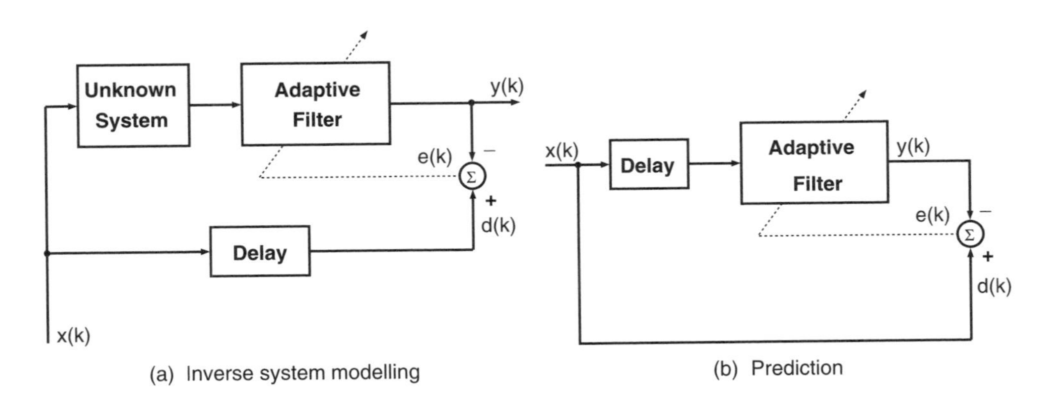

Figure 6.4 Adaptive filtering configurations

where the input in the filter memory (tap input vector) is $\mathbf{x}(k) = [x(k-1), \ldots, x(k-N)]^{\mathrm{T}}$ and the weight vector (vector of adaptive filter coefficients) is $\mathbf{w}(k) = [w_1(k), \ldots, w_N(k)]^{\mathrm{T}}$.

Wiener filter. Consider a linear feedforward filter with fixed coefficients $\mathbf{w}$, for which the optimisation task is to minimise the error criterion (cost function)

$$J(\mathbf{w}) = E\big[|e(k)|^2\big] = E\big[e(k)e^*(k)\big] = E\big[e_r^2(k) + e_i^2(k)\big] \tag{6.6}$$

This is a 'deterministic' function of the weight vector $\mathbf{w}$, and represents the estimated power of the output error. The aim is to find an 'optimal' vector with fixed coefficients $\mathbf{w}_{\mathrm{o}}$ which minimises $J(\mathbf{w})$. The Wiener filter provides a block solution which produces the best estimate of the desired signal in the mean square sense – for stationary input signals this solution is optimal in terms of second order statistics.

It is convenient to analyse the mean square error (Equation 6.6) using the description (Equation 6.2), where $e(k) = d(k) - \mathbf{w}^{\mathrm{H}}\mathbf{x}(k)$. Upon evaluating Equation (6.6), we have[1]

$$\begin{aligned} E\big[e(k)e^*(k)\big] &= E\left[\big(d(k) - \mathbf{w}^{\mathrm{H}}\mathbf{x}(k)\big)\big(d(k) - \mathbf{w}^{\mathrm{H}}\mathbf{x}(k)\big)^*\right] \\ &= E\big[d(k)d^*(k) - \mathbf{w}^{\mathrm{H}}\mathbf{x}(k)d^*(k) - \mathbf{w}^{\mathrm{T}}\mathbf{x}^*(k)d(k) + \mathbf{w}^{H}\mathbf{x}(k)\mathbf{w}^{T}\mathbf{x}^*(k)\big] \end{aligned} \tag{6.7}$$

Due to the linearity of the statistical expectation operator $E[\cdot]$, we have

$$\begin{aligned} E\big[e(k)e^*(k)\big] &= E\big[\,|d(k)|^2\,\big] - \mathbf{w}^{\mathrm{H}}E\big[\mathbf{x}(k)d^*(k)\big] - \mathbf{w}^{\mathrm{T}}E\big[\mathbf{x}^*(k)d(k)\big] + \mathbf{w}^{\mathrm{H}}E\big[\mathbf{x}(k)\mathbf{x}^{\mathrm{H}}(k)\big]\mathbf{w} \\ &= \sigma_d^2 - \mathbf{w}^{\mathrm{H}}\mathbf{p} - \mathbf{p}^{\mathrm{H}}\mathbf{w} + \mathbf{w}^{\mathrm{H}}\mathbf{R}\mathbf{w} \end{aligned} \tag{6.8}$$

where $\sigma_d^2 = E[|d(k)|^2]$ is the power of the desired response, $\mathbf{R} = E\big[\mathbf{x}(k)\mathbf{x}^{\mathrm{H}}(k)\big]$ denotes the complex valued input correlation matrix, $\mathbf{p} = E\big[d^*(k)\mathbf{x}(k)\big]$ is the crosscorrelation vector between the desired response and the input signal, subscripts $(\cdot)_r$ and $(\cdot)_i$ denote respectively the real and imaginary part of the complex number, $(\cdot)^*$ is the complex conjugation operator, and $(\cdot)^{\mathrm{H}}$ is the Hermitian transpose operator.

[1] Based on the identity $\mathbf{a}^{\mathrm{T}}\mathbf{b} = \mathbf{b}^{\mathrm{T}}\mathbf{a}$, we have $\mathbf{w}^{\mathrm{T}}\mathbf{x}^*(k) = \mathbf{x}^{\mathrm{H}}(k)\mathbf{w}$ and $\mathbf{w}^{\mathrm{T}}\mathbf{p}^* = \mathbf{p}^{\mathrm{H}}\mathbf{w}$.

We can solve for the optimal weight vector $\mathbf{w}_o$ by differentiating Equation (6.8) with respect to $\mathbf{w}$ and setting the result to zero, that is[2]

$$\nabla_{\mathbf{w}} J(\mathbf{w}) = 2\mathbf{R}\mathbf{w} - 2\mathbf{p} = \mathbf{0} \tag{6.9}$$

which gives the Wiener–Hopf solution[3]

$$\mathbf{w}_o = \arg\min_{\mathbf{w}} J(\mathbf{w}) = \mathbf{R}^{-1}\mathbf{p} \tag{6.10}$$

The minimum achievable mean square error is calculated by replacing the optimal weight vector (Equation 6.10) back into Equation (6.8), to give [308]

$$J_{\min} = J(\mathbf{w}_o) = \sigma_d^2 - \mathbf{p}^{\mathrm{H}}\mathbf{R}^{-1}\mathbf{p} \tag{6.11}$$

To emphasise that the cost function (Equation 6.6) is quadratic in $\mathbf{w}$ and hence has a global minimum $J_{\min}$ for $\mathbf{w} = \mathbf{w}_o$, we can rewrite Equation (6.8) as

$$J(\mathbf{w}) = J_{\min} + (\mathbf{w} - \mathbf{w}_o)^{\mathrm{H}}\, \mathbf{R}\, (\mathbf{w} - \mathbf{w}_o) \tag{6.12}$$

Notice that the mean square error $J(\mathbf{w})$ comprises two terms – the minimum achievable mean square error $J_{\min}$ and a term that is quadratic in the weight error vector $\mathbf{v} = \mathbf{w} - \mathbf{w}_o$, which for the Wiener solution $\mathbf{w} = \mathbf{w}_o$ vanishes. The weight error vector $\mathbf{v}$ is also called the *misalignment vector*, and plays an important role in the analysis of adaptive filtering algorithms.

For nonstationary data the optimal weight vector is time varying, and hence the Wiener solution is suboptimal.

6.2 The Complex Least Mean Square Algorithm

Due to the block nature of the Wiener solution and the requirement of stationarity of the input, together with a possibly prohibitively large correlation matrix, this algorithm is not suitable for real world real time applications. One way to mitigate this problem is to make use of the parabolic shape of the 'error surface' defined by $J(\mathbf{w}) = E[|e(k)|^2]$. As the convexity of the error surface guarantees the existence of the solution, we can reach the optimal solution $\mathbf{w} = \mathbf{w}_o$ recursively, by performing 'steepest descent' towards the minimum $J_{\min} = J(\mathbf{w}_o)$, that is

$$\mathbf{w}(k+1) = \mathbf{w}(k) - \mu \nabla_{\mathbf{w}} J(\mathbf{w}) \tag{6.13}$$

where μ is the stepsize, a small positive constant.

[2]The aim is to minimise the output error power, hence the cost function is a real function of complex variable, and does not have a derivative directly in $\mathbb{C}$. We can, however use the $\mathbb{CR}$ calculus to find $\nabla_{\mathbf{w}} J$, as explained in Chapter 5.
[3]For stationary inputs the correlation matrix $\mathbf{R}$ is almost always positive semidefinite, that is, $\mathbf{u}^{\mathrm{H}}\mathbf{R}\mathbf{u} \geq 0$ for every nonzero $\mathbf{u}$, and therefore nonsingular and invertible.

The complex least mean square (CLMS) algorithm, introduced in 1975 [307], performs 'stochastic gradient descent', whereby all the statistical quantities in the Wiener filtering problem are replaced by their instantaneous estimates, to give[4]

$$
\begin{aligned}
E\left[|e^2(k)|\right] &\rightarrow \frac{1}{2}|e^2(k)| \\
E\left[\mathbf{x}(k)\mathbf{x}^{\mathrm{H}}(k)\right] &\rightarrow \mathbf{x}(k)\mathbf{x}^{\mathrm{H}}(k) \\
E\left[\mathbf{x}(k)d(k)\right] &\rightarrow \mathbf{x}(k)d(k)
\end{aligned} \tag{6.14}
$$

The 'stochastic' cost function

$$J(k) = \frac{1}{2}|e(k)|^2 \tag{6.15}$$

is now time varying, and based on Equation (6.13) the weight vector update can be expressed as

$$\mathbf{w}(k+1) = \mathbf{w}(k) - \mu \nabla_{\mathbf{w}} J(k)_{|\mathbf{w}=\mathbf{w}(k)} \tag{6.16}$$

The gradient of the cost function with respect to the complex valued weight vector $\mathbf{w}(k) = \mathbf{w}_r(k) + \jmath\mathbf{w}_i(k)$ can be expressed as[5]

$$\nabla_{\mathbf{w}} J(k) = \nabla_{\mathbf{w}_r} J(k) + \jmath \nabla_{\mathbf{w}_i} J(k) = \frac{\partial J(k)}{\partial \mathbf{w}_r(k)} + \jmath \frac{\partial J(k)}{\partial \mathbf{w}_i(k)} \tag{6.17}$$

As the output error of the linear adaptive filter in Figure 6.2 is given by

$$e(k) = d(k) - \mathbf{x}^{\mathrm{T}}(k)\mathbf{w}(k)$$

the gradient with respect to the real part of the complex weight vector can be evaluated as

$$\nabla_{\mathbf{w}_r} J(k) = \frac{1}{2}\frac{\partial\left[e(k)e^*(k)\right]}{\partial \mathbf{w}_r(k)} = \frac{1}{2}e(k)\nabla_{\mathbf{w}_r} e^*(k) + \frac{1}{2}e^*(k)\nabla_{\mathbf{w}_r} e(k) \tag{6.18}$$

where

$$\nabla_{\mathbf{w}_r} e(k) = -\mathbf{x}(k) \qquad \nabla_{\mathbf{w}_r} e^*(k) = -\mathbf{x}^*(k)$$

The real part of the gradient of the cost function is then obtained as

$$\nabla_{\mathbf{w}_r} J(k) = -\frac{1}{2}e(k)\mathbf{x}^*(k) - \frac{1}{2}e^*(k)\mathbf{x}(k) \tag{6.19}$$

---

[4]For convenience, the cost function is scaled by $\frac{1}{2}$. This makes the weight updates easier to manipulate, and does not influence the result, as this factor will be cancelled after performing the derivative of the squared error term.

[5]We have already calculated this gradient in Section 5.4.4, using the $\mathbb{R}^*$-derivative – see Equation (5.42). In this Chapter, we provide a step by step derivation of CLMS – $\mathbb{CR}$ calculus will be used to simplify the derivation of learning algorithms for feedback filters in Chapters 7 and 15.

Similarly, the gradient of the cost function with respect to the imaginary part of the weight vector can be calculated as

$$\nabla_{\mathbf{w}_i} J(k) = \frac{\jmath}{2} e(k)\mathbf{x}^*(k) - \frac{\jmath}{2} e^*(k)\mathbf{x}(k) \tag{6.20}$$

By combining Equations (6.19) and (6.20) we obtain the gradient term in Equation (6.16) in the form

$$\nabla_{\mathbf{w}} J(k) = \nabla_{\mathbf{w}_r} J(k) + \jmath \nabla_{\mathbf{w}_i} J(k) = -e(k)\mathbf{x}^*(k)$$

Finally, the stochastic gradient adaptation for the weight vector can be expressed as

$$\mathbf{w}(k+1) = \mathbf{w}(k) + \mu e(k)\mathbf{x}^*(k), \qquad \mathbf{w}(0) = \mathbf{0} \tag{6.21}$$

This completes the derivation of the complex least mean square (CLMS) algorithm [307].

Two other frequently used formulations for the CLMS are

$$y = \mathbf{w}^{\mathrm{H}}(k)\mathbf{x}(k) \quad \rightarrow \quad \mathbf{w}(k+1) = \mathbf{w}(k) + \mu e^*(k)\mathbf{x}(k) \tag{6.22}$$

and

$$y = \mathbf{x}^{\mathrm{H}}(k)\mathbf{w}(k) \quad \rightarrow \quad \mathbf{w}(k+1) = \mathbf{w}(k) + \mu e(k)\mathbf{x}(k) \tag{6.23}$$

The above three formulations for CLMS are equivalent; forms (6.22) and (6.23) are sometimes more convenient for matrix manipulation.

6.2.1 Convergence of the CLMS Algorithm

To illustrate that for stationary signals the CLMS converges to the optimal Wiener solution, consider the expected value for the CLMS update

$$\mathbf{w}(k+1) = \mathbf{w}(k) + \mu E\big[e(k)\mathbf{x}^*(k)\big] \tag{6.24}$$

which has converged when $\mathbf{w}(k+1) = \mathbf{w}(k) = \mathbf{w}(\infty)$, that is, when the iteration (6.24) reaches its fixed point[6] [199, 256], and the weight update $\Delta\mathbf{w}(k) = \mu E[e(k)\mathbf{x}^*(k)] = \mathbf{0}$. This is achieved for[7]

$$\mathbf{0} = E\big[d^*(k)\mathbf{x}(k)\big] - E\big[\mathbf{x}(k)\mathbf{x}^{\mathrm{H}}(k)\big]\mathbf{w}(k) \quad \Longleftrightarrow \quad \mathbf{R}\mathbf{w} = \mathbf{p} \tag{6.25}$$

that is, for the Wiener solution. The condition $E[e(k)\mathbf{x}^*(k)] = \mathbf{0}$ is called the 'orthogonality condition' and states that the output error of the filter and the tap input vector are orthogonal ($e \perp \mathbf{x}$) when the filter has converged to the optimal solution.

[6] For more detail see Appendix P, Appendix O, and Appendix N.

[7] Upon applying the complex conjugation operator and setting $\mathbf{w}^* = \mathbf{w}$.

It is of greater interest, however, to analyse the evolution of the weights in time. As with any other estimation problem, we need to analyse the 'bias' and 'variance' of the estimator, that is:

- Convergence in the mean, to ascertain whether $\mathbf{w}(k) \rightarrow \mathbf{w}_o$ when $k \rightarrow \infty$;
- Convergence in the mean square, in order to establish whether the variance of the weight error vector $\mathbf{v}(k) = \mathbf{w}(k) - \mathbf{w}_o(k)$ approaches $J_{\min}$ as $k \rightarrow \infty$.

The analysis of convergence of linear adaptive filters is made mathematically tractable if we use so called *independence assumptions*, such that the filter coefficients are statistically independent of the data currently in filter memory, and $\{d(l), x(l)\}$ is independent of $\{d(k), x(k)\}$ for $k \neq l$.

Convergence in the mean. For convenience, the analysis will be based on the form (6.23). We can assume without loss in generality that the desired response

$$d(k) = \mathbf{x}^{\mathrm{H}}(k)\mathbf{w}_o + q(k) \tag{6.26}$$

where $q(k)$ is complex white Gaussian noise, with zero mean and variance σ_q^2, which is uncorrelated with $\mathbf{x}(k)$. Then, we have

$$\begin{aligned} e(k) &= \mathbf{x}^{\mathrm{H}}(k)\mathbf{w}_o + q(k) - \mathbf{x}^{\mathrm{H}}(k)\mathbf{w}(k) \\ \mathbf{w}(k+1) &= \mathbf{w}(k) + \mu\mathbf{x}(k)\mathbf{x}^{\mathrm{H}}\mathbf{w}_o - \mu\mathbf{x}(k)\mathbf{x}^{\mathrm{H}}(k)\mathbf{w}(k) + \mu q(k)\mathbf{x}(k) \end{aligned}$$

Subtracting the optimal weight vector $\mathbf{w}_o$ from both sides of the last equation, the weight error vector $\mathbf{v}(k) = \mathbf{w}(k) - \mathbf{w}_o$ can be expressed as[8]

$$\mathbf{v}(k+1) = \mathbf{v}(k) - \mu\mathbf{x}(k)\mathbf{x}^{\mathrm{H}}(k)\mathbf{v}(k) + \mu q(k)\mathbf{x}(k) \tag{6.27}$$

Applying the statistical expectation operator to both sides of Equation (6.27) and employing the independence assumptions, we have

$$E[\mathbf{v}(k+1)] = \big(\mathbf{I} - \mu E[\mathbf{x}(k)\mathbf{x}^{\mathrm{H}}(k)]\big)E[\mathbf{v}(k)] + \mu E[q(k)\mathbf{x}(k)] = \big(\mathbf{I} - \mu\mathbf{R}\big)E[\mathbf{v}(k)] \tag{6.28}$$

Unless the correlation matrix $\mathbf{R}$ is diagonal, there will be cross–coupling between the coefficients of the weight error vector. Since $\mathbf{R}$ is Hermitian and positive semidefinite, it can be rotated into a diagonal matrix by a unitary transformation[9]

$$\mathbf{R} = \mathbf{Q}\boldsymbol{\Lambda}\mathbf{Q}^{\mathrm{H}} \tag{6.29}$$

where $\boldsymbol{\Lambda} = diag\,(\lambda_1, \lambda_2, \ldots, \lambda_N)$ is a diagonal matrix comprising the real and positive eigenvalues of the correlation matrix, $\mathbf{Q}$ is the matrix of the corresponding eigenvectors, and $\lambda_1 \geq \lambda_2 \geq \cdots \geq \lambda_N$.

[8] The same result can be obtained from the original formulation (6.21) where $d(k) = \mathbf{x}^{\mathrm{T}}(k)\mathbf{w}_o + q(k)$, based on $\Delta\mathbf{w}(k) = \mu\mathbf{x}^*(k)e(k)$, and using the identity $\mathbf{x}^*(k)\mathbf{x}^{\mathrm{T}}(k) = \mathbf{x}(k)\mathbf{x}^{\mathrm{H}}(k)$.

[9] The eigenvectors may be chosen to be orthonormal in which case $\mathbf{Q}$ is unitary.

Rotating the weight error vector $\mathbf{v}(k)$ by the eigenmatrix $\mathbf{Q}$, that is, $\mathbf{v}'(k) = \mathbf{Q}\mathbf{v}(k)$, decouples the evolution of its coefficients. This rotation allows us to express the so called 'modes of convergence' solely in terms of the corresponding eigenvalues of the correlation matrix[10]

$$\mathbf{v}'(k+1) = (\mathbf{I} - \mu\mathbf{\Lambda})\,\mathbf{v}'(k) \tag{6.30}$$

Since $(\mathbf{I} - \mu\mathbf{\Lambda})$ is diagonal and the nth component of $\mathbf{v}'$ represents the projection of the vector $\mathbf{v}(k)$ onto the nth eigenvector of $\mathbf{R}$, every element of $\mathbf{v}'(k)$ evolves independently, and Equation (6.30) converges to zero if $|1 - \mu\lambda_n| < 1$. As the fastest mode of convergence corresponds to the maximum eigenvalue $\lambda_{\max}$, the condition for the convergence in the mean of the CLMS algorithm becomes[11]

$$0 < \mu < \frac{2}{\lambda_{\max}} \approx \frac{2}{\mathrm{tr}[\mathbf{R}]} \tag{6.31}$$

The trace of the input correlation matrix is equal to the product of the filter length and input signal power, and so an easier to estimate bound on the learning rate is given by

$$0 < \mu < \frac{2}{NE\left[|x(k)|^2|\right]} \tag{6.32}$$

Convergence in the mean square. To evaluate the mean square error, the CLMS must converge in the mean, and hence its learning rate must obey $0 < \mu < 2/\lambda_{\max}$. As the filter coefficients begin to converge in the mean, they start fluctuating around their optimum values, defined by $\mathbf{w}_{\mathrm{o}}$. This is due to the 'stochastic' gradient approximation used for the update of $\mathbf{w}(k)$, that is, the use of instantaneous estimates for the statistical moments within CLMS, as shown in Equation (6.14). As a result, the mean square error $\xi(k) = E[|e(k)|^2]$ exceeds the minimum mean square error $J_{\min}$ (6.11) by an amount referred to as the *excess mean square error*, denoted by $\xi_{\mathrm{EMSE}}(k)$, that is [110]

$$\xi(k) = J_{\min} + \xi_{\mathrm{EMSE}}(k) \tag{6.33}$$

The excess mean square error depends on second-order statistical properties of the desired response, input, and weight error vector. The plot showing time evolution of the mean square error is called the *learning curve*.

For convergence, it is of principal importance to preserve the asymptotic boundedness[12] of the mean square error $\xi(k)$. Based on Equation (6.12) and the signal model (6.26), the minimum mean square error is $J_{\min} = \sigma_q^2$, which gives[13]

$$\xi(k) = \sigma_q^2 + E\left[\mathbf{v}^{\mathrm{H}}(k)\mathbf{R}\mathbf{v}(k)\right] = \sigma_q^2 + \mathrm{tr}\left[\mathbf{R}\mathbf{K}(k)\right] \tag{6.34}$$

---

[10]Modes of convergence are defined as $v^n(k+1) = (1 - \mu\lambda_n)v^n(k),\ n = 1, \ldots, N$. As $v^n(k+1) = (1 - \lambda_n)^k v^n(0)$, we require $|1 - \mu\lambda_n| < 1$.

[11]Using the identity $\lambda_{\max} \leq \sum$(diagonal elements of $\mathbf{R}$) = $\mathrm{tr}[\mathbf{R}]$.

[12]For more detail on asymptotic stability, see Appendix O.

[13]Using the identity $E[\mathbf{v}^{\mathrm{H}}(k)\mathbf{R}\mathbf{v}(k)] = \mathrm{tr}[\mathbf{R}\mathbf{K}(k)] = \mathrm{tr}[\mathbf{K}(k)\mathbf{R}]$.

where the weight error correlation matrix $\mathbf{K}(k) = E[\mathbf{v}(k)\mathbf{v}^{\mathrm{H}}(k)]$ and $\xi_{\mathrm{EMSE}}(k) = \mathrm{tr}[\mathbf{RK}(k)]$. Since the correlation matrix $\mathbf{R}$ is bounded, the CLMS will converge in the mean square if the elements of $\mathbf{K}(k)$ remain bounded as $k \to \infty$. To analyse $\mathbf{K}(k)$ we can multiply Equation (6.27) by $\mathbf{v}^{H}(k)$, to give

$$\begin{aligned}\mathbf{v}(k+1)\mathbf{v}^{\mathrm{H}}(k+1) &= \mathbf{v}(k)\mathbf{v}^{\mathrm{H}}(k) - \mu\mathbf{x}(k)\mathbf{x}^{\mathrm{H}}(k)\mathbf{v}(k)\mathbf{v}^{\mathrm{H}}(k) - \mu\mathbf{v}(k)\mathbf{v}^{\mathrm{H}}(k)\mathbf{x}(k)\mathbf{x}^{\mathrm{H}}(k)\\ &\quad + \mu^2 q(k)q^*(k)\mathbf{x}(k)\mathbf{x}^{\mathrm{H}}(k) + \mu^2\mathbf{x}(k)\mathbf{x}^{\mathrm{H}}(k)\mathbf{v}(k)\mathbf{v}^{\mathrm{H}}(k)\mathbf{x}(k)\mathbf{x}^{\mathrm{H}}(k) + xt(k)\end{aligned} \tag{6.35}$$

where variable $xt(k)$ comprises the crossterms. Owing to the independence assumptions, the cross-terms vanish upon the application of the statistical expectation operator, and

$$\begin{aligned}E\left[\mathbf{x}(k)\mathbf{x}^{\mathrm{H}}(k)\mathbf{v}(k)\mathbf{v}^{\mathrm{H}}(k)\right] &= \mathbf{RK}(k)\\ E\left[\mathbf{v}(k)\mathbf{v}^{\mathrm{H}}(k)\mathbf{x}(k)\mathbf{x}^{\mathrm{H}}(k)\right] &= \mathbf{K}(k)\mathbf{R}\\ E\left[\mathbf{x}(k)\mathbf{x}^{\mathrm{H}}(k)\mathbf{v}(k)\mathbf{v}^{\mathrm{H}}(k)\mathbf{x}(k)\mathbf{x}^{\mathrm{H}}(k)\right] &= \mathbf{RK}(k)\mathbf{R} + \mathbf{R}\mathrm{tr}[\mathbf{RK}(k)]\\ E\left[|q(k)|^2\mathbf{x}(k)\mathbf{x}^{\mathrm{H}}(k)\right] &= \sigma_q^2\mathbf{R}\end{aligned} \tag{6.36}$$

to yield[14]

$$\mathbf{K}(k+1) = \mathbf{K}(k) - \mu\Big(\mathbf{RK}(k) + \mathbf{K}(k)\mathbf{R}\Big) + \mu^2\Big(\mathbf{RK}(k) + \mathrm{tr}[\mathbf{RK}(k)]\Big)\mathbf{R} + \mu^2\sigma_q^2\mathbf{R} \tag{6.37}$$

Evaluation of the excess mean square error based on Equation (6.37) is rather mathematically demanding, however, similarly to the analysis of convergence in the mean, since $\mathbf{R}$ is Hermitian and positive semidefinite, it can be rotated into a diagonal matrix by a unitary transformation[15] $\mathbf{R} = \mathbf{Q\Lambda Q}^{-1}$. Then, since $\mathbf{v}'(k) = \mathbf{Qv}(k)$, the rotated weight error correlation matrix becomes

$$\widetilde{\mathbf{K}}(k) = \mathbf{QK}(k)\mathbf{Q}^{\mathrm{H}} \tag{6.38}$$

To simplify the analysis of Equation (6.37), consider a white iid[16] input for which $\mathbf{R} = \sigma_x^2\mathbf{I}$, to yield

$$\widetilde{\mathbf{K}}(k+1) = \left(1 - \mu\sigma_x^2\right)^2\widetilde{\mathbf{K}}(k) + \mu^2\sigma_x^4\mathrm{tr}[\widetilde{\mathbf{K}}(k)]\mathbf{I} + \mu^2\sigma_q^2\sigma_x^2\mathbf{I} \tag{6.39}$$

Since $\widetilde{\mathbf{K}}(k)$ is a correlation matrix, for every element $\kappa_{mn}(k) \in \widetilde{\mathbf{K}}(k)$, we have

$$|\kappa_{mn}(k)|^2 \le \kappa_{mm}(k)\kappa_{nn}(k) \tag{6.40}$$

[14] We use the property that for zero mean, complex, jointly Gaussian x_1, x_2, x_3, x_4, the Gaussian Moment Factoring Theorem states that $E\left[x_1x_2^{\mathrm{H}}x_3x_4^{\mathrm{H}}\right] = x_1x_2^{\mathrm{H}} \cdot x_3x_4^{\mathrm{H}} + x_1x_4^{\mathrm{H}} \cdot x_2^{\mathrm{H}}x_3$. For a detailed analysis, we refer to [62, 76, 127].

[15] Here, $\mathbf{\Lambda} = \mathrm{diag}\,(\lambda_1, \lambda_2, \ldots, \lambda_N)$ is a diagonal matrix comprising the real and positive eigenvalues of the correlation matrix, $\mathbf{Q}$ is the matrix of the corresponding eigenvectors, and $\lambda_1 \ge \lambda_2 \ge \cdots \ge \lambda_N$.

[16] Independent identically distributed.

Thus, for convergence of Equation (6.39) it is sufficient to look only at the evolution of the diagonal elements of $\widetilde{\mathbf{K}}(k)$; the recursion for the calculation of these coefficients is provided in the classic result by Horowitz and Senne [127]. Consider a vector comprising the diagonal elements of $\widetilde{\mathbf{K}}(k)$, given by

$$\mathbf{s}(k) = [\kappa_{11}(k), \ldots, \kappa_{NN}(k)]^{\mathrm{T}}$$

Then, from Equation (6.39) we have

$$\mathbf{s}(k+1) = \left[\left(1 - \mu\sigma_x^2\right)^2 \mathbf{I} + \mu^2\sigma_x^4 \mathbf{1}\mathbf{1}^{\mathrm{T}}\right] \mathbf{s}(k) + \mu^2\sigma_q^2\sigma_x^2 \mathbf{1} \tag{6.41}$$

where $\mathbf{1}$ is an $N \times 1$ vector of ones. In the steady state, $\mathbf{s}(k+1) = \mathbf{s}(k) = \mathbf{s}(\infty)$, and Equation (6.41) can be expressed as

$$\mathbf{s}(\infty) = \frac{\mu\sigma_q^2 \mathbf{1}}{2\mathbf{I} - \mu\sigma_x^2\left(\mathbf{I} + \mathbf{1}\mathbf{1}^{\mathrm{T}}\right)} \tag{6.42}$$

We can use the matrix inversion lemma[17] [110] to find the inverse of the denominator of Equation (6.42), to give

$$\kappa_{pp}(\infty) = \mu\sigma_q^2 \frac{\frac{1}{2-\mu\sigma_x^2}}{1 - \mu\sum_{n=1}^{N}\frac{\sigma_x^2}{2-\mu\sigma_x^2}} \qquad p = 1, \ldots, N \tag{6.43}$$

The steady state excess mean square error $\xi_{\mathrm{EMSE}}(\infty) = \mathrm{tr}[\mathbf{R}\mathbf{K}(\infty)] = \mathrm{tr}[\mathbf{K}(\infty)\mathbf{R}]$ now becomes

$$\xi_{\mathrm{EMSE}}(\infty) = \sigma_x^2 \mathbf{s}^{\mathrm{T}}(\infty)\mathbf{1} = \sigma_q^2 \frac{\sum_{n=1}^{N}\frac{\mu\sigma_x^2}{2-\mu\sigma_x^2}}{1 - \sum_{n=1}^{N}\frac{\mu\sigma_x^2}{2-\mu\sigma_x^2}} \tag{6.44}$$

It is convenient to assess the performance of adaptive filters in terms of the misadjustment

$$\mathcal{M} = \frac{\xi_{\mathrm{EMSE}}(\infty)}{\xi_{\min}} = \frac{\xi_{\mathrm{EMSE}}(\infty)}{\sigma_q^2} \tag{6.45}$$

From Equations (6.34) and (6.43), for small μ the expression for misadjustment simplifies into

$$\mathcal{M} = \mu\frac{\frac{1}{2}\mathrm{tr}[\mathbf{R}]}{1 - \frac{1}{2}\mu\,\mathrm{tr}[\mathbf{R}]} = \mu\frac{\sigma_x^2 N}{2 - \mu\sigma_x^2 N} \approx \frac{1}{2}\mu\sigma_x^2 N \tag{6.46}$$

[17] The form of the matrix inversion lemma used states that for a positive definite $N \times N$ matrix $\mathbf{A}$, scalar a, and $N \times 1$ vector $\mathbf{a}$, we have

$$\left(\mathbf{A} + a\mathbf{a}\mathbf{a}^{\mathrm{H}}\right)^{-1} = \mathbf{A}^{-1} - \frac{a\mathbf{A}^{-1}\mathbf{a}\mathbf{a}^{\mathrm{H}}\mathbf{A}^{-1}}{1 + a\mathbf{a}^{\mathrm{H}}\mathbf{A}^{-1}\mathbf{a}}$$

For convergence in the mean square, the misadjustment must remain bounded and positive, that is $1 - \frac{1}{2}\mu \mathrm{tr}[\mathbf{R}] > 0$. Thus, the mean square error $\xi(k)$ converges asymptotically to $\xi(\infty) = J_{\min} = \sigma_q^2$ for

$$0 < \mu < \frac{2}{\mathrm{tr}[\mathbf{R}]} = \frac{2}{\sigma_x^2 N} \tag{6.47}$$

For mathematical tractability, the above bound on the learning rate has been derived for a white iid input. In practical applications this bound is considerably lower, and depends on the condition number of the input correlation matrix $\nu = \lambda_{\max}/\lambda_{\min}$, or equivalently on the flatness of the power spectrum of the input.

From Equation (6.46), as the misadjustment is directly proportional to the stepsize μ, the requirements of fast convergence and good steady state properties of an adaptive filtering algorithm are contradictory. An algorithm will have fast initial convergence for a large μ, whereas it will converge to the optimal solution for a small μ. For the same learning rate μ, shorter filters will exhibit lower misadjustment. Chapter 8 introduces adaptive filters with variable learning rates, which can cope better with the requirements of fast initial convergence and low misadjustment.

6.3 Nonlinear Feedforward Complex Adaptive Filters

Following on the derivation of the CLMS algorithm, we shall now introduce stochastic gradient learning algorithms for training feedforward nonlinear filters, for which a generic block diagram is shown in Figure 6.5. The nonlinearity $\Phi(\cdot)$ can be from any class addressed in Chapter 4.

To simplify the notation, the output of the filter in Figure 6.5, can be expressed as

$$\begin{aligned} y(k) = \Phi\big(\mathbf{x}^{\mathrm{T}}(k)\mathbf{w}(k)\big) &\rightarrow \Phi(k) = u\big(\sigma(k), \tau(k)\big) + \jmath v\big(\sigma(k), \tau(k)\big) = u(k) + \jmath v(k) \\ \mathbf{x}^{T}(k)\mathbf{w}(k) &\rightarrow \sigma(k) + \jmath\tau(k) \end{aligned} \tag{6.48}$$

6.3.1 Fully Complex Nonlinear Adaptive Filters

As fully complex nonlinearities are complex functions of complex variables, the Complex Nonlinear Gradient Descent (CNGD) algorithm for training this class of filters can be derived similarly to the derivation of CLMS. If $\Phi(k)$ is analytic on a region in $\mathbb{C}$, then based on the

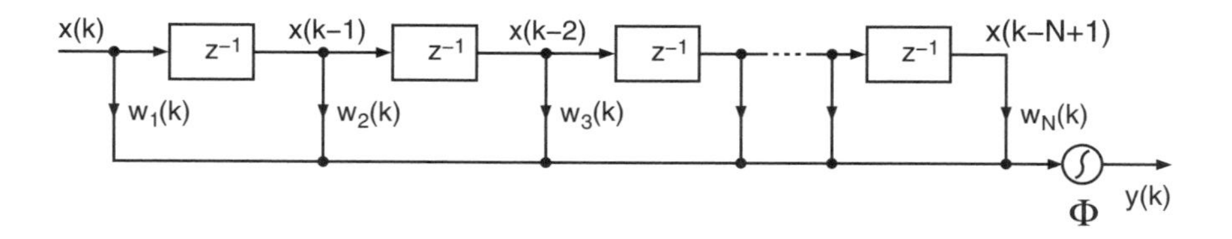

Figure 6.5 Nonlinear adaptive filter

cost function

$$J(k) = \frac{1}{2}|e(k)|^2 \tag{6.49}$$

we have

$$\nabla_{\mathbf{w}} J(k) = \frac{\partial J(k)}{\partial \mathbf{w}_r(k)} + \jmath \frac{\partial J(k)}{\partial \mathbf{w}_i(k)} = \nabla_{\mathbf{w}_r} J(k) + \jmath \nabla_{\mathbf{w}_i} J(k)$$

where

$$\begin{aligned}\nabla_{\mathbf{w}_r} J(k) &= \frac{1}{2} e(k) \nabla_{\mathbf{w}_r} e^*(k) + \frac{1}{2} e^*(k) \nabla_{\mathbf{w}_r} e(k) \\ \nabla_{\mathbf{w}_i} J(k) &= \frac{1}{2} e(k) \nabla_{\mathbf{w}_i} e^*(k) + \frac{1}{2} e^*(k) \nabla_{\mathbf{w}_i} e(k)\end{aligned} \tag{6.50}$$

Since the output error of the linear adaptive filter in Figure 6.2 can be expressed as

$$e(k) = d(k) - \Phi\big(\mathbf{x}^{\mathrm{T}}(k)\mathbf{w}(k)\big)$$

and the partial derivatives for the 'net input' $net(k) = \mathbf{x}^T(k)\mathbf{w}(k)$ from Equation (6.48) are

$$\begin{aligned}\frac{\partial \sigma(k)}{\partial \mathbf{w}_r(k)} &= \mathbf{x}_r(k) \quad \frac{\partial \sigma(k)}{\partial \mathbf{w}_i(k)} = -\mathbf{x}_i(k) \\ \frac{\partial \tau(k)}{\partial \mathbf{w}_r(k)} &= \mathbf{x}_i(k) \quad \frac{\partial \tau(k)}{\partial \mathbf{w}_i(k)} = \mathbf{x}_r(k)\end{aligned} \tag{6.51}$$

Thus, the gradients with respect to the real and parts of the weight vector in Equation (6.50) become

$$\frac{\partial J(k)}{\partial \mathbf{w}_r(k)} = -e_r(k)\left[u_\sigma(k)\mathbf{x}_r(k) + u_\tau(k)\mathbf{x}_i(k)\right] - e_i(k)\left[v_\sigma(k)\mathbf{x}_r(k) + v_\tau(k)\mathbf{x}_i(k)\right] \tag{6.52}$$

$$\frac{\partial J(k)}{\partial \mathbf{w}_i(k)} = -e_r(k)\left[-u_\sigma(k)\mathbf{x}_i(k) + u_\tau(k)\mathbf{x}_r(k)\right] - e_i(k)\left[-v_\sigma(k)\mathbf{x}_i(k) + v_\tau(k)\mathbf{x}_r(k)\right] \tag{6.53}$$

where $u_\sigma = \partial u/\partial \sigma,\ u_\tau = \partial u/\partial \tau,\ v_\sigma = \partial v/\partial \sigma,\ v_\tau = \partial v/\partial \tau$ exist and are bounded.

By employing the Cauchy–Riemann equations (see Chapter 5), we have

$$u_\sigma(k) = v_\tau(k) \qquad u_\tau(k) = -v_\sigma(k) \tag{6.54}$$

and the gradient $\nabla_{\mathbf{w}} J(k) = \nabla_{\mathbf{w}_r} J(k) + \jmath \nabla_{\mathbf{w}_i} J(k)$ simplifies into

$$\begin{aligned}\nabla_{\mathbf{w}} J(k) &= -\mathbf{x}^*(k)\Big[e_r(k)\big(u_\sigma(k) - \jmath v_\sigma(k)\big) + e_i(k)\big(v_\sigma(k) + \jmath u_\sigma(k)\big)\Big] \\ &= -\mathbf{x}^*(k)\big[\Phi'^*(k)e_r(k) + \jmath\Phi'^*(k)e_i(k)\big] \\ &= -\mathbf{x}^*(k)\Phi'^*(k)e(k)\end{aligned} \tag{6.55}$$

where, for convenience, $\Phi'(\mathbf{x}^{\mathrm{T}}(k)\mathbf{w}(k)) = \Phi'(k)$.

Finally, the weight update for the complex nonlinear gradient descent (CNGD) algorithm for training fully complex nonlinear adaptive filters is given by

$$\mathbf{w}(k+1) = \mathbf{w}(k) + \mu e(k)\Phi'^{*}(k)\mathbf{x}^{*}(k), \qquad \mathbf{w}(0) = \mathbf{0} \tag{6.56}$$

The nonlinear nature of the filter is reflected in the first derivative of a fully complex nonlinearity within the update. This weight update has the same generic form as CLMS (Equation 6.21) and hence it degenerates into CLMS for a linear Φ. Properties of the class of fully complex nonlinear functions are given in Chapter 4 and Appendix E, whereas their representation as Möbius transformations is addressed in Chapter 11.

6.3.2 Derivation of CNGD using $\mathbb{CR}$ calculus

The complex nonlinear gradient descent algorithm can be alternatively derived using the $\mathbb{CR}$ calculus. From Equation (5.40), the gradient of the cost function is calculated along the conjugate direction of the weights, that is $\nabla_{\mathbf{w}} J = 2\partial J(k)/\partial \mathbf{w}^{*}(k)$.

For most nonlinear activation functions used in this work[18]

$$\partial\Phi^{*}/\partial net = \partial\Phi/\partial net^{*} = \Phi'^{*} \qquad \partial net^{*}(k)/\partial \mathbf{w}^{*}(k) = \mathbf{x}^{*}(k) \tag{6.57}$$

Thus $\nabla_{\mathbf{w}} J(k) = -e(k)\Phi'^{*}(k)\mathbf{x}^{*}(k)$ and we obtain the algorithm (6.56).

Convergence of feedforward nonlinear adaptive filters. Because of the effects of nonlinearity, it is difficult to derive directly the conditions for convergence of fully complex adaptive feedforward filters. However, for a contractive Φ, an approximate analysis can be conducted based on the contraction mapping theorem (see Appendix P) [190], and the convergence analysis in Section 6.2.1. We shall next analyse the mean weight vector convergence for fully complex feedforward adaptive filters.

We can express, without loss in generality, the desired response as

$$d(k) = \Phi(\mathbf{x}^{T}(k)\mathbf{w}_{o}) + q(k) \tag{6.58}$$

where $q(k)$ is complex Gaussian noise with variance σ_q^2, uncorrelated with $\mathbf{x}(k)$. The instantaneous output error and the weight update can now be expressed as

$$\begin{aligned} e(k) &= \Phi\big(\mathbf{x}^{T}(k)\mathbf{w}_{o}\big) + q(k) - \Phi\big(\mathbf{x}^{T}(k)\mathbf{w}(k)\big) \\ \mathbf{w}(k+1) &= \mathbf{w}(k) + \mu q(k)\Phi'^{*}(k)\mathbf{x}^{*}(k) - \mu\Phi'^{*}(k)\mathbf{x}^{*}(k)\Big[\Phi\big(\mathbf{x}^{T}(k)\mathbf{w}(k)\big) - \Phi\big(\mathbf{x}^{T}(k)\mathbf{w}_{o}\big)\Big] \end{aligned} \tag{6.59}$$

If Φ is a contraction,[19] then the term in the square brackets can be approximated as (recall that $\mathbb{C}$ is not an ordered field – see Appendix A)

$$|\Phi\big(\mathbf{x}^{T}(k)\mathbf{w}(k)\big) - \Phi\big(\mathbf{x}^{T}(k)\mathbf{w}_{o}\big)| \leq \gamma|\mathbf{x}^{T}(k)(\mathbf{w}(k) - \mathbf{w}_{o})| = \beta|\mathbf{x}^{T}(k)\mathbf{v}(k)| \tag{6.60}$$

[18] In addition, for the class of elementary transcendental functions, which are typical fully complex nonlinear activation functions used in this work, we have $(\Phi')^{*} = (\Phi^{*})' = \Phi'^{*}$.

[19] By the contraction mapping theorem $|\Phi(b) - \Phi(a)| \leq \gamma|b - a|,\ \gamma < 1,\ a, b \in S \subset \mathbb{C}$ (see Appendix P).

Subtract the optimal weight vector $\mathbf{w}_o$ from both sides of Equation (6.59) and for simplicity ignore the modulus in Equation (6.60) to obtain the recursion for the weight error vector

$$\mathbf{v}(k+1) = \mathbf{v}(k) - \mu\beta\Phi'^{*}(k)\mathbf{x}^{*}(k)\mathbf{x}^{T}(k)\mathbf{v}(k) + \mu q(k)\Phi'^{*}(k)\mathbf{x}^{*}(k) \tag{6.61}$$

For Φ a contraction, its first derivative $|\Phi'(k)| < 1$ and can be replaced by a constant $\gamma < 1$. Upon applying the statistical expectation operator and using the independence assumptions $(q \perp \mathbf{x})$

$$E\big[|\mathbf{v}(k+1)|\big] \leq \big|\mathbf{I} - \mu\beta\gamma E[\mathbf{x}(k)\mathbf{x}^{H}(k)]\big| E\big[|\mathbf{v}(k)|\big] \tag{6.62}$$

Similarly to Equation (6.32), the modes of convergence for Equation (6.62) are dominated by the largest eigenvalue of $\mathbf{R} = E[\mathbf{x}(k)\mathbf{x}^{H}(k)]$, and for a white iid input the bound on the learning rate which preserves convergence in the mean is

$$0 < \mu < \frac{2}{\alpha \parallel \mathbf{x}(k) \parallel_2^2} \tag{6.63}$$

where α is a positive parameter derived from the first derivative of Φ.

6.3.3 Split-complex Approach

This class of filters has the same general architecture as fully complex feedforward adaptive filters shown in Figure 6.5. As has been shown in Chapter 4, we differentiate between the real–imaginary and amplitude–phase split-complex approaches. The former is suitable for signals which exhibit symmetry around the real and imaginary axes, and the latter is best suited for rotational processes.

Real–imaginary split-complex approach (RISC). Split-complex nonlinear activation functions were originally introduced for complex neural networks employed for binary classification [166] and nonlinear equalisation in communications [27]. The output of a feedforward nonlinear adaptive filter with a RISC nonlinearity is given by

$$y(k) = \Phi(\mathbf{x}^{T}(k)\mathbf{w}(k)) = \Phi\big(net(k)\big) = \sigma\big(net_r(k)\big) + \jmath\sigma\big(net_i(k)\big) \tag{6.64}$$

where the real and imaginary parts of the complex net input $net(k)$ are processed separately by real valued sigmoid functions σ. Figure 6.6 shows one such nonlinearity – a hyperbolic tangent split complex activation function.

As the gradients with respect to the real and imaginary parts of the output are calculated independently – in the same way as in real valued nonlinear adaptive filters, the weight update in the RISC approach becomes

$$\mathbf{w}(k+1) = \mathbf{w}(k) + \mu\Big(e_r(k)\sigma'\big(net_r(k)\big) + \jmath e_i(k)\sigma'\big(net_i(k)\big)\Big)\mathbf{x}^{*}(k) \tag{6.65}$$

Amplitude–phase split-complex approach (APSC). Examples include filters with the output nonlinearity proposed by Georgiou and Koutsougeras [88], given by

$$\Phi(z) = \frac{z}{c + |z|/r} \tag{6.66}$$

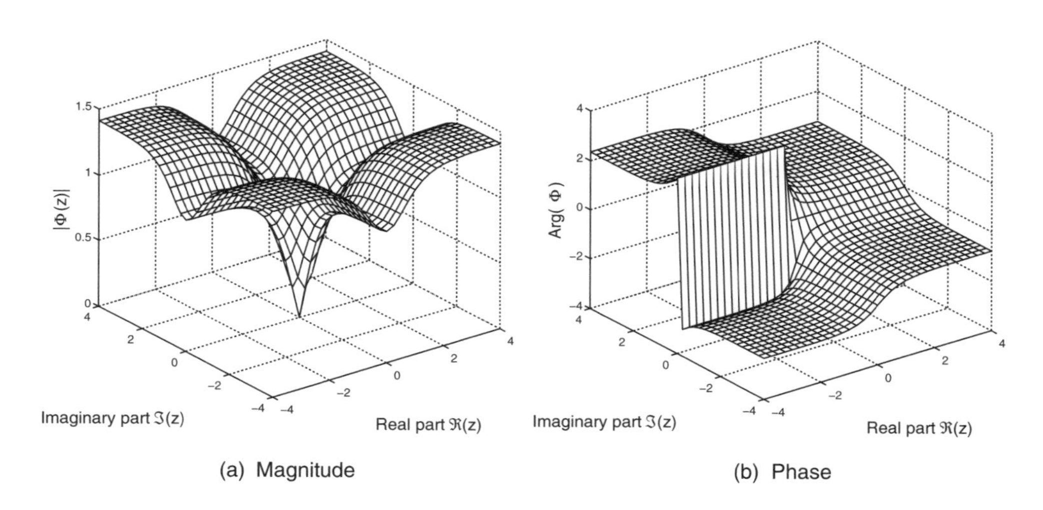

Figure 6.6 Real–imaginary split-complex function $\Phi(z) = \tanh(u) + \jmath \tanh(v)$

and filters based on the output nonlinearity introduced by Hirose [118], given by

$$\Phi(z) = \tanh(|z|/m)e^{\jmath \arg(z)} \tag{6.67}$$

where c, r, and m are real positive constants. The filter output for the class of APSC nonlinearities has a general form

$$y(k) = \Phi\big(\mathbf{x}^{\mathrm{T}}(k)\mathbf{w}(k)\big) = \Phi\big(net(k)\big) = \sigma\big(|net(k)|\big)e^{\jmath \arg(net(k))} \tag{6.68}$$

for which the update becomes (for $e^{\jmath\varphi} = e^{\jmath \arg(net(k))}$)

$$\mathbf{w}(k+1) = \mathbf{w}(k) + \mu\mathbf{x}^*(k)e^{\jmath\varphi}\left[\Re\left\{e(k)\sigma'(|net(k)|)e^{\jmath\varphi}\right\} + \jmath\Im\left\{\frac{1}{|net(k)|}e(k)\sigma(|net(k)|)e^{\jmath\varphi}\right\}\right] \tag{6.69}$$

where the symbols $\Re\{\cdot\}$ and $\Im\{\cdot\}$ denote respectively the real and imaginary part of a complex number.

Boundendess vs differentiability. The output of split-complex nonlinear adaptive filters is bounded by virtue of the saturation type real valued nonlinearities within the activation functions. However, split-complex functions are not differentiable in the complex sense, and hence the stochastic gradient learning algorithms for RISC and APSC adaptive filters do not have the same generic form as CLMS.

6.3.4 Dual Univariate Adaptive Filtering Approach (DUAF)

The dual univariate approach deals with complex data by splitting the input signal into its real and imaginary parts and treating them as independent real valued quantities [156, 213], as shown in Figure 6.7. Dual univariate filters therefore perform suboptimally when dealing with complex valued signals with rich nonlinear behaviour and coupling between the real and

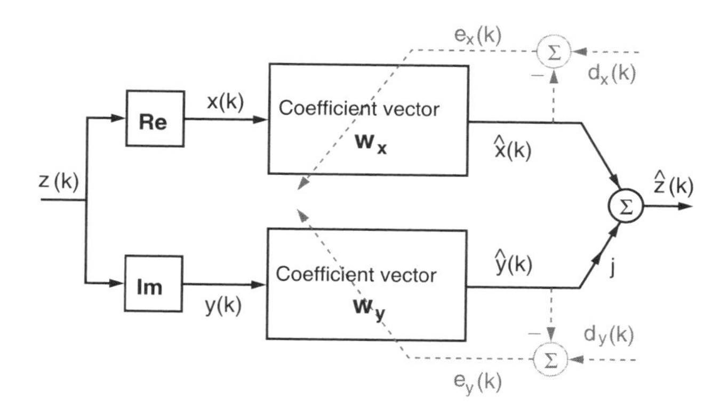

Figure 6.7 Dual univariate real adaptive filter

imaginary parts. However, they are simple and fast, and can be used for processes for which the real and imaginary part are not heavily correlated.

Any real valued adaptive filter can be used for the x and y channel, these can be linear, nonlinear, feedforward or recurrent [190], and the two channels do not have to share the same filter. In the simplest case, the weight vectors $\mathbf{w}_x(k)$ and $\mathbf{w}_y(k)$ are updated using standard LMS, to give

$$\begin{aligned}\mathbf{w}_x(k+1) &= \mathbf{w}_x(k) + \mu e_x(k)\mathbf{x}(k)\\ \mathbf{w}_y(k+1) &= \mathbf{w}_y(k) + \mu e_y(k)\mathbf{y}(k)\end{aligned} \tag{6.70}$$

where the corresponding instantaneous output errors are

$$e_x(k) = d_x(k) - \hat{x}(k) \qquad\qquad e_y(k) = d_y(k) - \hat{y}(k) \tag{6.71}$$

6.4 Normalisation of Learning Algorithms

The weight update $\Delta\mathbf{w}(k)$ within stochastic gradient learning algorithms is proportional to the input vector $\mathbf{x}(k)$, which causes gradient noise amplification for large magnitudes of the input. For instance, within the CLMS algorithm, we have $\Delta\mathbf{w}(k) = \mu e(k)\mathbf{x}^*(k)$. Learning algorithms can be made independent of the input signal power by normalising their updates, as shown below.

Normalised CLMS (NCLMS). Observe from Equation (6.32) that for the convergence of CLMS in the mean square, the bound on the stepsize is given by

$$0 < \mu < \frac{2}{\sigma_x^2 N}$$

and that the input signal power σ_x^2 can be estimated as[20]

$$\sigma_x^2 = E\big[|x(k)|^2\big] \approx \frac{1}{N}\sum_{n=1}^{N}|x(k-n)|^2 = \frac{1}{N}\mathbf{x}^{\mathrm{H}}(k)\mathbf{x}(k)$$

which leads to the following bound on the stepsize

$$0 < \mu < \frac{2}{\mathbf{x}^{\mathrm{H}}(k)\mathbf{x}(k)} \tag{6.72}$$

This stepsize can be incorporated into the CLMS, to give the normalised CLMS (NCLMS) algorithm

$$\mathbf{w}(k+1) = \mathbf{w}(k) + \frac{\eta}{\|\mathbf{x}(k)\|_2^2}e(k)\mathbf{x}^*(k) \tag{6.73}$$

where $0 < \eta < 2$. The effective stepsize

$$\mu(k) = \frac{\eta}{\|\mathbf{x}(k)\|_2^2} \tag{6.74}$$

is time varying, it alters the magnitude, but not the direction of the estimated gradient vector and greatly reduces the gradient noise, resulting in faster convergence as compared with CLMS [110].

Another, albeit approximate, way to perform normalisation of stochastic gradient learning algorithms for feedforward adaptive filters is to find the stepsize which minimises the error $e(k+1)$ based on its Taylor series expansion [187, 190, 279], given by

$$e(k+1) = e(k) + \sum_{n=1}^{N}\frac{\partial e(k)}{\partial w_n(k)}\Delta w_n(k) + \sum_{n=1}^{N}\sum_{m=1}^{N}\frac{\partial e^2(k)}{\partial w_n(k)\partial w_m(k)}\Delta w_n(k)\Delta w_m(k) + \cdots \tag{6.75}$$

For the CLMS (in the prediction setting $\mathbf{x}(k) = [x(k-1), \ldots, x(k-N]^{\mathrm{T}})$ the second- and higher-order terms in Equation (6.75) vanish, and

$$\frac{\partial e(k)}{\partial w_n(k)} = -x(k-n) \qquad\qquad \Delta w_n(k) = \mu e(k)x^*(k-n)$$

which gives

$$e(k+1) = e(k)\big[1 - \mu \|\mathbf{x}(k)\|_2^2\big]$$

The error $e(k+1)$ is zero for $e(k) = 0$ (trivial solution), and for

$$\mu(k) = \frac{1}{\|\mathbf{x}(k)\|_2^2} \tag{6.76}$$

that is, the stepsize of NCLMS.

[20]For convenience, we consider the prediction setting, where $\mathbf{x}(k) = [x(k-1), \ldots, x(k-N)]^{\mathrm{T}}$.

Normalised CNGD. Similarly to NCLMS, the CNGD algorithm for fully complex nonlinear adaptive filters can be normalised based on the Taylor series expansion (6.75), where

$$\frac{\partial e(k)}{\partial w_n(k)} = -\Phi'(k)x(k-n) \qquad \Delta w_n(k) = \mu e(k)\Phi'^*(k)x^*(k-n)$$

This gives the following 'optimal' learning rate for the complex nonlinear gradient descent algorithm

$$\mu(k) = \frac{1}{|\Phi'(k)|^2 \parallel \mathbf{x}(k) \parallel_2^2} \tag{6.77}$$

Finally, the normalised complex nonlinear gradient descent (NCNGD) algorithm can be expressed as

$$\mathbf{w}(k+1) = \mathbf{w}(k) + \frac{\eta}{|\Phi'(k)|^2 \parallel \mathbf{x}(k) \parallel_2^2} e(k)\Phi'^*(k)\mathbf{x}^*(k) \tag{6.78}$$

This alternative derivation of optimal stepsizes is only approximate – strictly speaking the normalisation should be based on the minimisation of the *a posteriori* error[21] $d(k)-\mathbf{x}^{\mathrm{T}}(k)\mathbf{w}(k+1)$ [61, 248].

6.5 Performance of Feedforward Nonlinear Adaptive Filters

Simulations have been conducted to illustrate:

- performance comparison between the linear CLMS and the fully complex, split complex, and dual univariate approaches;
- learning curves, showing convergence of learning algorithms when processing linear and nonlinear signals;
- performance comparison within the class of fully complex filters, for all the elementary transcendental functions used as nonlinearities.

Simulations were performed in a one step ahead prediction setting, on a linear AR(4) signal given by

$$y(k) = 1.79y(k-1) - 1.85y(k-2) + 1.27y(k-3) - 0.41y(k-4) + n(k) \tag{6.79}$$

where $n(k)$ is zero mean complex valued white Gaussian noise with variance $\sigma_n^2 = 1$, together with benchmark nonlinear Lorenz and Ikeda series, and a segment of real world wind data (speed and direction) recorded by an ultrasonic anemometer.[22] The Lorenz, Ikeda, and wind signals were made complex valued by convenience of representation, as explained in Section 2.2; for instance, for the complex wind vector we have $\mathbf{v}(k) = v(k)e^{\jmath\phi(k)}$, where v denotes the wind speed and $\phi(k)$ direction.

[21] For more detail see Chapter 10 and Appendix M.

[22] These datasets are described in detail in Chapters 8 and 13.

Table 6.1 Prediction performance (in R_p [dB]) for the fully complex (FCAF), real–imaginary split-complex (RISC), linear CLMS, and dual univariate (DUAF) adaptive filters

	AR(4)	Lorenz	Ikeda	Wind
FACF	2.5222	15.2475	1.2364	10.2217
RISC	2.4969	14.4656	1.1535	9.8896
CLMS	2.5224	14.5456	1.1967	9.8939
DUAF	1.6528	11.7529	–0.3072	8.7353

The quantitative performance criterion was the prediction gain

$$R_p = 10 \log \frac{\sigma_y^2}{\sigma_e^2} [dB] \tag{6.80}$$

For the AR(4) and Ikeda series, the results were averaged over 200 independent simulations, whereas for the complex Lorenz and wind signal single trial simulations were performed. In all cases, the filter tap length was $N = 4$, the signals were standardised to zero mean and maximum magnitude $|x|_{\max} = 0.8$, and the learning rate was $\mu = 0.01$.

Table 6.1 shows the prediction performance for all the feedforward adaptive filtering architectures considered. The data were 1000 samples long, and R_p was calculated over the last 100 samples. The fully complex filter used the complex tanh as the output nonlinearity and exhibited the best performance for nonlinear signals, whereas for the linear AR(4) process its performance was similar to that of CLMS. The performances of the split complex filter, which used the real tanh, and the CLMS were similar. The dual univariate approach had the worst performance, as by design, it did not take into account the correlation between the real and imaginary channel of complex quantities.

Figure 6.8 shows learning curves for the linear CLMS, nonlinear CNGD, DUAF, and normalised CLMS (NCLMS). This was achieved by averaging 200 independent simulations in a

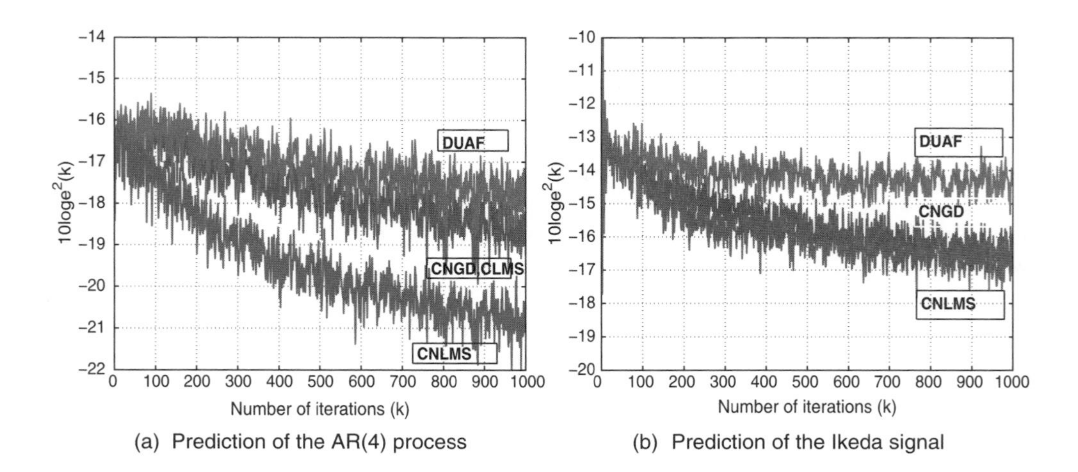

(a) Prediction of the AR(4) process

(b) Prediction of the Ikeda signal

Figure 6.8 Learning curves for one step ahead prediction of a linear and nonlinear signal

Table 6.2 Prediction performance (in R_p [dB]) for the fully complex filter and all elementary transcendental functions

	tan	sin	arctan	arcsin	tanh	sinh	arctanh	arcsinh
AR(4)	4.8968	4.9318	4.9057	4.9149	4.8955	4.9052	4.8820	4.8855
Lorenz	15.4061	16.9497	16.7381	16.0620	16.9510	16.0966	15.3699	16.9214
Ikeda	2.1739	2.3037	2.3361	2.2226	2.3383	2.2332	2.1846	2.3187
Wind	11.2484	11.7029	10.7832	12.0004	11.2454	11.9136	11.4867	11.5850

one step ahead prediction setting, for the linear AR(4) model (6.79) and nonlinear Ikeda map. Figure 6.8(a) illustrates that, for the linear signal, the fully complex nonlinear filter trained by CNGD performed similarly to CLMS, DUAF exhibited slightly inferior performance, whereas the NCLMS outperformed all the other algorithms. For the nonlinear Ikeda map, however, the CNGD had considerable performance advantage over DUAF, and approached the performance of NCLMS, as shown in Figure 6.8(b).

In the second set of simulations, only the fully complex nonlinearities were considered and the prediction performances were evaluated over the last 1000 samples of 6000 sample long datasets, and for all the elementary transcendental functions, as shown in Table 6.2. As the filters exhibited similar performances for all the fully complex nonlinearities, for convenience, we will most frequently use the tanh function.

6.6 Summary: Choice of a Nonlinear Adaptive Filter

The linear adaptive filter shown in Figure 6.2 is a simple and powerful adaptive filtering architecture. Its output is a linear combination of the inputs in filter memory and the set of adaptive filter coefficients (filter weights). However, due to its linear nature, this filter may perform suboptimally when processing nonlinear signals. To this end, we have introduced a class of nonlinear adaptive filters which comprise the standard linear adaptive filter and the output nonlinearity, as shown in Figure 6.5. This structure, called a nonlinear adaptive filter or a dynamical perceptron, is also a basic building block for neural networks. Properties and convergence of linear and nonlinear complex adaptive filters have been analysed, and their performance has been illustrated for both signals complex by design and by convenience of representation.

To summarise:

- The complex least mean square (CLMS) algorithm has been introduced for training complex linear adaptive filters, and its convergence in the mean and in the mean square has been addressed.
- Two classes of complex nonlinear adaptive filters have been introduced – fully complex and split-complex, together with the dual univariate approach.
- When selecting a nonlinear function at the output of a complex nonlinear adaptive filter, we need to choose between differentiability (fully complex) and boundedness (split-complex); this choice depends on the application – split-complex filters are more commonly used in neural networks for classification, whereas fully complex filters are a more natural choice in nonlinear adaptive filtering.

- Nonlinear adaptive filters based on fully complex nonlinearities retain the same generic form in their updates as linear adaptive filters, and provide the most consistent performance.
- When using fully complex filters we often need to standardise inputs; to allow for more freedom in the processing of general complex signals, Chapter 9 introduces nonlinear adaptive filters with an adaptive amplitude of nonlinearity.
- For best performance, the learning rate should be large at the beginning of the adaptation – for fast convergence, and small in the steady state – for good misadjustment. These are contradictory requirements, and to deal with these issues filters with a variable stepsize are introduced in Chapter 8.
- Learning algorithms in this chapter have been introduced based on standard complex statistics and are optimal when processing circular signals, that is, signals with rotation invariant distributions. However, when processing noncircular data (see Chapter 12), it is more appropriate to use so called widely linear models, which take into account both the covariance $C = E[\mathbf{x}(k)\mathbf{x}^{\mathrm{H}}(k)]$ and pseudocovariance $\mathbb{P} = E[\mathbf{x}(k)\mathbf{x}^{\mathrm{T}}(k)]$ functions. One such adaptive filtering algorithm is the augmented CLMS (ACLMS), introduced in Chapter 13.

Chapter 7 extends the class of feedforward nonlinear adaptive filters to allow feedback. Feedback architectures are very useful when modelling systems with long impulse responses which would require long feedforward filters.

7

Adaptive Filters with Feedback

Real world signals are, in general, nonlinear, nonstationary, and generated by systems with long impulse responses. For their efficient modelling, then, it is natural to consider architectures with:

- nonlinearity, in order to cater for the possibly nonlinear signal natures;
- feedback, to make it possible to model systems with long impulse responses and memory, together with processes with long time dependencies;
- adaptive coefficient update, to be able to cope with nonstationarity.

This chapter introduces training algorithms for complex valued adaptive filters with feedback, as a natural extension of standard feedforward adaptive filters, introduced in Chapter 6. Our emphasis is on nonlinear adaptive filters, and so learning algorithms will be developed for temporal problems.

Due to the necessity for online adaptive mode of operation, the learning algorithms will be developed based on direct gradient calculation,[1] and for the following architectures:

- Linear Infinite Impulse Response (IIR) adaptive filters. These are standard linear adaptive filters, equipped with feedback – this makes them perfectly suited for adaptive filtering applications when memory is a main concern, for instance, to represent Autoregressive Moving Average (ARMA) models.
- Recurrent perceptrons and Recurrent Neural Networks (RNN). A recurrent perceptron can be considered an IIR filter equipped with output nonlinearity (see Chapter 4), hence making it suitable to represent Nonlinear ARMA (NARMA) models. An RNN is a generalisation of a recurrent perceptron, it has multiple output and hidden neurons, is fully connected, and can have both local and global feedback [190].

The analysis is performed using $\mathbb{CR}$ calculus; this greatly simplifies the derivation of learning algorithms and allows us to develop generic expressions for the update of the sensitivity terms within the algorithms.

[1]No backpropagation networks will be considered.

Complex Valued Nonlinear Adaptive Filters: Noncircularity, Widely Linear and Neural Models
Danilo P. Mandic and Vanessa Su Lee Goh

7.1 Training of IIR Adaptive Filters

An infinite impulse response (IIR) adaptive filter (for convenience, the architecture is presented in the prediction configuration) is shown in Figure 7.1. It consists of a feedforward tap input delay line comprising the external input $\mathbf{x}(k) = [x(k-1), \ldots, x(k-M)]^{\mathrm{T}}$, delayed feedback $\mathbf{y}(k) = [y(k-1), \ldots, y(k-N)]^{\mathrm{T}}$, and the corresponding adaptive filter coefficients $\mathbf{b}(k) = [b_1(k), \ldots, b_M(k)]^{\mathrm{T}}$ and $\mathbf{a}(k) = [a_1(k), \ldots, a_N(k)]^{\mathrm{T}}$, all complex valued. The IIR adaptive filtering architecture can be used within all the four standard adaptive filtering configurations addressed in Section 6.1, that is, system identification, inverse system modelling, noise cancellation, and prediction.[2]

The output $y(k)$ of the adaptive IIR filter in Figure 7.1 is given by[3]

$$y(k) = \sum_{n=1}^{N} a_n(k)y(k-n) + \sum_{m=1}^{M} b_m(k)x(k-m) = \mathbf{a}^{\mathrm{T}}(k)\mathbf{y}(k) + \mathbf{b}^{\mathrm{T}}(k)\mathbf{x}(k) \tag{7.1}$$

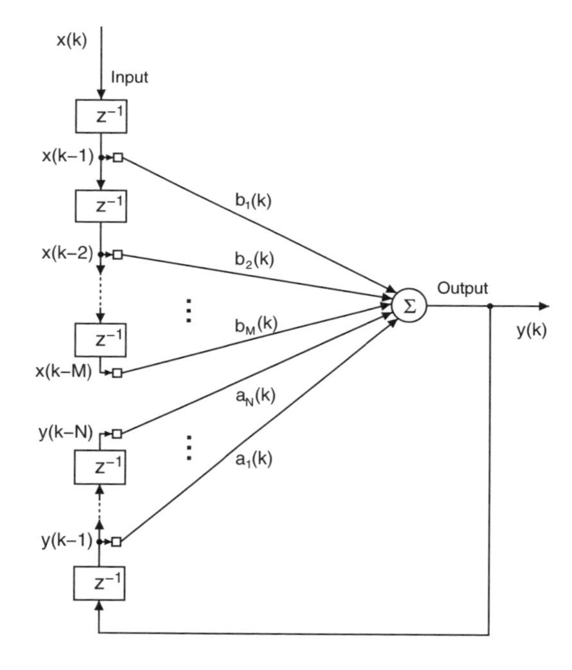

Figure 7.1 An adaptive infinite impulse response (IIR) filter (in the prediction setting)

[2] A comprehensive account of adaptive IIR filters can be found in the work by Regalia, Shynk, and Johnson [136, 248, 274], whereas a recursive algorithm for the training of complex valued adaptive IIR filters was introduced by Shynk [275]. The derivation of an *output–error* learning algorithm for complex valued adaptive IIR filters presented in this Chapter is based on [190, 275].

[3] This equation is written for the prediction setting. Depending on the application, we may also have the term $b_0(k)x(k)$, and the second sum would start from $m = 0$. The adaptive IIR filter in a system identification configuration is also referred to as a Model Reference Adaptive System (MRAS) in the control literature.

and reflects the autoregressive moving average (ARMA) nature of this architecture. The output of an IIR filter in Equation (7.1) can be expressed in a more compact form as

$$y(k) = \frac{B(k, z^{-1})}{1 - A(k, z^{-1})} x(k) \tag{7.2}$$

The input–output relation for IIR filters can have alternative forms corresponding to those in Equations (6.1–6.3) for linear adaptive FIR filters. For consistency, and without loss in generality, we will develop learning algorithms based on the input–output relationship (Equation 7.1).

7.1.1 Coefficient Update for Linear Adaptive IIR Filters

The adaptive IIR architecture can be used as a generic adaptive filter in Figure 6.1, and the output error and cost function can be defined in the same way as those for FIR filters, that is

$$\begin{aligned} e(k) &= d(k) - y(k) \\ J(k) &= \frac{1}{2}|e(k)|^2 = \frac{1}{2}e(k)e^*(k) \end{aligned} \tag{7.3}$$

In the stochastic gradient descent setting, we wish to update the filter weights

$$\mathbf{w}(k) = [b_1(k), \ldots, b_M(k), a_1(k), \ldots, a_N(k)]^{\mathrm{T}} \tag{7.4}$$

recursively, based on

$$\mathbf{w}(k+1) = \mathbf{w}(k) - \mu \nabla_{\mathbf{w}} J(k)_{|\mathbf{w}=\mathbf{w}(k)} \tag{7.5}$$

Function $J(k)$ is a real function of complex variable, and its gradient is calculated using $\mathbb{R}$-derivatives,[4] as shown in Section 5.3.1. This allows us to write the above update in an expanded form

$$\begin{aligned} \mathbf{w}^r(k+1) &= \mathbf{w}^r(k) - \mu \nabla_{\mathbf{w}^r} J(k) \\ \mathbf{w}^i(k+1) &= \mathbf{w}^i(k) - \mu \nabla_{\mathbf{w}^i} J(k) \end{aligned} \tag{7.6}$$

where the overall weight vector of the IIR filter, and the gradients with respect to the real and imaginary parts of the weights are given by

$$\begin{aligned} \mathbf{w}(k) &= \mathbf{w}^r(k) + \jmath \mathbf{w}^i(k) \\ \nabla_{\mathbf{w}^r} J(k) &= \left[\frac{\partial J(k)}{\partial b_1^r(k)}, \ldots, \frac{\partial J(k)}{\partial b_M^r(k)}, \frac{\partial J(k)}{\partial a_1^r(k)}, \ldots, \frac{\partial J(k)}{\partial a_N^r(k)}\right]^{\mathrm{T}} \\ \nabla_{\mathbf{w}^i} J(k) &= \left[\frac{\partial J(k)}{\partial b_1^i(k)}, \ldots, \frac{\partial J(k)}{\partial b_M^i(k)}, \frac{\partial J(k)}{\partial a_1^i(k)}, \ldots, \frac{\partial J(k)}{\partial a_N^i(k)}\right]^{\mathrm{T}} \end{aligned} \tag{7.7}$$

[4] The gradient of the cost function is calculated as $\nabla_{\mathbf{w}} J = \partial J/\partial \mathbf{w}^r + \jmath \partial J/\partial \mathbf{w}^i$.

The 'compound' gradient

$$\nabla_{\mathbf{w}} J(k) = \nabla_{\mathbf{w}^r} J(k) + \jmath \nabla_{\mathbf{w}^i} J(k) \tag{7.8}$$

can now be expressed as[5]

$$\nabla_{\mathbf{w}} J(k) = -\frac{1}{2} e^*(k)\big[\nabla_{\mathbf{w}^r} y(k) + \jmath \nabla_{\mathbf{w}^i} y(k)\big] - \frac{1}{2} e(k)\big[\nabla_{\mathbf{w}^r} y^*(k) + \jmath \nabla_{\mathbf{w}^i} y^*(k)\big] \tag{7.9}$$

Denote 'sensitivities'

$$\begin{aligned} \boldsymbol{\pi}^{\circ}(k) &= \frac{1}{2}\big[\nabla_{\mathbf{w}^r} y(k) + \jmath \nabla_{\mathbf{w}^i} y(k)\big] \\ \boldsymbol{\pi}^{\star}(k) &= \frac{1}{2}\big[\nabla_{\mathbf{w}^r} y^*(k) + \jmath \nabla_{\mathbf{w}^i} y^*(k)\big] \end{aligned} \tag{7.10}$$

Due to the feedback in the IIR filtering architecture, these sensitivities are not straightforward to compute. We shall now derive a recursive expression for $\boldsymbol{\pi}^{\star}(k)$ and show that $\boldsymbol{\pi}^{\circ}(k) = \mathbf{0}$.

Calculation of $\boldsymbol{\pi}^{\circ}(k)$. The elements of the vector $\boldsymbol{\pi}^{\circ}(k) = [\pi_1^{\circ}(k), \ldots, \pi_{N+M}^{\circ}(k)]^T$ are effectively the $\mathbb{R}^*$-derivatives[6] (Equation 5.12). As the filter output $y(k)$ is a standard complex function of complex variable, its $\mathbb{R}^*$-derivatives are zero (see Section 5.3.2) and thus $\boldsymbol{\pi}^{\circ}(k) = \mathbf{0}$. Alternatively, it is straightforward to show that for every $w_n(k) \in \mathbf{w}(k)$

$$\pi_n^{\circ}(k) = \frac{1}{2}\left[\frac{\partial y^r(k)}{\partial w_n^r(k)} + \jmath \frac{\partial y^i(k)}{\partial w_n^r(k)} + \jmath \frac{\partial y^r(k)}{\partial w_n^i(k)} - \frac{\partial y^i(k)}{\partial w_n^i(k)}\right] = 0 \tag{7.11}$$

as the output $y(k)$ is a complex function of complex variable (7.1), and admits the use of the Cauchy–Riemann equations

$$\frac{\partial y^r(k)}{\partial w_n^r(k)} = \frac{\partial y^i(k)}{\partial w_i^r(k)} \quad \text{and} \quad \frac{\partial y^i(k)}{\partial w_n^r(k)} = -\frac{\partial y^r(k)}{\partial w_n^i(k)}$$

to give $\pi_n^{\circ}(k) = 0$ and $\boldsymbol{\pi}^{\circ}(k) = \mathbf{0}$.

Calculation of $\boldsymbol{\pi}^{\star}(k)$. From Equation (7.10), the sensitivities $\pi_n^{\star}(k) \in \boldsymbol{\pi}^{\star}(k)$ are given by

$$\pi_n^{\star}(k) = \frac{1}{2}\left[\frac{\partial y^r(k)}{\partial w_n^r(k)} - \jmath \frac{\partial y^i(k)}{\partial w_n^r(k)} + \jmath \frac{\partial y^r(k)}{\partial w_n^i(k)} + \frac{\partial y^i(k)}{\partial w_n^i(k)}\right] = \frac{\partial y^r(k)}{\partial w_n^r(k)} - \jmath \frac{\partial y^i(k)}{\partial w_n^r(k)} = \frac{\partial y^*(k)}{\partial w_n(k)} \tag{7.12}$$

[5]The gradient with respect to the real part of the weight vector $\nabla_{\mathbf{w}^r} J(k) = -\frac{1}{2}\left[e^*(k)\nabla_{\mathbf{w}^r} y(k) + e(k)\nabla_{\mathbf{w}^r} y^*(k)\right]$, whereas for the gradient with respect to the imaginary part of the weight vector we have $\nabla_{\mathbf{w}^i} J(k) = -\frac{1}{2}\left[e^*(k)\nabla_{\mathbf{w}^i} y(k) + e(k)\nabla_{\mathbf{w}^i} y^*(k)\right]$, and thus $J = -e^*(k)\boldsymbol{\pi}^{\circ}(k) - e(k)\boldsymbol{\pi}^{\star}(k)$.

[6]Recall that the $\mathbb{R}$-derivative is given by $\partial f/\partial z = \frac{1}{2}\left[\frac{\partial f}{\partial x} - \jmath \frac{\partial f}{\partial y}\right]$, the $\mathbb{R}^*$-derivative by $\partial f/\partial z^* = \frac{1}{2}\left[(\partial f/\partial x) + \jmath(\partial f/\partial y)\right]$, and that for a holomorphic function the $\mathbb{R}$-derivative is equal to the standard complex derivative, whereas the $\mathbb{R}^*$-derivative vanishes.

which gives

$$\pi^\star(k) = \left[\frac{\partial y^*(k)}{\partial b_1(k)}, \ldots, \frac{\partial y^*(k)}{\partial b_M(k)}, \frac{\partial y^*(k)}{\partial a_1(k)}, \ldots, \frac{\partial y^*(k)}{\partial a_N(k)}\right]^{\mathrm{T}} \tag{7.13}$$

In this compact form, the sensitivites are now standard complex partial derivatives, as both $y^*(k)$ and $w_n(k)$ are complex functions of complex variable. It is, however, convenient to evaluate them starting from Equation (7.10), that is, based on Wirtinger calculus,[7] to give

$$\begin{aligned}
\frac{\partial y^*(k)}{\partial b_m^r(k)} &= x^*(k-m) + \sum_{l=1}^{N} a_l^*(k) \frac{\partial y^*(k-l)}{\partial b_m^r(k)} \\
\frac{\partial y^*(k)}{\partial b_m^i(k)} &= -\jmath x^*(k-m) + \sum_{l=1}^{N} a_l^*(k) \frac{\partial y^*(k-l)}{\partial b_m^i(k)} \\
\frac{\partial y^*(k)}{\partial a_n^r(k)} &= y^*(k-n) + \sum_{l=1}^{N} a_l^*(k) \frac{\partial y^*(k-l)}{\partial a_n^r(k)} \\
\frac{\partial y^*(k)}{\partial a_n^i(k)} &= -\jmath y^*(k-n) + \sum_{l=1}^{N} a_l^*(k) \frac{\partial y^*(k-l)}{\partial a_n^i(k)}
\end{aligned} \tag{7.14}$$

The above expressions for the sensitivites are with respect to the past values of the output $y^*(k-1), \ldots, y^*(k-N)$ and the current values of the filter coefficients $a_n(k)$ and $b_m(k)$, and do not admit recursive calculation. It is, however, reasonable to assume that for a sufficiently small learning rate μ, the filter coefficients will exhibit little variation,[8] that is

$$\mathbf{w}(k-1) \approx \mathbf{w}(k-2) \approx \cdots \approx \mathbf{w}(k-N) \approx \mathbf{w}(k-N) \tag{7.15}$$

Using this assumption, we have

$$\begin{aligned}
\frac{\partial y^*(k)}{\partial b_m^r(k)} &= x^*(k-m) + \sum_{l=1}^{N} a_l^*(k) \frac{\partial y^*(k-l)}{\partial b_m^r(k-l)} \\
\frac{\partial y^*(k)}{\partial b_m^i(k)} &= -\jmath x^*(k-m) + \sum_{l=1}^{N} a_l^*(k) \frac{\partial y^*(k-l)}{\partial b_m^i(k-l)} \\
\frac{\partial y^*(k)}{\partial a_n^r(k)} &= y^*(k-n) + \sum_{l=1}^{N} a_l^*(k) \frac{\partial y^*(k-l)}{\partial a_n^r(k-l)} \\
\frac{\partial y^*(k)}{\partial a_n^i(k)} &= -\jmath y^*(k-n) + \sum_{l=1}^{N} a_l^*(k) \frac{\partial y^*(k-l)}{\partial a_n^i(k-l)}
\end{aligned} \tag{7.16}$$

[7] Starting from $y^*(k) = \sum_{m=1}^{M} b_m^*(k) x^*(k-m) + \sum_{n=1}^{N} a_n^*(k) y^*(k-n)$.

[8] This approximation is particularly good for a small feedback order N, for more detail see [190, 248, 274, 293].

which enables a recursive update of the sensitivities as, for instance, terms $\partial y^*(k-l)/\partial b^r_m(k-l)$ can be considered delayed versions of the term $\partial y^*(k)/\partial b^r_m(k)$, and thus $\pi^\star_{b_m(k)}(k) = x^*(k-m) + \sum_{l=1}^N a^*_l(k)\pi^\star_{b_m(k)}(k-m)$.

For compactness, we can define the overall input vector to an adaptive IIR filter as a concatenation of the external inputs and feedback, that is[9]

$$\mathbf{u}(k) = \left[x(k-1), \ldots, x(k-M), y(k-1), \ldots, y(k-N)\right]^{\mathrm{T}} \tag{7.17}$$

The recursive update for the sensitivities is now obtained in the form

$$\pi^\star_n(k) = u^*_n(k) + \sum_{l=1}^{N} w^*_{l+M}(k)\pi^\star_n(k-l), \qquad \pi^\star_n(0) = 0, \; n = 1, \ldots, M+N \tag{7.18}$$

and a recursive algorithm for training linear complex valued IIR adaptive filters becomes

$$\mathbf{w}(k+1) = \mathbf{w}(k) + \mu e(k)\boldsymbol{\pi}^\star(k) \tag{7.19}$$

The weight update is along the direction of the the conjugate gradient of the cost function. This also corresponds to the direction providing the maximum rate of gradient change, and conforms with the result obtained by the $\mathbb{CR}$ calculus in Section 5.4.4. The learning algorithm (Equation 7.19) has the same generic form as that for FIR filters, and simplifies into the CLMS when the feedback is removed. It also simplifies into the corresponding algorithm for real valued adaptive IIR filters [190] when all the inputs are real valued.

7.1.2 Training of IIR filters with Reduced Computational Complexity

Computation of the sensitivities involves significant computational complexity. However, the expression for sensitivities allows for significant simplification [275, 293]. Denote

$$\begin{aligned} \theta(k) &= \pi^\star_{a_1}(k) = \frac{\partial y^*(k)}{\partial a_1(k)} \\ \psi(k) &= \pi^\star_{b_1}(k) = \frac{\partial y^*(k)}{\partial b_1(k)} \end{aligned} \tag{7.20}$$

The assumption (7.15) allows us to write

$$\begin{aligned} \theta(k-n) &\approx \pi^\star_{a_n}(k) \qquad n = 1, \ldots, N \\ \psi(k-m) &\approx \pi^\star_{b_m}(k) \qquad m = 1, \ldots, M \end{aligned}$$

and

$$\boldsymbol{\pi}^\star(k) = [\psi(k-1), \ldots, \psi(k-M), \theta(k-1), \ldots, \theta(k-N)]^{\mathrm{T}} \tag{7.21}$$

---

[9]Thus, for instance, $u_2(k) = x(k-2)$ or $u_{M+1}(k) = y(k-1)$, which correspond to the weights $w_2(k) = b_2(k)$ and $w_{M+1}(k) = a_1(k)$.

This new vector of sensitivities offers much reduced computational complexity and memory requirements, and can replace the standard sensitivities in the update of the IIR adaptive filter (Equation 7.19).

7.2 Nonlinear Adaptive IIR Filters: Recurrent Perceptron

The recurrent perceptron can be viewed as an adaptive IIR filter equipped with output nonlinearity, as shown in Figure 7.2. This architecture may also have an optional bias input, set to $(1 + \jmath)$, revealing its link with recurrent neural networks, as it can be regarded as a recurrent neural network with one neuron (for more detail see [190] and Chapter 3). It is convenient to consider the bias input as an $(M + 1)$th external input with constant value $(1 + \jmath)$, and an associated filter weight $b_{M+1}(k)$.

The output of a recurrent perceptron is thus given by

$$y(k) = \Phi\left(\sum_{n=1}^{N} a_n(k)y(k-n) + \sum_{m=1}^{M} b_m(k)x(k-m) + (1+\jmath)b_{M+1}(k)\right) = \Phi\big(net(k)\big) \tag{7.22}$$

indicating its suitability to represent Nonlinear ARMA (NARMA) models. Notice that the net input has the same functional expression as the output of an IIR filter, that is, $net(k) = y_{IIR}(k)$. The overall input vector to a recurrent perceptron and the weight vector are given respectively by

$$\begin{aligned} \mathbf{u}(k) &= \big[x(k-1), \ldots, x(k-M), 1+\jmath, y(k-1), \ldots, y(k-N)\big]^{\mathrm{T}} \\ \mathbf{w}(k) &= \big[b_1(k), \ldots, b_{M+1}(k), a_1(k), \ldots, a_N(k)\big]^{\mathrm{T}} \end{aligned} \tag{7.23}$$

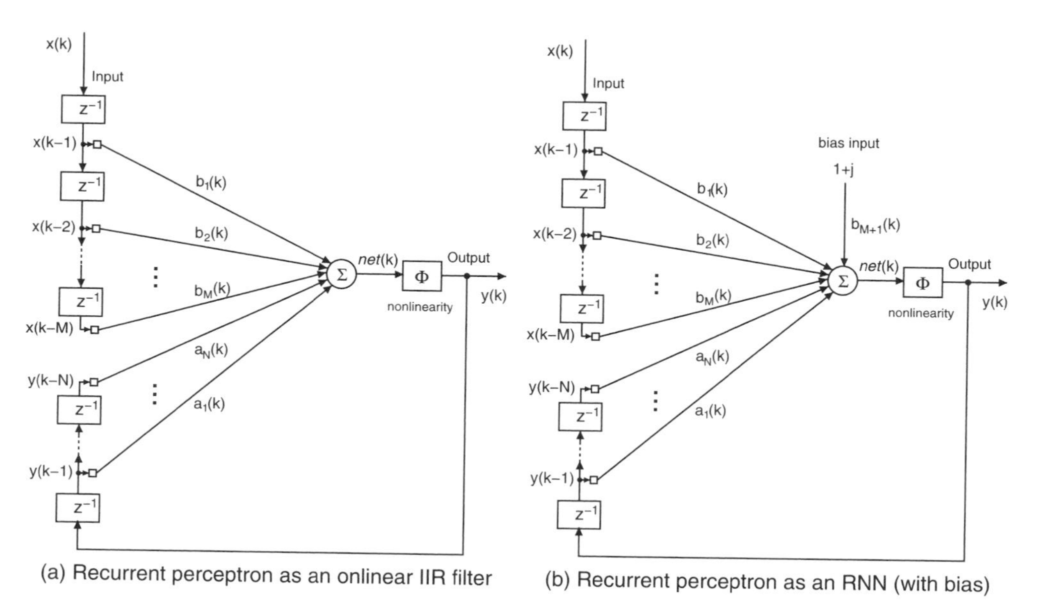

Figure 7.2 A recurrent perceptron (nonlinear IIR filter) in the prediction setting

The expressions for the cost function and weight update are the same as those in Equations (7.3–7.7) for the IIR case, giving

$$\nabla_{\mathbf{w}} J(k) = -e^*(k)\boldsymbol{\pi}^\circ(k) - e(k)\boldsymbol{\pi}^\star(k) \tag{7.24}$$

To evaluate the gradient of the cost function we therefore first need to calculate the sensitivities of the output with respect to the filter coefficients, given by

$$\begin{aligned}\boldsymbol{\pi}^\circ(k) &= \frac{1}{2}\left[\nabla_{\mathbf{w}^r} y(k) + \jmath\nabla_{\mathbf{w}^i} y(k)\right]\\ \boldsymbol{\pi}^\star(k) &= \frac{1}{2}\left[\nabla_{\mathbf{w}^r} y^*(k) + \jmath\nabla_{\mathbf{w}^i} y^*(k)\right]\end{aligned} \tag{7.25}$$

Calculation of $\boldsymbol{\pi}^\circ(k)$. From the analysis of the corresponding term for the linear IIR filter, the sensitivities $\pi_n^\circ(k) = 0,\ n = 1, \ldots, N + M + 1$, since they represent $\mathbb{R}^*$-derivatives of a holomorphic complex valued function $y(k)$. Alternatively, we can use the Cauchy–Riemann equations $\partial y^r(k)/\partial w_n^r(k) = \partial y^i(k)/\partial w_i^r(k)$ and $\partial y^i(k)/\partial w_n^r(k) = -\partial y^r(k)/\partial w_n^i(k)$, to show that[10]

$$\pi_n^\circ(k) = \frac{1}{2}\left[\frac{\partial y^r(k)}{\partial w_n^r(k)} + \jmath\frac{\partial y^i(k)}{\partial w_n^r(k)} + \jmath\frac{\partial y^r(k)}{\partial w_n^i(k)} - \frac{\partial y^i(k)}{\partial w_n^i(k)}\right] = 0 \tag{7.26}$$

Calculation of $\boldsymbol{\pi}^\star(k)$. Although these have the same form as the corresponding sensitivities for the IIR filter, the evaluation of terms $\partial y^*(k)/\partial w_n(k)$ needs some attention, as $y = \Phi(net(k))$ is a nonlinear function of the net input. Let us first convert the $\mathbb{R}^*$-derivative of $y^*(k)$ in Equation (7.25) into the corresponding $\mathbb{C}$-derivative, that is

$$\pi_n^\star(k) = \frac{1}{2}\left[\frac{\partial y^r(k)}{\partial w_n^r(k)} - \jmath\frac{\partial y^i(k)}{\partial w_n^r(k)} + \jmath\frac{\partial y^r(k)}{\partial w_n^i(k)} + \frac{\partial y^i(k)}{\partial w_n^i(k)}\right] = \frac{\partial y^r(k)}{\partial w_n^r(k)} - \jmath\frac{\partial y^i(k)}{\partial w_n^r(k)} = \frac{\partial y^*(k)}{\partial w_n(k)} \tag{7.27}$$

A direct evaluation of $\partial y^*(k)/\partial w_n(k) = \partial\Phi^*(net(k))/\partial w_n(k)$ is not straightforward. However, as explained in Chapter 4, most nonlinear activation functions belong the the class of elementary transcendental functions, for which

$$\Phi^*(z) = \Phi(z^*) \tag{7.28}$$

Thus, for instance for the complex $tanh(z) = (e^z - e^{-z})/(e^z + e^{-z})$ function, we have[11]

$$\tanh^*(z) = \frac{e^{x-\jmath y} - e^{-x+\jmath y}}{e^{x-\jmath y} + e^{-x+\jmath y}} = \frac{e^{z^*} - e^{-z^*}}{e^{z^*} + e^{-z^*}} = \tanh(z^*) \tag{7.29}$$

[10] Similarly to the calculation for the term $\pi_n^\star$, and following the corresponding calculations for the linear IIR filter, we can show that $\pi_n^\circ(k) = \Phi'(net(k))\left(\sum_{l=1}^{N} w_{l+M+1}(k)\pi_n^\circ(k-l)\right)$. This is an unforced difference equation, initialised with zero, and π_n° vanishes. Alternatively, as this recursion does not have a driving term, it will decay to zero. See also Appendix N and Appendix O.

[11] We can also write $\Phi^*(z) = \Phi(z^*)$ for any function for which the Taylor series expansion has real coefficients, see [167] and Section 4.1.

Now, the sensitivities $\pi_n^\star(k)$ can be expressed as (bearing in mind that $(y^*(net) = y(net^*))$

$$\frac{\partial y^*(k)}{\partial w_n(k)} = \frac{\partial y(k)}{\partial net^*(k)} \frac{\partial net^*(k)}{\partial w_n(k)} \tag{7.30}$$

The term $\partial net^*(k)/\partial w_n(k)$ can be calculated in the same way as the conjugate sensitivities for the linear IIR filter in Equations (7.13–7.18), to give

$$\frac{\partial y(k)}{\partial net^*(k)} = \Phi'^*\big(net(k)\big)$$

$$\frac{\partial net^*(k)}{\partial w_n(k)} = u_n^*(k) + \sum_{l=1}^{N} w_{l+M+1}^*(k)\pi_n^\star(k-l)$$

$$\pi_n^\star(k) = \Phi'^*(net(k))\left(u_n^*(k) + \sum_{l=1}^{N} w_{l+M+1}^*(k)\pi_n^\star(k-l)\right) \tag{7.31}$$

The weight update for a recursive algorithm for the training of nonlinear IIR filters (recurrent perceptron) is therefore given by

$$\mathbf{w}(k+1) = \mathbf{w}(k) + \mu e(k)\boldsymbol{\pi}^\star(k) \tag{7.32}$$

This algorithm simplifies into the corresponding algorithm for complex IIR filters (Equation 7.19) upon the removal of the nonlinearity, and into the CNGD algorithm for nonlinear feedforward filters when no feedback is present.

7.3 Training of Recurrent Neural Networks

Figure 7.3 shows a Williams–Zipser fully connected recurrent neural network (FCRNN), which consists of N neurons, with M external inputs and a bias input, set to $(1 + \jmath)$. The network has two distinct layers – the external input/feedback layer and a layer of processing elements (neurons). We usually have the teaching signal available for only some of the outputs. In Figure 7.3, $y_1(k), \ldots, y_L(k)$ correspond to *output neurons*, whereas $y_{L+1}(k), \ldots, y_N(k)$ correspond to *hidden neurons*.

Let $y_n(k)$ denote the complex valued output of each neuron $n = 1, \ldots, N$ at time index k, and $\mathbf{x}(k)$ the $(M \times 1)$ external complex valued input vector. The overall input to the network $\mathbf{I}(k)$ represents a concatenation of the input and output vectors and the bias $(1 + \jmath)$, and is given by[12]

$$\begin{aligned}\mathbf{I}(k) &= [x(k-1), \ldots, x(k-M), 1+\jmath, y_1(k-1), \ldots, y_N(k-1)]^{\mathrm{T}} \\ &= [I_1(k), \ldots, I_{M+N+1}(k)]^{\mathrm{T}} = \mathbf{I}^r(k) + \jmath\mathbf{I}^i(k)\end{aligned} \tag{7.33}$$

Since the RNN in Figure 7.3 is fully connected, the weights connecting the input layer and every neuron y_n, $(n = 1, \ldots, N)$ form an $(M + N + 1) \times 1$ dimensional weight vector

[12]Based on Figure 7.3, in the prediction setting considered here, $x_1(k) = x(k-1), \ldots, x_M(k) = x(k-M)$.

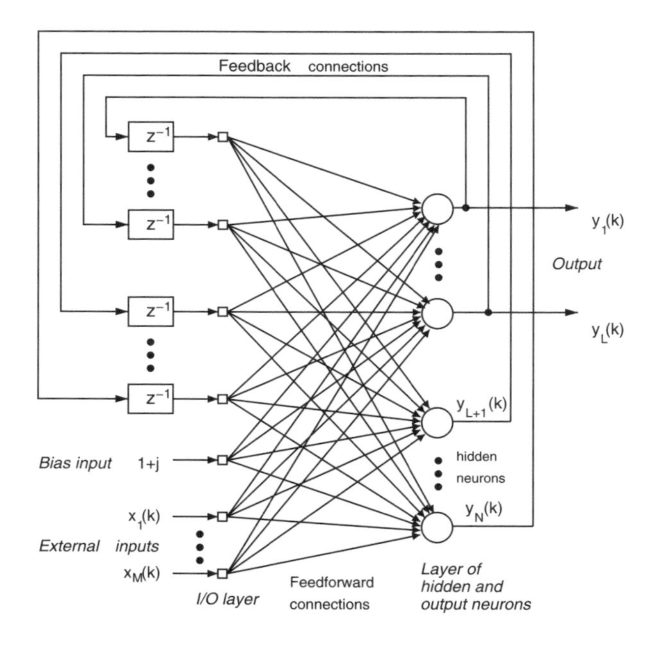

Figure 7.3 A fully connected recurrent neural network (FCRNN)

$\mathbf{w}_n = \left[w_{n,1}, \ldots, w_{n,M+N+1}\right]^{\mathrm{T}}$. It is convenient to combine all the weights in an RNN within a complex valued weight matrix $\mathbf{W}(k) = [\mathbf{w}_1(k), \ldots, \mathbf{w}_N(k)]$.

The output of every neuron in the network can be expressed as

$$y_n(k) = \Phi\big(net_n(k)\big) = y_n^r(k) + \jmath y_n^i(k), \quad n = 1, \ldots, N \tag{7.34}$$

where

$$net_n(k) = \sum_{q=1}^{M+N+1} w_{n,q}(k) I_q(k) \tag{7.35}$$

is the net input at the nth neuron at time instant k and $\Phi(\cdot)$ is a fully complex nonlinear activation function of a neuron (see Chapter 4 and Appendix E).

The cost function for the training of complex recurrent networks is of the same form as that for complex valued FIR filters, and for a networks with L output neurons, it is given by[13]

$$J(k) = \frac{1}{2}\sum_{l=1}^{L} |e_l(k)|^2 = \frac{1}{2}\sum_{l=1}^{L} e_l(k) e_l^*(k) \tag{7.36}$$

[13] In the case when only the first neuron is considered as the output neuron, as in simple nonlinear adaptive filtering applications, the cost function becomes the standard $J(k) = \frac{1}{2}e(k)e^*(k)$.

A direct stochastic gradient learning algorithm for training RNNs is called the Real Time Recurrent Learning (RTRL) [93, 190, 311]. Using stochastic gradient adaptation, we have $(n = 1, \dots, N,\ q = 1, \dots, M + N + 1)$

$$w_{n,q}(k+1) = w_{n,q}(k) + \Delta w_{n,q}(k) = w_{n,q}(k) - \mu \nabla_{w_{n,q}} J(k)_{|w_{n,q}=w_{n,q}(k)} \tag{7.37}$$

where μ is the learning rate. Since the errors can be calculated only for the output neurons $y_1(k), \dots, y_L(k)$, following the same procedure as for the recurrent perceptron, we have

$$\begin{aligned} \nabla_{w_{n,q}} J(k) &= \frac{\partial J(k)}{\partial w^r_{n,q}(k)} + \jmath \frac{\partial J(k)}{\partial w^i_{n,q}(k)} \\ &= -\sum_{l=1}^{L} \left[e_l(k) \pi^{\star l}_{n,q}(k) + e^*_l(k) \pi^{\circ l}_{n,q}(k) \right] \end{aligned} \tag{7.38}$$

where for every output neuron $y_l,\ l = 1, \dots, L$ we have the sensitivities

$$\begin{aligned} \boldsymbol{\pi}^{\circ l}(k) &= \frac{1}{2} \left[\nabla_{\mathbf{w}^r_l} y_l(k) + \jmath \nabla_{\mathbf{w}^i_l} y_l(k) \right] \\ \boldsymbol{\pi}^{\star l}(k) &= \frac{1}{2} \left[\nabla_{\mathbf{w}^r_l} y^*(k) + \jmath \nabla_{\mathbf{w}^i_l} y^*(k) \right] \end{aligned} \tag{7.39}$$

Following the analysis in Equations (7.26–7.31), we have $\boldsymbol{\pi}^{\circ l}(k) = \mathbf{0},\ l = 1, \dots, L$ whereas the recursive update for the sensitivities $\pi^{\star l}_{n,q}(k)$ becomes[14]

$$\pi^{\star l}_{n,q}(k) = \Phi'^*_l \left(net_l(k) \right) \left[\sum_{p=1}^{N} w^*_{l,M+1+p}(k) \pi^{\star p}_{n,q}(k-p) + \delta_{nl} I^*_q(k) \right], \quad \pi^{\star p}_{n,q}(0) = 0 \tag{7.40}$$

where

$$\delta_{nl} = \begin{cases} 1, & n = l \\ 0, & n \neq l \end{cases} \tag{7.41}$$

is the Kronecker delta.

Finally, the complex real time recurrent learning (CRTRL) algorithm for the training of complex valued recurrent neural networks performs the weight update given by [93]

$$w_{n,q}(k+1) = w_{n,q}(k) + \mu \sum_{l=1}^{L} e_l(k) \pi^{\star l}_{n,q}(k) \tag{7.42}$$

This degenerates to the recursive algorithm for the training of a complex recurrent perceptron, in the case of an RNN with one neuron.

[14] We use the symbol Φ_l to highlight that every neuron within an RNN can have different nonlinearity.

7.3.1 Other Learning Algorithms and Computational Complexity

The CRTRL algorithm has been developed for temporal problems, however, it can be applied to RNNs operating in different settings, for instance, for classification. The CRTRL algorithm is rather computationally demanding, and its complexity is $\mathcal{O}(N^4)$, however, the required memory size remains constant and will not increase with the length of the training sequence, thus making it suitable for real time processing. Computational complexity can be significantly reduced by the approximation (7.21), and without significant loss in performance.

The memory requirements and computational complexity of complex valued neural networks differ from those of real valued neural networks. A complex addition is roughly equivalent to two real additions, whereas a complex multiplication can be represented as four real multiplications and two additions or three multiplications and five additions.[15]

Other frequently used learning algorithms for training RNNs include the backpropagation through time (BPTT) and recurrent backpropagation (RBP) algorithms [111], and their variants.

7.4 Simulation Examples

To provide insight into the performance of feedback adaptive filtering architectures, two sets of simulations were conducted:

- performance comparison between a complex RNN trained with the CRTRL with a fully complex nonlinearity (FCRTRL) and a CRTRL with a split-complex nonlinearity (SCRTRL);[16]
- performance comparison between an RNN, IIR, and FIR filter trained respectively by CRTRL, recursive algorithm for adaptive IIR filters, and CGND and CLMS.

The test signals included a complex stable linear $AR(4)$ process given by

$$r(k) = 1.79r(k-1) - 1.85r(k-2) + 1.27r(k-3) - 0.41r(k-4) + n(k) \tag{7.43}$$

with doubly white[17] complex white Gaussian noise (CWGN) $n(k) \sim \mathcal{N}(\mathbf{0}, \mathbf{1})$ as the driving input, and the complex benchmark nonlinear input signal given by

$$x(k) = \frac{x(k-1)}{1 + x^2(k-1)} + r^3(k) \tag{7.44}$$

together with a segment of a real world wind signal, made complex by convenience of representation. The quantitative measure of performance was the standard prediction gain $R_p = 10 \log \sigma_y^2 / \sigma_e^2$.

[15] There are two ways to perform a complex multiplication: (i) $(a + \jmath b)(c + \jmath d) = (ac - bd) + \jmath(bc + ad)$; and (ii) $(a + \jmath b)(c + \jmath d) = a(c + d) - d(a + b) + \jmath\,(a\,(c + d) + c\,(b - a))$.

[16] The split-complex RTRL (SCRTRL) can be derived similarly to the derivation of split-complex nonlinear feedforward filters in Section 6.3.3.

[17] The real and imaginary components of CWGN are mutually independent sequences having equal variances so that $\sigma^2 = \sigma_r^2 + \sigma_i^2$.

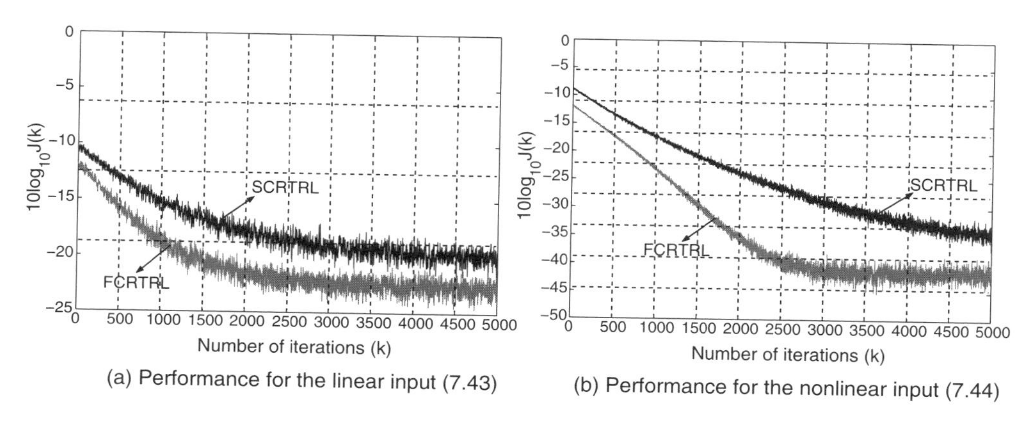

Figure 7.4 Performance of fully complex and split complex CRTRL for one step ahead prediction of a linear and nonlinear benchmark signal

Fully complex vs split-complex CRTRL. In the first set of simulations, the nonlinearity at the neuron was chosen to be the logistic sigmoid function, given by

$$\Phi(x) = \frac{1}{1 + e^{-\beta x}} \tag{7.45}$$

the slope was chosen to be $\beta = 1$ and learning rate $\mu = 0.01$. The complex RNN had $N = 2$ neurons with $L = 1$ output neuron, and a tap input length of $M = 4$. Figures 7.4(a) and 7.4(b) show respectively the learning curves (averaged over 100 independent trials) for the FCRTRL and SCRTRL performing adaptive prediction of complex coloured (7.43) and nonlinear (7.44) signals. The fully complex approach showed superior performance for both the inputs; after 5000 iterations the performance improvement over SCRTRL was roughly 3 [dB] for the linear and 6 [dB] for the nonlinear signal. The performances of FCRTRL and SCRTRL were next compared on single trial complex wind data. Figures 7.5(a) and 7.5(b) illustrate the tracking

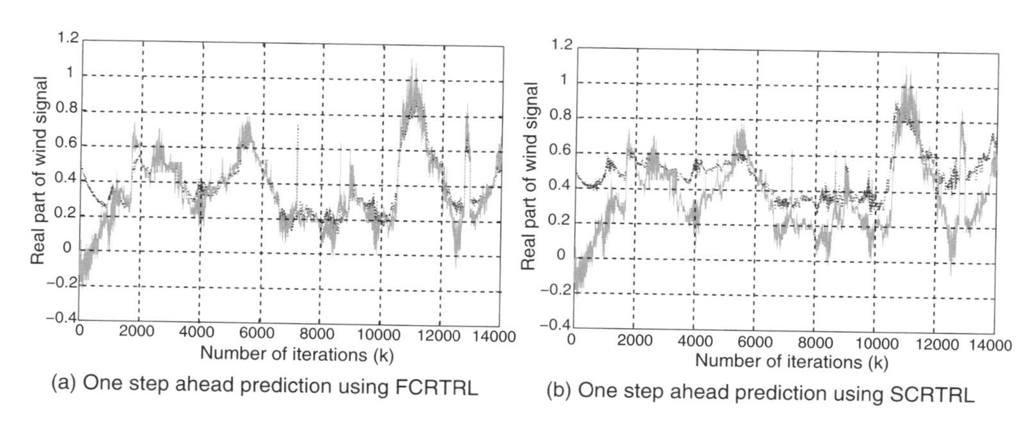

Figure 7.5 Performance comparison between FCRTRL and SCRTRL for the one step ahead prediction of a complex wind signal. Solid line: original wind signal; dotted line: predicted signal

Table 7.1 Prediction gains R_p for linear and nonlinear adaptive filters

Signal	Nonlinear N1 (7.44)	AR4(7.43)	Wind
R_p (dB) for linear FIR filter – CLMS	1.87	2.99	4.89
R_p (dB) for nonlinear FIR filter – CNGD	2.50	3.10	5.11
R_p (dB) for linear adaptive IIR filter	2.91	3.50	5.25
R_p (dB) for fully connected RNN – CRTRL	3.76	3.54	6.12

abilities of FCRTRL and SCRTRL; the FCRTRL achieved much improved prediction performance over the SCRTRL.

Performance of RNN vs other architectures. For the simulations, the nonlinearity at the neurons was chosen to be the fully complex tanh function

$$\Phi(x) = \frac{e^{\beta x} - e^{-\beta x}}{e^{\beta x} + e^{-\beta x}} \tag{7.46}$$

with slope $\beta = 1$. The complex recurrent neural network (Figure 7.3) had $N = 3$ neurons, with $L = 1$ output neuron. For the IIR filter (Figure 7.1), the order of feedback was $N = 3$. In all cases, the length of the external tap input delay line was $M = 5$. The signals were standardised to zero mean and maximum magnitude $|x|_{\max} = 0.8$, and the learning rate was $\mu = 0.1$. The performances of FIR and IIR filters (both linear and nonlinear) and RNNs used as nonlinear adaptive filters were compared, based on the complex valued LMS (CLMS), complex valued NGD (CNGD), recursive algorithm for adaptive IIR filters, and CRTRL. Table 7.1 comprises the performances for the complex linear AR(4) process (7.43), complex benchmark nonlinear signal $N1$ (7.44), and a segment of complex wind signal.

For the nonlinear benchmark signal $N1$, it is expected that the nonlinear RNN would outperform the linear IIR and FIR filters. This is confirmed in Table 7.1, as the CRTRL algorithm exhibited the best prediction gain for the nonlinear signal, followed by the linear adaptive IIR filter, nonlinear FIR filter (CNGD), and the linear FIR filter trained by CLMS. Similar observations can be made for the simulations on the real world complex wind signal.

To summarise:

- Feedback adaptive filtering architectures have potential advantages over their feedforward counterparts, as they are naturally suited to model systems with memory, signals with long term correlations, and long impulse responses.
- The linear infinite impulse response (IIR) adaptive filtering architecture has been introduced, and it is shown that it is suitable for the representation of autoregressive moving average (ARMA) models.
- It has been shown that, due to the presence of feedback, a recursive learning algorithm for training such IIR filters is rather computationally complex, however, it also needs very few feedforward and feedback tap input delay elements to model processes for which standard finite impulse response (FIR) filters would require a large filter length.

- A recurrent perceptron is introduced both as a nonlinear feedback adaptive filter (IIR filter with output nonlinearity) and as a recurrent neural network (RNN) with only one neuron; it has also been shown that it can represent nonlinear ARMA (NARMA) models.
- A fully complex recurrent neural network has been introduced, and a direct gradient algorithm for its training, called the complex valued real time recurrent learning (CRTRL), has been rigorously derived for fully complex nonlinear activation functions of neurons.
- It has been shown that the use of the $\mathbb{CR}$ calculus allows us to unify and greatly simplify the derivation of learning algorithms, and that the learning algorithms for IIR filters, recurrent perceptron, and RNNs have the same generic form.
- The performance of complex RNNs has been evaluated against IIR filters and both linear and nonlinear FIR filters; it has been shown that they offer improved modelling capabilities and performance, especially for nonlinear signals with rich dynamics. As real world signals are typically nonlinear, nonstationary and with long correlations, and systems have long impulse responses, nonlinear feedback models emerge as a natural processing framework, computational complexity permitting.

8

Filters with an Adaptive Stepsize

Standard adaptive filtering algorithms typically experience convergence problems when processing nonlinear and nonstationary signals, and signals with large dynamical ranges and ill-conditioned tap input autocorrelation matrices. To help circumvent this problem, it is convenient to employ Variable Stepsize (VSS) algorithms, these are based on two simultaneous unconstrained optimisation procedures – one used for the weight vector update and the other for the update of the adaptive learning rate.

Consider a stochastic gradient weight update for a nonlinear adaptive finite impulse response filter (Figure 8.1), given by

$$\mathbf{w}(k+1) = \mathbf{w}(k) + \mu(k)e(k)\Phi'^{*}(k)\mathbf{x}^{*}(k), \quad \Phi'^{*}(k) = \Phi'^{*}\left(\mathbf{x}^{\mathrm{T}}(k)\mathbf{w}(k)\right) \tag{8.1}$$

The parameter μ is a (possibly adaptive) stepsize and is critical to the convergence. To introduce a variable stepsize into this class of algorithms, we shall extend the corresponding VSS approaches for real valued adaptive filters, for which a number of gradient adaptive stepsize LMS algorithms have been developed [13, 24, 180, 207]. These include:

- Gradient adaptive stepsize (GASS) algorithms based on the minimisation of $\nabla_{\mu} J$, such as the algorithms introduced by Benveniste [24], Ang and Farhang [13] and Mathews and Xie [207] (see Appendix K).
- Algorithms based upon a regularisation of the normalised LMS (NLMS) update, where the regularisation factor $\varepsilon(k)$ is made adaptive. One such approach is the generalised normalised gradient descent (GNGD) algorithm [180] and its variants [47].
- Heuristic approaches, based for instance on imposing hard constraints on the lower and upper bounds for the stepsize, superimposing regression on the stepsize sequence [2], reducing the computational complexity by employing sign algorithms [68], and employing graded updates [210].

This chapter introduces gradient adaptive stepsize algorithms for both linear and nonlinear complex valued adaptive filters [97]. It is shown that this improves the speed of convergence and stability of such filters, albeit at a cost of increased computational complexity. Analysis

Complex Valued Nonlinear Adaptive Filters: Noncircularity, Widely Linear and Neural Models
Danilo P. Mandic and Vanessa Su Lee Goh

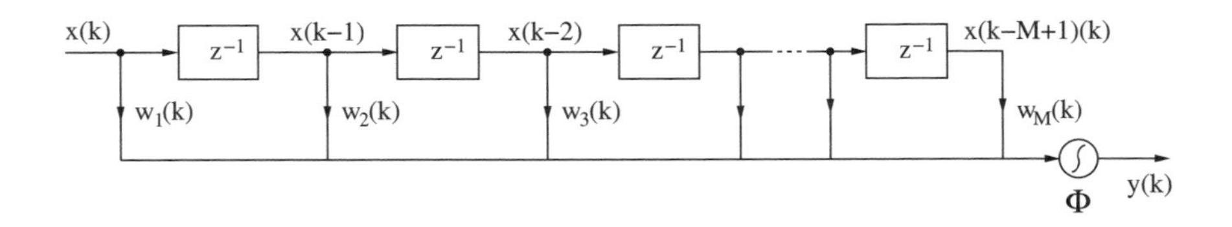

Figure 8.1 A nonlinear FIR filter with nonlinearity Φ

is supported by simulations on both benchmark and real world nonlinear and nonstationary data.

8.1 Benveniste Type Variable Stepsize Algorithms

To introduce an adaptive stepsize into nonlinear FIR filters (Figure 8.1) in $\mathbb{C}$, based on the weight update (Equation 8.1) we may perform a standard stochastic gradient adaptation, given by

$$\mu(k) = \mu(k-1) - \rho \nabla_{\mu} J(k)_{|\mu=\mu(k-1)} \tag{8.2}$$

where parameter ρ is the stepsize. Based on the standard cost function $J(k) = \frac{1}{2}|e(k)|^2 = \frac{1}{2}e(k)e^*(k)$, the gradient $\nabla_{\mu} J(k)$ is calculated from

$$\nabla_{\mu} J(k) = \frac{1}{2}\left[e(k)\frac{\partial e^*(k)}{\partial \mu(k-1)} + e^*(k)\frac{\partial e(k)}{\partial \mu(k-1)}\right] \tag{8.3}$$

The main issue in the derivation of the gradient adaptive stepsize algorithms is the calculation of the partial derivatives $\partial e^*(k)/\partial \mu(k-1)$ and $\partial e(k)/\partial \mu(k-1)$ from Equation (8.3). For instance, to calculate the term $\partial e^*(k)/\partial \mu(k-1)$, we need to evaluate

$$\frac{\partial e^*(k)}{\partial \mu(k-1)} = \frac{\partial e_r(k)}{\partial \mu(k-1)} - \jmath \frac{\partial e_i(k)}{\partial \mu(k-1)} \tag{8.4}$$

For a fully complex analytic nonlinear activation function $\Phi(net(k)) = \Phi(k)$, by using the Cauchy–Riemann equations (see Chapter 5), this yields

$$\frac{\partial e^*(k)}{\partial \mu(k-1)} = -\mathbf{x}^H(k)\Phi'^*(k)\frac{\partial \mathbf{w}^*(k)}{\partial \mu(k-1)} \tag{8.5}$$

By a similar calculation, the second gradient term in Equation (8.3) $\partial e(k)/\partial \mu(k-1)$ becomes

$$\frac{\partial e(k)}{\partial \mu(k-1)} = -\mathbf{x}^T(k)\Phi'(k)\frac{\partial \mathbf{w}(k)}{\partial \mu(k-1)} \tag{8.6}$$

For simplicity, denote[1]

$$\boldsymbol{\psi}(k) = \frac{\partial \mathbf{w}(k)}{\partial \mu(k-1)} \approx \frac{\partial \mathbf{w}(k)}{\partial \mu(k)}$$

Similarly to gradient adaptive stepsize (GASS) real valued adaptive filters given in Appendix K, we introduce three algorithms within the class of algorithms based on $\partial J/\partial \mu$ based on different ways to evaluate of the 'sensitivity' term $\boldsymbol{\psi}(k)$, (for more detail see [97]).

Benveniste type VSS algorithm (BVSS) [24]. Based on the weight update (Equation 8.1), calculate

$$\begin{aligned}\boldsymbol{\psi}(k) &= \boldsymbol{\psi}(k-1) + \frac{\partial \mu(k-1)}{\partial \mu(k-1)} e(k-1)\Phi'^{*}(k-1)\mathbf{x}^{*}(k-1) \\ &+ \mu(k-1)\frac{\partial e(k-1)}{\partial \mu(k-1)}\Phi'^{*}(k-1)\mathbf{x}^{*}(k-1) + \mu(k-1)e(k-1)\frac{\partial \Phi'^{*}(k-1)}{\partial \mu(k-1)}\mathbf{x}^{*}(k-1)\end{aligned}$$

The last term can be neglected, to give (compare with Equation K.9 in Appendix K)

$$\begin{aligned}\boldsymbol{\psi}(k) &= \left[\mathbf{I} - \mu(k-1)\left|\Phi'(k-1)\right|^{2}\mathbf{x}^{*}(k-1)\mathbf{x}^{\mathrm{T}}(k-1)\right]\boldsymbol{\psi}(k-1) + e(k-1)\Phi'^{*}(k-1)\mathbf{x}^{*}(k-1) \\ &= \boldsymbol{\Lambda}(k)\boldsymbol{\psi}(k-1) + e(k-1)\Phi'^{*}(k-1)\mathbf{x}^{*}(k-1)\end{aligned} \tag{8.7}$$

This approach effectively performs filtering of noisy instantaneous gradients $e(k-1)\Phi'^{*}(k-1)\mathbf{x}^{*}(k-1)$, where the term within the square brackets, $\boldsymbol{\Lambda}(k)$, determines the properties of such a filter. Based on the stepsize adaptation in Equation (8.2), the BVSS algorithm becomes

$$\mu(k) = \mu(k-1) + \frac{\rho}{2}\left[e(k)\Phi'^{*}(k)\mathbf{x}^{H}(k)\boldsymbol{\psi}^{*}(k) + e^{*}(k)\Phi'(k)\mathbf{x}^{\mathrm{T}}(k)\boldsymbol{\psi}(k)\right] \tag{8.8}$$

Farhang–Ang type VSS algorithm (FVSS) [13]. To simplify the BVSS update, replace the time varying term in the square brackets in Equation (8.7) by a constant $0 < \alpha < 1$. With this simplification, the sensitivity term $\boldsymbol{\psi}(k)$ becomes

$$\boldsymbol{\psi}(k) = \alpha\boldsymbol{\psi}(k-1) + e(k-1)\Phi'^{*}(k-1)\mathbf{x}^{*}(k-1) \tag{8.9}$$

This way, the noisy instantaneous gradients $e(k-1)\Phi'^{*}(k-1)\mathbf{x}^{*}(k-1)$ are filtered with a low pass filter with a fixed coefficient α.

Mathews type VSS algorithm (MVSS) [207]. By setting $\alpha = 0$, calculation of the sensitivity term in Equation (8.9) can be further simplified, to obtain the MVSS algorithm, for which

$$\boldsymbol{\psi}(k) = e(k-1)\Phi'^{*}(k-1)\mathbf{x}^{*}(k-1) \tag{8.10}$$

is based only on the noisy instantaneous estimates of the gradient from Equation (8.6).

[1] For more details on GASS approaches for real valued filters, see Appendix K.

Table 8.1 Gradients within the VSS algorithms and their complexities

Algorithm	Sensitivity term $\boldsymbol{\psi}(k) = \partial \mathbf{w}(k)/\partial \mu(k)$	Multiplications
BVSS (8.7)	$\boldsymbol{\psi}(k) = \boldsymbol{\Lambda}(k)\boldsymbol{\psi}(k-1) + e(k-1)\Phi'^{*}(k-1)\mathbf{x}^{*}(k-1)$	12
FVSS (8.9)	$\boldsymbol{\psi}(k) = \alpha\boldsymbol{\psi}(k-1) + e(k-1)\Phi'^{*}(k-1)\mathbf{x}^{*}(k-1), \quad 0 < \alpha < 1$	9
MVSS (8.10)	$\boldsymbol{\psi}(k) = e(k-1)\Phi'^{*}(k-1)\mathbf{x}^{*}(k-1)$	8

Table 8.2 Gradients within the VSS algorithms for linear adaptive filters

Algorithm	Sensitivity term $\boldsymbol{\psi}(k) = \partial \mathbf{w}(k)/\partial \mu(k)$
BVSS	$\boldsymbol{\psi}(k) = \boldsymbol{\Lambda}(k)\boldsymbol{\psi}(k-1) + e(k-1)\mathbf{x}^{*}(k-1)$
FVSS	$\boldsymbol{\psi}(k) = \alpha\boldsymbol{\psi}(k-1) + e(k-1)\mathbf{x}^{*}(k-1), \quad 0 < \alpha < 1$
MVSS	$\boldsymbol{\psi}(k) = e(k-1)\mathbf{x}^{*}(k-1)$

Table 8.1 summarizes the three VSS algorithms and their associated computational complexities. These VSS algorithms have been derived for nonlinear adaptive filters in $\mathbb{C}$, their counterparts for linear adaptive filters can be obtained by removing the terms associated with nonlinearity Φ. Thus, for instance, for MVSS we have

$$\begin{aligned} \boldsymbol{\psi}(k) &= e(k-1)\mathbf{x}^{*}(k-1) \\ \mu(k) &= \mu(k-1) + \frac{\rho}{2}\left[e(k)\mathbf{x}^{H}(k)e^{*}(k-1)\mathbf{x}(k-1) + e^{*}(k)\mathbf{x}^{T}(k)e(k-1)\mathbf{x}^{*}(k-1)\right] \end{aligned} \tag{8.11}$$

Weight gradients for the GASS updates for linear adaptive filters[2] are given in Table 8.2.

8.2 Complex Valued GNGD Algorithms

Based on the standard weight update (Equation 8.1), adaptive learning rates $\eta(k)$ for the CLMS, normalised CLMS (CNLMS), and normalised complex nonlinear gradient descent (CNNGD) algorithms are given by

$$\eta(k) = \begin{cases} \mu & \text{for CLMS,} \\ \dfrac{\mu}{\|\mathbf{x}(k)\|_2^2 + \varepsilon} & \text{for CNLMS,} \\ \dfrac{\mu}{|\Phi'(k)|^2 \, \|\mathbf{x}(k)\|_2^2 + \varepsilon} & \text{for CNNGD} \end{cases} \tag{8.12}$$

where ε is a regularisation parameter used to prevent divergence for close to zero inputs.

[2]In BVSS for linear filters, $\boldsymbol{\Lambda}(k) = \mathbf{I} - \mu(k-1)\mathbf{x}^{*}(k-1)\mathbf{x}^{\mathrm{T}}(k-1)$.

The class of GASS algorithms performs 'linear' updates which depend upon estimators of $\partial J(k)/\partial\mu$. This is achieved by making the stepsize (amplification factor) within the weight update gradient adaptive. These algorithms are based on two coupled unconstrained optimisation procedures (for weights and the stepsize), which can affect their robustness to the initial values of the parameters (especially in the case of MVSS).

Another approach would be to introduce additional robustness and stability based on the Generalised Normalised Gradient Descent (GNGD) algorithm [180], which performs a 'non-linear' update of the learning rate by making the regularisation term ε in the denominator of the CNLMS stepsize gradient adaptive. For simplicity, we shall first derive the complex GNGD (CGNGD) algorithm for linear adaptive filters in $\mathbb{C}$. Following the approach from [180], consider a regularised CNLMS update

$$\begin{aligned}\mathbf{w}(k+1) &= \mathbf{w}(k) + \eta(k)e(k)\mathbf{x}^*(k)\\ \eta(k) &= \frac{\mu}{\|\mathbf{x}(k)\|_2^2 + \varepsilon(k)}\end{aligned} \tag{8.13}$$

where $\varepsilon(k)$ is an adaptive regularisation parameter. Based on the cost function

$$J(k) = \frac{1}{2}e(k)e^*(k) = \frac{1}{2}\,|e(k)|^2$$

we shall perform stochastic gradient adaptation of the adaptive regularisation parameter

$$\varepsilon(k) = \varepsilon(k-1) - \rho\nabla_\varepsilon\, J(k)|_{\varepsilon=\varepsilon(k-1)} \tag{8.14}$$

The gradient $\nabla_\varepsilon J(k)|_{\varepsilon=\varepsilon(k-1)}$ can be evaluated as

$$\nabla_\varepsilon J(k)|_{\varepsilon=\varepsilon(k-1)} = \frac{1}{2}\left[e(k)\frac{\partial e^*(k)}{\partial\varepsilon(k-1)} + e^*(k)\frac{\partial e(k)}{\partial\varepsilon(k-1)}\right] \tag{8.15}$$

where the corresponding partial derivatives[3] are

$$\begin{aligned}\frac{\partial e_r(k)}{\partial\varepsilon(k-1)} &= \frac{\partial e_r(k)}{\partial\mathbf{w}_r(k)}\frac{\partial\mathbf{w}_r(k)}{\partial\eta(k-1)}\frac{\partial\eta(k-1)}{\partial\varepsilon(k-1)} + \frac{\partial e_r(k)}{\partial\mathbf{w}_i(k)}\frac{\partial\mathbf{w}_i(k)}{\partial\eta(k-1)}\frac{\partial\eta(k-1)}{\partial\varepsilon(k-1)}\\ \frac{\partial e_i(k)}{\partial\varepsilon(k-1)} &= \frac{\partial e_i(k)}{\partial\mathbf{w}_r(k)}\frac{\partial\mathbf{w}_r(k)}{\partial\eta(k-1)}\frac{\partial\eta(k-1)}{\partial\varepsilon(k-1)} + \frac{\partial e_i(k)}{\partial\mathbf{w}_i(k)}\frac{\partial\mathbf{w}_i(k)}{\partial\eta(k-1)}\frac{\partial\eta(k-1)}{\partial\varepsilon(k-1)}\end{aligned} \tag{8.16}$$

---

[3] Since $\partial e(k)/\partial\varepsilon(k-1) = \partial e_r(k)/\partial\varepsilon(k-1) + \jmath\partial e_i(k)/\partial\varepsilon(k-1)$ and $\partial e^*(k)/\partial\varepsilon(k-1) = \partial e_r(k)/\partial\varepsilon(k-1) - \jmath\partial e_i(k)/\partial\varepsilon(k-1)$.

The complex generalised normalised gradient descent (CGNGD) algorithm can now be summarised as[4]

$$
\begin{aligned}
\mathbf{w}(k+1) &= \mathbf{w}(k) + \eta(k)e(k)\mathbf{x}^*(k) \\
\eta(k) &= \frac{\mu}{\|\mathbf{x}(k)\|_2^2 + \varepsilon(k)} \\
\varepsilon(k) &= \varepsilon(k-1) - \rho\mu \frac{\Re\left\{e(k)e^*(k-1)\mathbf{x}^H(k)\mathbf{x}(k-1)\right\}}{\left[\|\mathbf{x}(k-1)\|^2 + \varepsilon(k-1)\right]^2}
\end{aligned}
\tag{8.17}
$$

where $\Re\{\cdot\}$ denotes the real part of a complex quantity.

8.2.1 Complex GNGD for Nonlinear Filters (CFANNGD)

To introduce the GNGD stepsize adaptation into the class of complex valued nonlinear FIR filters (dynamical perceptron from Figure 8.1 for which the stepsize is given in Equation 8.12), the gradient

$$
\nabla_\varepsilon J(k)|_{\varepsilon=\varepsilon(k-1)} = \frac{\partial J(k)}{\partial \varepsilon(k-1)} = \frac{1}{2}\left[e(k)\frac{\partial e^*(k)}{\partial \varepsilon(k-1)} + e^*(k)\frac{\partial e(k)}{\partial \varepsilon(k-1)}\right]
\tag{8.18}
$$

is calculated similarly to that in the CGNGD. After expanding the partial derivatives of the error similarly to those in Equation (8.16), and using the Cauchy–Riemann equations, that is

$$
\begin{aligned}
\frac{\partial e^*(k)}{\partial \varepsilon(k-1)} &= -\mathbf{x}^{\mathrm{H}}(k)\frac{\partial \mathbf{w}^*(k)}{\partial \varepsilon(k-1)}\Phi'^*\big(net(k)\big) \\
\frac{\partial e(k)}{\partial \varepsilon(k-1)} &= -\mathbf{x}^{\mathrm{T}}(k)\frac{\partial \mathbf{w}(k)}{\partial \varepsilon(k-1)}\Phi'\big(net(k)\big)
\end{aligned}
\tag{8.19}
$$

the weight gradient with respect to the adaptive regularisation parameter $\varepsilon(k)$ is obtained in the form (for a detailed derivation see Appendix L)

$$
\frac{\partial \mathbf{w}(k)}{\partial \varepsilon(k-1)} = -\frac{e(k)\Phi'^*\big(net(k)\big)\mathbf{x}^*(k)}{\left[|\Phi'(net(k-1))|^2\,\|\mathbf{x}(k-1)\|_2^2 + \varepsilon(k-1)\right]^2}
\tag{8.20}
$$

The variant of CGNGD for nonlinear adaptive filters is termed the complex valued fully adaptive normalised nonlinear gradient descent (CFANNGD), and can be summarised as [106, 108]

$$
\begin{aligned}
e(k) &= d(k) - \Phi\big(net(k)\big) = d(k) - \Phi(k) \\
net(k) &= \mathbf{x}^{\mathrm{T}}(k)\mathbf{w}(k)
\end{aligned}
$$

[4] For a full derivation see Appendix L.

$$\eta(k) = \frac{1}{|\Phi'(k)|^2 \, \|\mathbf{x}(k)\|_2^2 + \varepsilon(k)}$$

$$\varepsilon(k) = \varepsilon(k-1) - \rho\mu \frac{\Re\left\{e(k)e^*(k-1)\Phi'^*(k)\Phi'(k-1)\mathbf{x}^{\mathrm{H}}(k)\mathbf{x}(k-1)\right\}}{\left[|\Phi'(net(k-1))|^2 \, \|\mathbf{x}(k-1)\|_2^2 + \varepsilon(k-1)\right]^2} \tag{8.21}$$

8.3 Simulation Examples

In all the experiments, the order of the adaptive FIR filters (both linear and nonlinear) was chosen to be $M = 4$, with the slope of the nonlinear complex tanh function $\beta = 1$. Simulations were undertaken by averaging 200 independent trials on the prediction of both complex valued benchmark coloured and nonlinear signals, as well as single trial real-life signals (complex valued radar and two sets of wind data), given below.

L1. Linear complex stable $AR(4)$ process, given by

$$y(k) = 1.79y(k-1) - 1.85y(k-2) + 1.27y(k-3) - 0.41y(k-4) + n(k) \tag{8.22}$$

N1. Complex nonlinear benchmark signal [216]

$$y(k) = \frac{y^2(k-1)(y(k-1)+2.5)}{1 + y^2(k-1) + y^2(k-2)} + n(k-1) \tag{8.23}$$

N2. Complex nonlinear benchmark signal [216]

$$y(k) = \frac{y(k-1)}{1 + y^2(k-1)} + n^3(k) \tag{8.24}$$

Wind. The wind measurements were sampled at 50 Hz for a interval of 1 h. The measurements were recorded at 1 and 17 m above ground level. The measurements at different heights yielded different wind dynamics. The wind vector can be expressed in the complex domain $\mathbb{C}$ as $v(t)e^{\jmath\theta(t)} = v_{East}(t) + \jmath v_{North}(t)$. Here, the two wind components, the speed v and direction θ, which are of different natures, are modelled as a single quantity in a complex representation space.

IPIX radar. The samples correspond to the sea clutter signal captured by the McMaster IPIX Radar at OHGR (Osborne Head Gunnery Range), Dartmouth, Nova Scotia, Canada, on a cliff facing the Atlantic Ocean. The dataset used is referred to as 'high sea state data' (hi.zip)[5] and contains magnitude and phase of the sea clutter signal collected in November 1993.

The driving noise $n(k)$ in Equations (8.22 – 8.24) was complex doubly white noise (CDWN) with zero mean and unit variance, that is $n(k) = n_r(k) + \jmath n_i(k)$ where $n_r(k), n_i(k) \sim \mathcal{N}(0, 1)$.

[5]Publicly available from http://soma.ece.mcmaster.ca/ipix/dartmouth/datasets.html.

Table 8.3 Prediction gain R_p for standard CNGD and VSS CNGD algorithms

Algorithm	Parameters	L1	N1	N2	Wind (1m)	Wind (17m)	Radar
CNGD	$\mu(0) = 0.05$	5.010	3.22	1.91	7.89	9.55	10.15
BVSS	$\mu(0) = 0.05, \rho = 0.0002$	6.61	4.65	5.09	17.55	14.41	15.51
FVSS	$\alpha = 0.95, \mu(0) = 0.05, \rho = 0.0002$	6.27	4.12	5.09	17.26	14.00	14.31
MVSS	$\mu(0) = 0.05, \rho = 0.0002$	4.01	2.90	1.88	8.10	9.78	13.14

The real and imaginary components of CDWN were therefore mutually statistically independent sequences and with equal variances so, that $\sigma_n^2 = \sigma_{n_r}^2 + \sigma_{n_i}^2$.

The measurement used to assess the performance was the prediction gain, given by [114]

$$R_p = 10 \log_{10} \left(\frac{\sigma_x^2}{\hat{\sigma}_e^2} \right) \text{dB} \tag{8.25}$$

where σ_x^2 denotes the variance of the input signal $x(k)$, and $\hat{\sigma}_e^2$ denotes the estimated variance of the forward prediction error $e(k)$.

Table 8.3 compares prediction gains of a nonlinear adaptive filter trained by the CNGD and the GASS VSS algorithms. In conformance with the analysis, in all the cases, the BVSS had best performance, followed by the FVSS, whereas MVSS and standard CNGD usually had similar performances. Improvement in the performance when using the VSS class of algorithms is especially noticeable for signals made complex by convenience of representation (columns 6–8 in Table 8.3).

Figure 8.2 illustrates the convergence of the class of GASS algorithms for prediction of the coloured signal (Equation 8.22). The BVSS and FVSS exhibited fastest convergence, as illustrated in Figure 8.2(a), whereas MVSS had similar behaviour to CNGD. Figure 8.2(b) illustrates the evolution of the corresponding stepsizes – BVSS showed its robust nature by having smoothest convergence of its stepsize, whereas the MVSS, due to its instantaneous

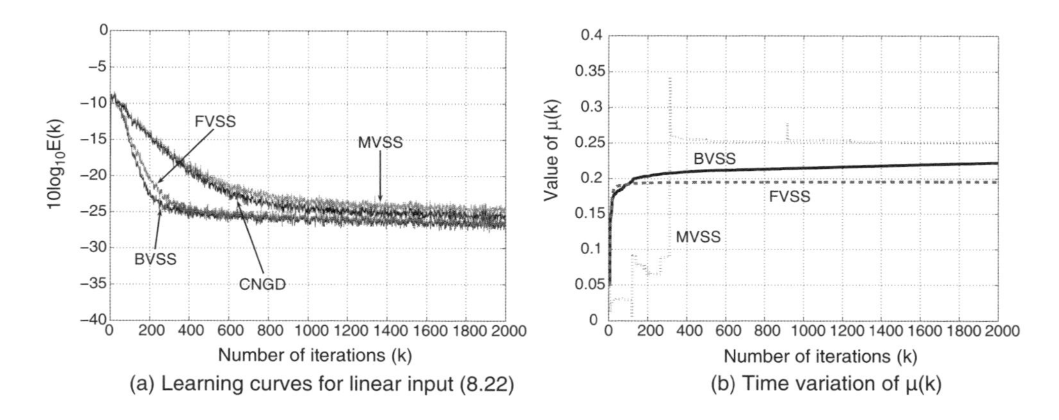

Figure 8.2 Performance of GASS algorithms for the linear signal (Equation 8.22)

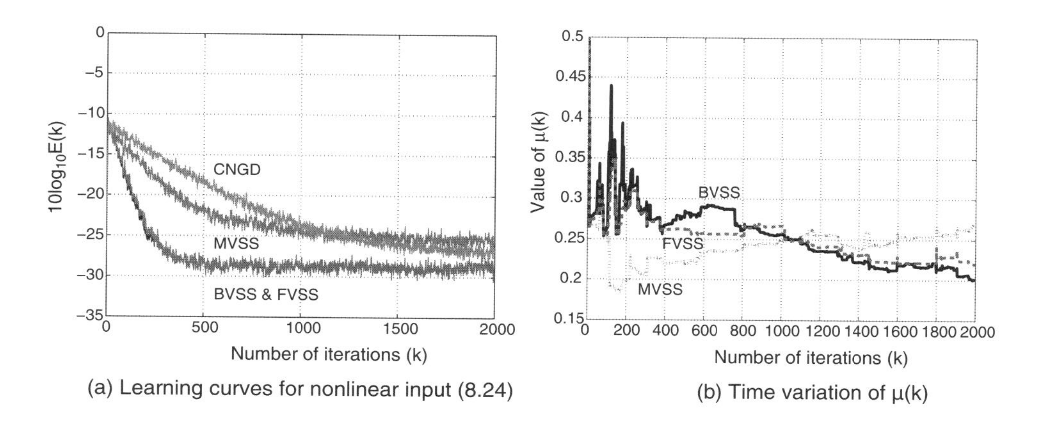

Figure 8.3 Performance of GASS algorithms for the nonlinear signal (Equation 8.24)

estimates of the weight gradient, experienced problems with outliers and instability (e.g. around sample number 350). This is also reflected in the values of the prediction gain in the third column in Table 8.3.

In the next set of experiments, simulations from Figure 8.2 were repeated for the nonlinear input (Equation 8.24). Figure 8.3(a) shows that BVSS and FVSS had fastest convergence, and their performance was similar. In the steady state, the BVSS and FVSS converged to the same solution as CNGD. The initial convergence of MVSS was faster, however, its steady state error was larger than that of CNGD. This also illustrates the sensitivity of MVSS to the choice of initial parameters and its relative instability as compared to BVSS and FVSS. Figure 8.3(b) illustrates the evolution of the stepsize parameter for the learning curves from Figure 8.3(a). Again, the BVSS and FVSS stepsizes settled whereas the MVSS stepsize was fluctuating (see also prediction gains for signals $N1$ and $N2$ in Table 8.3).

To illustrate the robustness of the BVSS algorithm to variations in its parameters, Figure 8.4(a) shows the variation of the prediction gain R_p (8.25) for a range of initial values of $\mu(0)$ and $\rho(0)$, for prediction of a wind signal. Owing to the rigorous derivation of BVSS, the choice of initial values did not have significant effect on the performance. This robustness is less pronounced for the simplified versions – the FVSS and MVSS algorithms. Figure 8.4(b) provides insight into the behaviour of the corresponding learning rates.

Figure 8.5 compares the performances of CGNGD and CNLMS;[6] the initial values used for the CGNGD simulations were $\rho = 0.15$ and $\varepsilon(0) = 0.1$. Figure 8.5(a) illustrates the CGNGD exhibiting faster convergence than CNLMS and similar steady state performance, for relatively small values of the learning rate μ. To illustrate the excellent stability of CGNGD, the critical condition of nearly vanishing inputs (for which NLMS diverges) was generated by setting the value of the stepsize close to the NLMS stability bound ($\mu = 1.9$) – the CGNGD converged faster and had better steady state properties. Figure 8.5(b) shows learning curves for a similar experiment on the nonlinear signal (8.24). The stepsize was set to $\mu = 2$, for which the CGNGD was stable and converged, whereas CNLMS diverged.

[6]The performance of CFANNGD is illustrated in Figure 9.4 of Chapter 9.

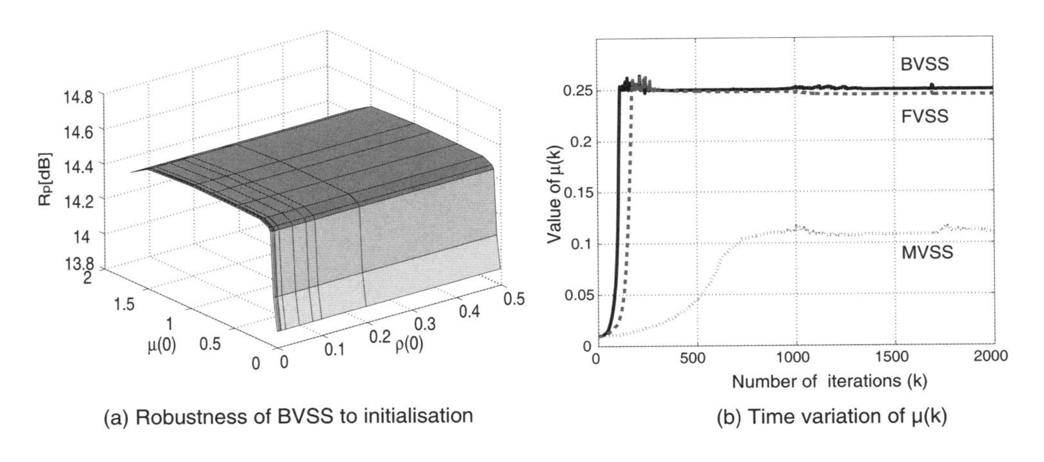

(a) Robustness of BVSS to initialisation

(b) Time variation of μ(k)

Figure 8.4 Performance of the BVSS, FVSS, and MVSS algorithms on prediction of wind signal (17 m). (a) Dependence on initial values of the parameters; (b) evolution of $\mu(k)$

To summarise:

- It has been recognised that a fixed stepsize, which governs the speed of convergence and steady state error of stochastic gradient algorithms, is not an optimal choice for nonstationary environments and signals with ill–conditioned tap input correlation matrix. To that end several variable stepsize (VSS) algorithms have been proposed.
- Ideally, we want an algorithm for which the speed of convergence is fast and the steady state error is small when operating in a stationary environment, whereas in a nonstationary environment the algorithm should change the learning rate according to the dynamics of the input signal, so as to achieve as good a performance as possible.

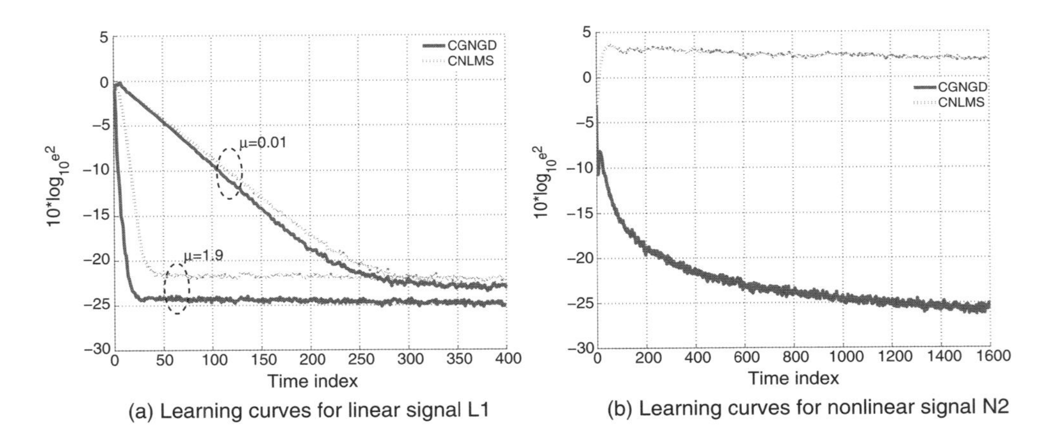

(a) Learning curves for linear signal L1

(b) Learning curves for nonlinear signal N2

Figure 8.5 Learning curves for CNLMS and CGNGD for the linear signal L1 (Equation 8.22) and the nonlinear signal N2 (Equation 8.24). Left: $\mu = 0.01$ for top curves and $\mu = 1.9$ for bottom curves; right: $\mu = 2$

- Two classes of VSS algorithms have been introduced: those based on stochastic gradient adaptation of the stepsize parameter (GASS) and those based on making the regularisation parameters within the CNLMS algorithm gradient adaptive (CGNGD).
- The analysis has been conducted for nonlinear finite impulse response adaptive filters (dynamical perceptron) with a fully complex activation function. The linear counterparts of the BVSS, GVSS, MVSS, and CFANNGD algorithms are obtained by removing the terms within the updates which contain the nonlinearity Φ.
- The adaptive stepsize algorithms derived in this chapter can be extended to recurrent neural networks in a generic way. The corresponding algorithms for infinite impulse filters (IIR) in $\mathbb{C}$ are obtained by removing the effects of nonlinearity Φ, whereas by removing feedback we arrive at the algorithms for nonlinear FIR filters.
- Chapter 9 illustrates how to deal with signals with large dynamical ranges in a different way–by making the magnitude of the complex nonlinear activation function gradient adaptive.

9

Filters with an Adaptive Amplitude of Nonlinearity

Real world signals are typically nonstationary and nonlinear, and therefore their dynamical range is not known beforehand. Hence, nonlinear adaptive filters based on a fixed nonlinearity may be inappropriate for some applications.[1] One way to circumvent this problem is by *dynamical range reduction*, described below. Another way to match the dynamics of the input with the nonlinearity within the filter is to equip nonlinear adaptive filters with an adaptive amplitude of the nonlinearity [92, 105, 108, 294]. This also helps to circumvent problems that arise from dynamical range reduction by estimate subtractions, such as the accumulation of errors. This chapter introduces learning algorithms with an adaptive amplitude of nonlinearity for both the feedforward and feedback nonlinear adaptive filters.

9.1 Dynamical Range Reduction

Dynamical range reduction is a method whereby a preprocessor dynamically transforms the range of the external input process so as to make it fit the range of filter nonlinearity. In the output stage, the filter is then equipped with a post-processor which performs dynamical range extension, in order to recover the original range of the process in hand.

There are several ways to perform dynamical range reduction [170]:

- *Range reduction by estimate subtraction* is shown in Figure 9.1 and includes
  - Range reduction by differencing, which is based on simple differentiation given by

$$x_d(k) = x(k) - x(k-1) \tag{9.1}$$

[1]This problem is usually alleviated by standardising the input to a certain range and variance (see Appendix G), however, this requires prior knowledge about the process in hand.

Complex Valued Nonlinear Adaptive Filters: Noncircularity, Widely Linear and Neural Models
Danilo P. Mandic and Vanessa Su Lee Goh

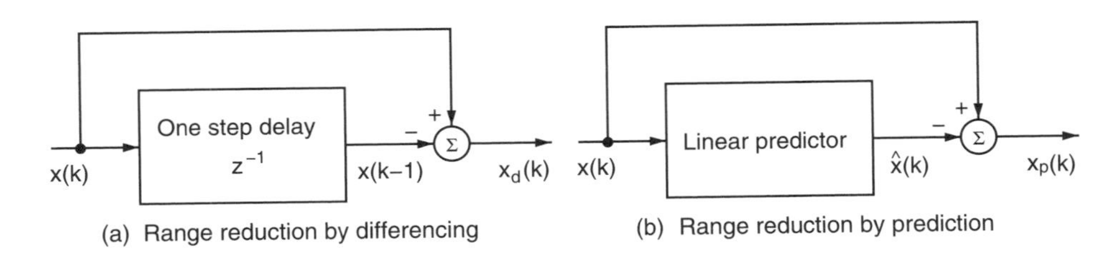

Figure 9.1 Schemes for dynamical range reduction based on estimate subtraction

This method is very simple, but requires careful initialisation.

- Range reduction by linear prediction, whereby the signal with a reduced range $x_p(k)$ is generated as

$$x_p(k) = x(k) - \hat{x}(k) \tag{9.2}$$

where $\hat{x}(k)$ denotes the prediction of $x(k)$.

The method of range reduction by differencing is, in fact, a special case of range reduction by prediction, where the prediction is based on a so called 'persistent' estimate, that is, $\hat{x}(k) = x(k-1)$ [91].

- *Range reduction by homomorphic transformation* comprises both the dynamic range reduction and extension [226, 260]. A two-layer homomorphic neural network (HNN) architecture is shown in Figure 9.2 (the homomorphic layer is within the frame). The layer of $\log(\cdot)$ functions reduces the dynamical range of the input, the so modified inputs are first combined adaptively in a linear fashion and their range is then extended by a layer of exponential functions and a linear output layer.

The remainder of this chapter will discuss another method, the use of filters with *an adaptive amplitude of nonlinearity*.

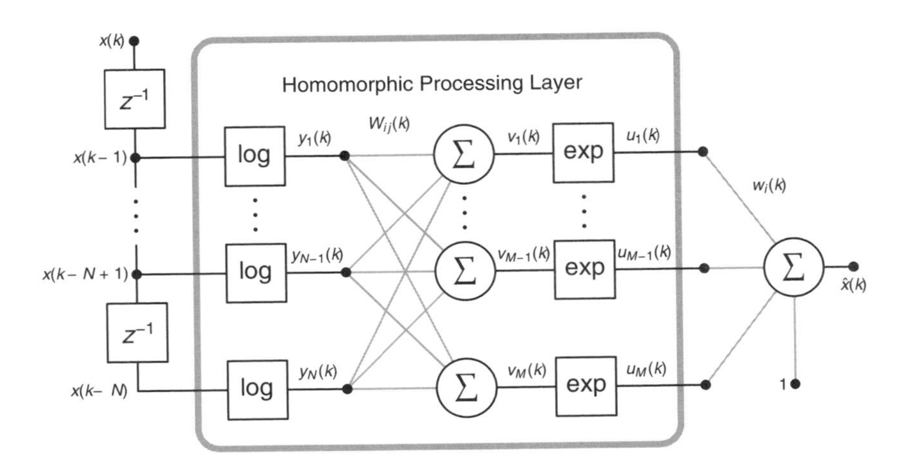

Figure 9.2 Homomorphic neural network

9.2 FIR Adaptive Filters with an Adaptive Nonlinearity

To make nonlinear adaptive FIR filters in $\mathbb{C}$ (see Figure 8.1) suitable for the processing of signals with unknown dynamics, the amplitude of the analytic nonlinear activation function can be made adaptive according to [294]

$$\Phi\big(net(k)\big) = \Phi\big(\mathbf{x}^{\mathrm{T}}(k)\mathbf{w}(k)\big) = \lambda(k)\overline{\Phi}\big(\mathbf{x}^{\mathrm{T}}(k)\mathbf{w}(k)\big) = \lambda(k)\big(\overline{u}(k) + \jmath\overline{v}(k)\big), \quad \lambda \in \mathbb{R}^{+} \tag{9.3}$$

where $\lambda(k)$ denotes an adaptive amplitude of the nonlinearity $\Phi\left(\mathbf{x}^{\mathrm{T}}(k)\mathbf{w}(k)\right)$, and $\overline{\Phi}\left(\mathbf{x}^{\mathrm{T}}(k)\mathbf{w}(k)\right)$ is the activation function with unit amplitude. For the logistic sigmoid function we have

$$\Phi\big(net(k), \eta, \lambda(k)\big) = \frac{\lambda(k)}{1 + e^{-\beta net(k)}} \tag{9.4}$$

where $net(k) \in \mathbb{C}$. Thus, for $\lambda(k) = 1$ we have $\Phi(k) = \overline{\Phi}(k)$. A stochastic gradient update for the adaptive amplitude from (9.3) is given by

$$\lambda(k+1) = \lambda(k) - \rho\nabla_{\lambda} J(k)_{|\lambda=\lambda(k)} \tag{9.5}$$

where

$$\nabla_{\lambda} J(k)_{|\lambda=\lambda(k)} = \frac{1}{2}\left[e^{*}(k)\frac{\partial e(k)}{\partial\lambda(k)} + e(k)\frac{\partial e^{*}(k)}{\partial\lambda(k)}\right] \tag{9.6}$$

is the gradient of the cost function with respect to the amplitude of the activation function $\lambda(k)$, and $\rho \in \mathbb{R}$ the stepsize of the algorithm, a small constant. Noticing that both $\lambda(k)$ and $J(k) = \frac{1}{2}e(k)e^{*}(k)$ are real valued, we have

$$\begin{aligned}\frac{\partial e(k)}{\partial\lambda(k)} &= \frac{\partial e_r(k)}{\partial\lambda(k)} + \jmath\frac{\partial e_i(k)}{\partial\lambda(k)} = \frac{\partial\big[d_r(k) - \lambda(k)\overline{u}(k)\big]}{\partial\lambda(k)} + \jmath\frac{\partial\big[d_i(k) - \lambda(k)\overline{v}(k)\big]}{\partial\lambda(k)} \\ &= -\overline{u}(k) - \jmath\overline{v}(k) = -\overline{\Phi}(k)\end{aligned} \tag{9.7}$$

Similarly to the derivation of $\partial e(k)/\partial\lambda(k)$, the second term $\partial e^{*}(k)/\partial\lambda(k)$ from Equation (9.6) is given by

$$\frac{\partial e^{*}(k)}{\partial\lambda(k)} = -\big(\overline{u}(k) - \jmath\overline{v}(k)\big) = -\overline{\Phi}^{*}(k) \tag{9.8}$$

Since the adaptive amplitude of nonlinearity $\lambda(k) \in \mathbb{R}^{+}$, its update has the form

$$\lambda(k+1) = \lambda(k) + \frac{\rho}{2}\left|e^{*}(k)\overline{\Phi}\big(\mathbf{x}^{\mathrm{T}}(k)\mathbf{w}(k)\big) + e(k)\overline{\Phi}^{*}\big(\mathbf{x}^{\mathrm{T}}(k)\mathbf{w}(k)\big)\right| \tag{9.9}$$

This adaptive amplitude of nonlinearity can be also used in conjunction with any of the standard stochastic gradient learning algorithms. For instance, a normalised CNGD algorithm equipped with an adaptive amplitude of nonlinearity can be expressed as

$$\begin{aligned}
e(k) &= d(k) - \Phi\big(\mathbf{x}^{\mathrm{T}}(k)\mathbf{w}(k)\big) \\
\Phi\big(\mathbf{x}^{\mathrm{T}}(k)\mathbf{w}(k)\big) &= \lambda(k)\overline{\Phi}\big(\mathbf{x}^{\mathrm{T}}(k)\mathbf{w}(k)\big) \\
\mathbf{w}(k+1) &= \mathbf{w}(k) + \eta(k)e(k)\Phi'^{*}\big(\mathbf{x}^{\mathrm{T}}(k)\mathbf{w}(k)\big)x^{*}(k) \\
\eta(k) &= \frac{1}{\big|\Phi'\big(\mathbf{x}^{T}(k)\mathbf{w}(k)\big)\big|^{2}\,\|\mathbf{x}(k)\|_{2}^{2} + \varepsilon} \\
\lambda(k+1) &= \lambda(k) + \frac{\rho}{2}\Big|e^{*}(k)\overline{\Phi}\big(\mathbf{x}^{T}(k)\mathbf{w}(k)\big) + e(k)\overline{\Phi}^{*}\big(\mathbf{x}^{T}(k)\mathbf{w}(k)\big)\Big|
\end{aligned}$$

9.3 Recurrent Neural Networks with Trainable Amplitude of Activation Functions

Assume that every neuron $l = 1, \ldots, N$ in an RNN is equipped with a nonlinearity for which the amplitude is made adaptive, that is

$$y_l(k) = \Phi\big(net_l(k)\big) = \lambda_l(k)\overline{\Phi}\big(net_l(k)\big), \quad l = 1, \ldots, N \tag{9.10}$$

where[2]

$$net_l(k) = \mathbf{I}^{\mathrm{T}}(k)\mathbf{w}_l(k) \tag{9.11}$$

The symbol $\lambda_l(k)$ denotes the adaptive amplitude of the nonlinearity at the lth neuron, whereas $\overline{\Phi}\big(net_l(k)\big)$ denotes the nonlinearity with a unit amplitude. Thus, if $\lambda_l = 1$ it follows that $\Phi\big(net_l(k)\big) = \overline{\Phi}\big(net_l(k)\big)$. For an RNN with N output neurons, the update for the gradient adaptive amplitude at the lth neuron is calculated based on [294]

$$\lambda_l(k+1) = \lambda_l(k) - \rho\nabla_{\lambda_l(k)}J(k), \quad l = 1, \ldots, N \tag{9.12}$$

where $\nabla_{\lambda_l(k)}J(k)$ denotes the gradient of the cost function with respect to the amplitude of the activation function λ, ρ is the stepsize, and the cost function is given by

$$J(k) = \frac{1}{2}\sum_{l=1}^{N}|e_l|^2(k) \tag{9.13}$$

From Equation (9.13), the gradient $\nabla_{\lambda_l(k)}J(k)$ can be obtained as

$$\nabla_{\lambda_l(k)}J(k) = \frac{\partial J(k)}{\partial\lambda_l(k)} = \sum_{l=1}^{N} e_l(k)\frac{\partial e_l(k)}{\partial\lambda_l(k)} = -\sum_{l=1}^{N} e_l(k)\frac{\partial y_l(k)}{\partial\lambda_l(k)} \tag{9.14}$$

[2]The overall input to the network $\mathbf{I}(k)$ represents the concatenation of the feedback vector $\mathbf{y}(k)$, external input $\mathbf{x}(k)$, and bias input $(1 + \jmath)$, as shown in Section 7.3.

where

$$\frac{\partial y_l(k)}{\partial \lambda_l(k)} = \overline{\Phi}\big(\mathbf{I}^{\mathrm{T}}(k)\mathbf{w}_l(k)\big) + \lambda_l(k)\overline{\Phi}'\big(\mathbf{I}^{\mathrm{T}}(k)\mathbf{w}_l(k)\big) \times \frac{\partial\left[\mathbf{I}^{\mathrm{T}}(k)\mathbf{w}_l(k)\right]}{\partial \lambda_l(k)}$$

$$= \overline{\Phi}\big(\underbrace{\mathbf{I}^{\mathrm{T}}(k)\mathbf{w}_l(k)}_{net_l(k)}\big) + \lambda_l(k)\overline{\Phi}'\big(\underbrace{\mathbf{I}^{\mathrm{T}}(k)\mathbf{w}_l(k)}_{net_l(k)}\big) \times \frac{\partial}{\partial \lambda_l(k)}\Big(\sum_{n=1}^{p+N+1} w_{l,n}(k)I_n(k)\Big) \quad (9.15)$$

Since $\partial\lambda_l(k-1)/\partial\lambda_l(k) = 0$, the second term in Equation (9.15) vanishes, to give

$$\nabla_{\lambda_l(k)} J(k) = \frac{\partial J(k)}{\partial \lambda_l(k)} = -\sum_{l=1}^{N} e_l(k)\overline{\Phi}\big(net_l(k)\big) \quad (9.16)$$

We next consider the following three cases:

- Case 1: *Common adaptive nonlinearity for all the neurons*. In this case $\lambda_l(k)=\lambda(k)$ for all $l = 1, \ldots, N$ and Equation (9.12) becomes [92]

$$\lambda(k+1) = \lambda(k) + \rho\left|\sum_{l=1}^{N} e_l(k)\overline{\Phi}\big(net_l(k)\big)\right| \quad (9.17)$$

- Case 2: *Common adaptive nonlinearity for each layer*. Since a fully connected recurrent network has two layers, assume that the output layer consists of M output neurons and the hidden layer contains the remaining $(N - M)$ neurons. This way, we have [90]

$$y_l(k) = \Phi\big(net_l(k)\big) = \begin{cases} \lambda_1(k)\overline{\Phi}\big(net_l(k)\big), & l = 1, \ldots, M \\ \lambda_2(k)\overline{\Phi}\big(net_l(k)\big), & l = M+1, \ldots, N \end{cases} \quad (9.18)$$

  where

$$\lambda_1(k+1) = \lambda_1(k) + \rho\left|\sum_{l=1}^{M} e_l(k)\overline{\Phi}\big(net_l(k)\big)\right|, \quad l = 1, \ldots, M$$

$$\lambda_2(k+1) = \lambda_2(k) + \rho\left|\sum_{l=M+1}^{N} e_l(k)\overline{\Phi}\big(net_l(k)\big)\right|, \quad l = M+1, \ldots, N \quad (9.19)$$

- Case 3: *Different $\lambda_l(k)$ for each neuron in the network*. This is the most general case where every neuron is equipped with an adaptive amplitude of nonlinearity, that is

$$y_l(k) = \Phi\big(net_l(k)\big) = \lambda_l(k)\overline{\Phi}\big(net_l(k)\big)$$

$$\lambda_l(k+1) = \lambda_l(k) + \rho\left|e_l(k)\overline{\Phi}\big(net_l(k)\big)\right|, \quad l = 1, \ldots, N \quad (9.20)$$

9.4 Simulation Results

In all the experiments, the amplitudes of the inputs were scaled to be within the range [0,0.1] and the nonlinearity was chosen to be the complex logistic sigmoid function, given by

$$\Phi\left(\mathbf{x}^{\mathrm{T}}(k)\mathbf{w}(k), \beta, \lambda(k)\right) = \frac{\lambda(k)}{1 + e^{-\beta\mathbf{x}^{\mathrm{T}}(k)\mathbf{w}(k)}} \tag{9.21}$$

with slope $\beta = 1$.

Simulations were undertaken by averaging 200 independent trials on prediction of both complex valued benchmark (coloured, Equation 8.22) and nonlinear, (Equation 8.23) signals as well as real world wind signals.[3] The measurement used to assess the performance was the prediction gain $R_p = 10log_{10}(\sigma_x^2/\hat{\sigma}_e^2)(dB)$. Simulations are provided for both the FIR and recurrent nonlinear adaptive filters.

Nonlinear FIR filter. In all the experiments, the order of the nonlinear adaptive filter was chosen to be $L = 4$, learning rate $\mu = 0.001$, and an initial amplitude $\lambda(0) = 0.1$. The stepsize for the adaptive amplitude was chosen to be $\rho = 0.01$. The best choice of the regularisation factor ε for the CNNGD and CFANNGD[4] algorithms was $\varepsilon = 0.15$. Figure 9.3 illustrates the dependence of the prediction gain R_p on initialisation of the adaptive amplitude $\lambda(0)$ and regularisation factor $\varepsilon(0)$, whereas Figure 9.4 shows convergence curves for the prediction of nonlinear input (Equation 8.23). From Figure 9.4, the algorithms equipped with an adaptive amplitude of nonlinearity converged faster and had better steady state properties. Figure 9.5(a) shows the behaviour of $\lambda(k)$ on prediction of a synthetic complex valued nonlinear input (16.3),

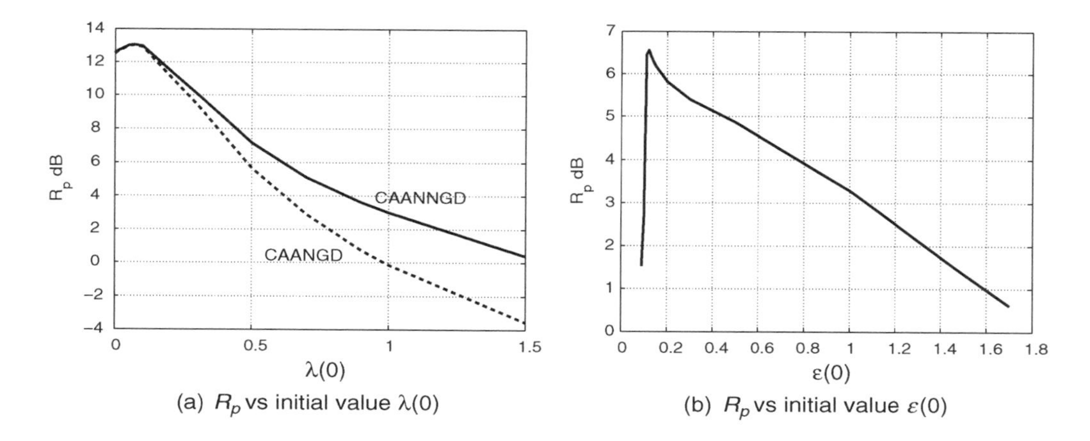

Figure 9.3 Prediction gain as a function of the initial amplitude of nonlinearity $\lambda(k)$ and the initial value of the regularisation parameter $\varepsilon(k)$, for prediction of nonlinear signal (Equation 8.23)

[3]Publicly available from *"http://mesonet.agron.iastate.edu/"*.

[4]See Section 8.2.1 in Chapter 8 for reference to these algorithms.

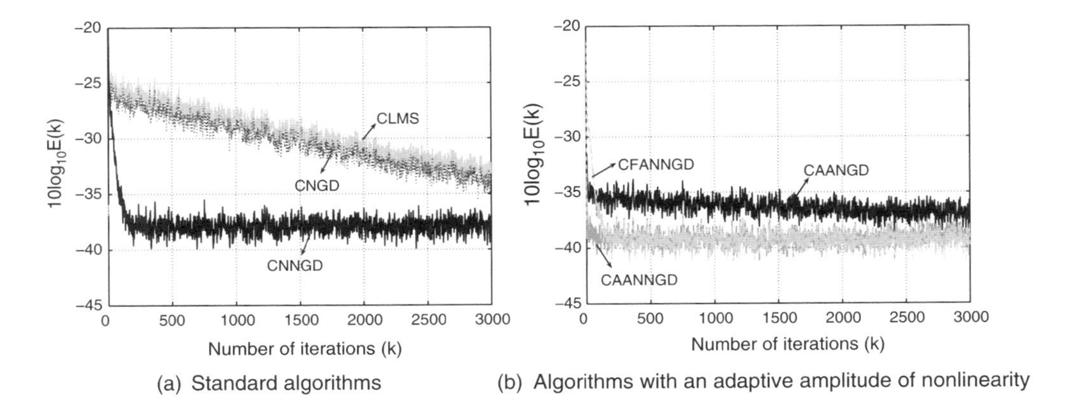

(a) Standard algorithms (b) Algorithms with an adaptive amplitude of nonlinearity

Figure 9.4 Comparison of standard algorithms with algorithms equipped with an adaptive amplitude of nonlinearity (CAANGD and CAANNGD) and those with a gradient adaptive stepsize (CFANNGD) for prediction of nonlinear signal (Equation 8.23)

the amplitude of the nonlinearity clearly adapts according to the dynamics of the input, leading to improved performance.

Complex RNNs. In all the experiments, the RNN had $N = 5$ neurons with the length of the external tap input $L = 7$ and first order feedback, learning rate $\mu = 0.1$, and an initial amplitude of nonlinearities $\lambda(0) = 1$. The stepsize for the adaptive amplitude was chosen to be $\rho = 0.15$. Figure 9.5(b) shows that that the amplitude $\lambda(k)$ within the AACRTRL followed the dynamics of the input, leading to improved performance. Figure 9.6 shows learning curves for the CRTRL and AACRTRL algorithms for both the coloured and nonlinear input. In both cases, the AACRTRL algorithm, equipped with an adaptive amplitude of nonlinearity, outperformed the CRTRL algorithm and exhibited faster convergence.

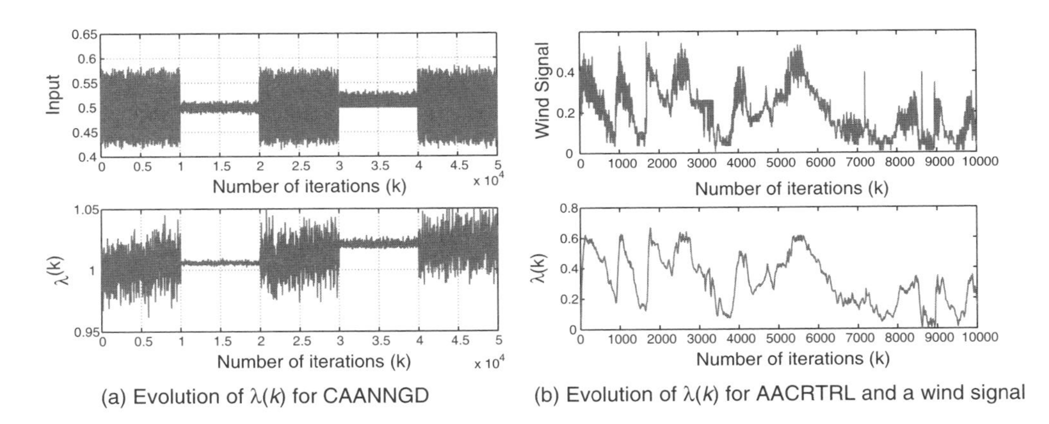

(a) Evolution of $\lambda(k)$ for CAANNGD (b) Evolution of $\lambda(k)$ for AACRTRL and a wind signal

Figure 9.5 Evolution of the adaptive amplitude $\lambda(k)$ for two classes of complex nonlinear adaptive filters

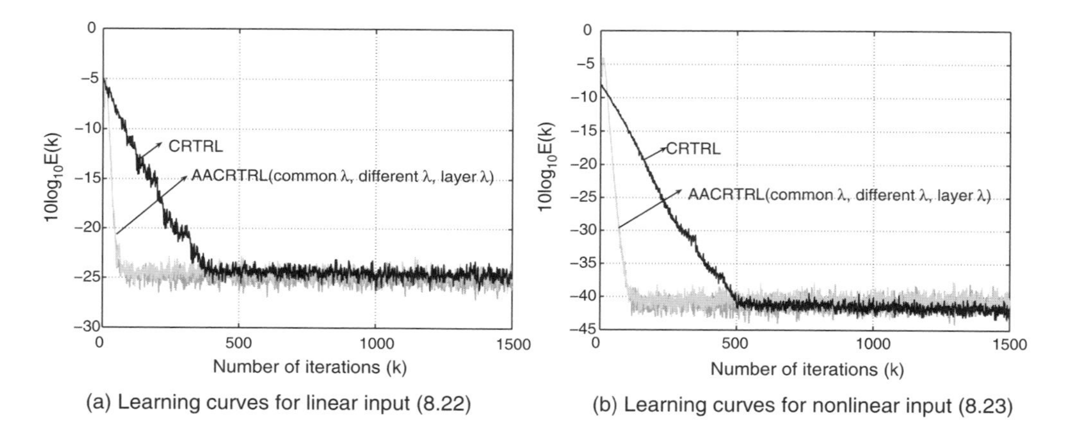

Figure 9.6 Comparison between the standard CRTRL algorithm and the AACRTRL algorithm for the three cases of adaptive amplitude:- common adaptive nonlinearity for the whole network; common adaptive nonlinearity for every network layer; individual adaptive amplitude of nonlinearity for every neuron

The quantitative performance results for the RNNs equipped with an adaptive amplitude of nonlinearity are summarised in Table 9.1. The case with an individual adaptive amplitude of nonlinearity at every neuron performed marginally best, but was also most demanding in terms of computational complexity.

To summarise:

- Performance of nonlinear adaptive filters depends on the range of the input; to circumvent this problem input signals are usually first standardised (see Appendix G).
- Standardisation of input is not realistic in the online adaptive mode of operation, and methods for dynamical range reduction based on the subtraction of a signal estimate from its original value are commonly employed. These methods, however, suffer from the accumulation of differencing and integration errors.
- Alternatively, dynamical range reduction and extension can be achieved based on a hormomorphic transform [226, 260]; this method operates in an automated manner, however the log–exp transformation may be quite sensitive.

Table 9.1 Prediction gain R_p for there CRTRL based adaptive amplitude algorithms

Different learning algorithms	Nonlinear N1 (8.23)	Linear L1 (8.22)
CRTRL with fixed amplitude of nonlinearity	4.012	3.845
CRTRL with common adaptive nonlinearity	5.562	4.724
CRTRL with 'layer-by-layer' adaptive nonlinearity	5.771	4.837
CRTRL with neuron-by-neuron adaptive nonlinearity	5.723	4.891

- To make nonlinear adaptive filters in $\mathbb{C}$ more efficient in nonstationary environments and for signals with large dynamics, the nonlinearities within such filters are equipped with an adaptive amplitude.
- Three different algorithms with an adaptive amplitude of nonlinearity are considered: a common nonlinear activation function with a common adaptive amplitude for all the neurons; a common nonlinearity with a common adaptive amplitude per network layer; individual adaptive amplitude of nonlinearity for every neuron in the network.
- It is shown that this class of algorithms has the potential to outperform the standard algorithms for a variety of benchmark and real world signals. This is achieved at little expense in terms of computational complexity.

10

Data-reusing Algorithms for Complex Valued Adaptive Filters

The class of data-reusing (DR) algorithms is based on *a posteriori* error adaptation and exhibits in general better convergence than standard *a priori* error based algorithms [190]. This is achieved in a fixed point iteration-like fashion by reusing the external input data while performing *a posteriori* weight update iterations.[1] By combining the recursive mode of learning based on the *a priori* output error and iterative mode of learning based on the *a posteriori* errors, data-reusing algorithms effectively operate between the time instants[2] k and $(k + 1)$. In this chapter, we introduce the class of data-reusing algorithms for complex valued adaptive filters, both feedforward and recurrent. The error bounds and convergence conditions are provided for both the case of contractive and of expansive complex activation functions.[3]

10.1 The Data-reusing Complex Valued Least Mean Square (DRCLMS) Algorithm

Data-reusing (DR) algorithms are modifications of standard algorithms whereby at every discrete time instant, k, after the 'recursive' standard update, the available desired response $d(k)$ and input vector $\mathbf{x}(k)$ are reused in order to refine the estimate of the filter coefficients (weights) [261]. Such updates are known as *a posteriori* updates.[4] Following the approach from [254],

[1] For more detail on *a posteriori* mode of learning, see Appendix M.
[2] Another class of algorithms which operate this way are fractional delay filters, see Chapter 11.
[3] A meromorphic complex valued activation function Φ is a contraction if $|\Phi(a + b)| < |\Phi(a) + \Phi(b)|$. Function Φ is an expansion if $|\Phi(a + b)| > |\Phi(a) + \Phi(b)|$. For more detail see Appendix P.
[4] For real valued adaptive filters this is elaborated in Appendix M.

Complex Valued Nonlinear Adaptive Filters: Noncircularity, Widely Linear and Neural Models
Danilo P. Mandic and Vanessa Su Lee Goh

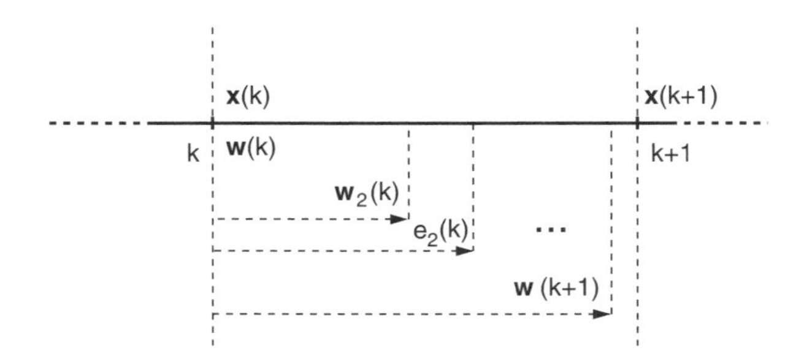

Figure 10.1 Time alignment within the data-reusing (*a posteriori*) approach

the weight update in the data-reusing complex valued LMS (CLMS) algorithm can be written as

$$\mathbf{w}_{t+1}(k) = \mathbf{w}_t(k) + \mu e_t(k)\mathbf{x}^*(k) \tag{10.1}$$

$$e_t(k) = d(k) - \mathbf{x}^{\mathrm{T}}(k)\mathbf{w}_t(k), \quad t = 1, \dots, L \tag{10.2}$$

where $\mathbf{w}_1(k) = \mathbf{w}(k),\ \mathbf{w}_{L+1}(k) = \mathbf{w}(k+1)$ and t represents the order of data-reuse iteration. For $L = 1$, Equations (10.1) and (10.2) boil down to the standard CLMS algorithm, given in Chapter 6. Time alignment for the *a priori* and *a posteriori* mode of operation is shown in Figure 10.1. In conformance with the analysis in [254], we can express the final DR weight update from (10.1) as

$$\begin{aligned}
\mathbf{w}(k+1) = \mathbf{w}_{L+1}(k) &= \mathbf{w}_L(k) + \mu e_L(k)\mathbf{x}^*(k) \\
&= \mathbf{w}_{L-1}(k) + \mu\big(e_{L-1}(k) + e_L(k)\big)\mathbf{x}^*(k) \\
&= \mathbf{w}(k) + \mu\sum_{t=1}^{L} e_t(k)\mathbf{x}^*(k).
\end{aligned} \tag{10.3}$$

To establish relationship between the *a priori* the *a posteriori* error (see also Appendix M), consider $t = 2$, that is

$$\begin{aligned}
e_2(k) &= d(k) - \mathbf{x}^{\mathrm{T}}(k)\mathbf{w}_2(k) \\
&= d(k) - \mathbf{x}^{\mathrm{T}}(k)\left[\mathbf{w}_1(k) + \mu e_1(k)\mathbf{x}^*(k)\right] \\
&= e_1(k)\left[1 - \mu\mathbf{x}^{\mathrm{T}}(k)\mathbf{x}^*(k)\right]
\end{aligned} \tag{10.4}$$

Consequently, the tth DR error can be expressed as

$$e_t(k) = e(k)\left[1 - \mu\mathbf{x}^{\mathrm{T}}(k)\mathbf{x}^*(k)\right]^{t-1}, \quad t = 1, \dots, L \tag{10.5}$$

and the total error after L data reusing iterations $\sum_{t=1}^{L} e_t(k)$ is given by

$$\sum_{t=1}^{L} e_t(k) = \sum_{t=1}^{L} e(k)\left[1 - \mu\mathbf{x}^T(k)\mathbf{x}^*(k)\right]^{t-1}$$
$$= \frac{e(k)\left[1 - \left(1 - \mu\mathbf{x}^T(k)\mathbf{x}^*(k)\right)^L\right]}{\mu\mathbf{x}^T(k)\mathbf{x}^*(k)}. \tag{10.6}$$

Finally, the DR weight update for L iterations of the DRCLMS algorithm becomes [135, 254, 261]

$$\mathbf{w}(k+1) = \mathbf{w}(k) + \frac{1 - \left[1 - \mu\mathbf{x}^{\mathrm{T}}(k)\mathbf{x}^*(k)\right]^L}{\mathbf{x}^{\mathrm{T}}(k)\mathbf{x}^*(k)} e(k)\mathbf{x}^*(k). \tag{10.7}$$

More detail on the class of *a posteriori* and data-reusing algorithms and a geometric interpretation of their convergence can be found in Appendix M and Appendix P.

Simulations. Figure 10.2 shows the learning curves for CLMS and DRCLMS with $L = 1$, $L = 3$, and $L = 10$. For convenient visualisation, the learning rate was chosen to be $\mu = 0.001$. There were 100 independent trials averaged on the prediction of complex valued coloured (Equation 8.22) and nonlinear (Equation 8.23) input. The speed of convergence improved with the order of DR iterations, approaching the NCLMS algorithm in the limit (for $L \to \infty$), as illustrated in Appendix M.

10.2 Data-reusing Complex Nonlinear Adaptive Filters

We now extend the class of DR algorithms to complex nonlinear adaptive filters realised as a dynamical perceptron. The principle of data-reusing relies on the updated weight vector

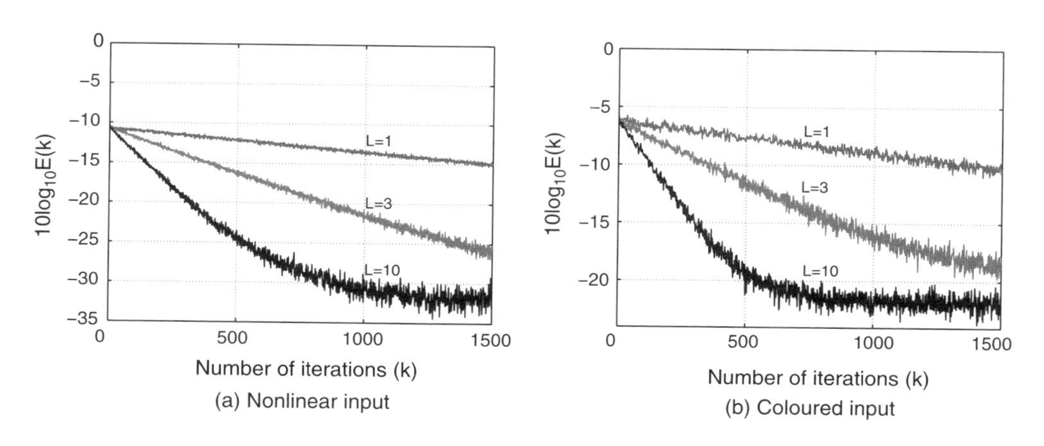

Figure 10.2 Performance of CLMS and data-reusing CLMS (DRCLMS) with $L = 1$, 3, and 10 for prediction of a coloured input (Equation 8.22) and nonlinear input (Equation 8.23)

$\mathbf{w}(k+1)$ being available before the next input vector $\mathbf{x}(k+1)$ (as shown in Figure 10.1). Similarly to the derivation of the DRCLMS algorithm, a complex data-reusing nonlinear gradient descent (CDRNGD) algorithm can be expressed as[5]

$$e_t(k) = d(k) - \Phi\left(\mathbf{x}^{\mathrm{T}}(k)\mathbf{w}_t(k)\right)$$

$$\mathbf{w}_{t+1}(k) = \mathbf{w}_t(k) + \mu\Phi'^{*}\left(\mathbf{x}^{\mathrm{T}}(k)\mathbf{w}_t(k)\right)e_t(k)\mathbf{x}^{*}(k) \tag{10.8}$$

where

$$\mathbf{w}_1(k) = \mathbf{w}(k), \quad \mathbf{w}_{L+1}(k) = \mathbf{w}(k+1) \tag{10.9}$$

and L denotes the number of data reusing iterations. For $L = 1$ the DRCNGD algorithm reduces to the standard complex nonlinear gradient descent (CNGD) algorithm. From the analysis for the linear filter in Section 10.1 it follows that the final DR weight update becomes

$$\begin{aligned}
\mathbf{w}(k+1) &= \mathbf{w}_{L+1}(k)\\
&= \mathbf{w}_L(k) + \mu\Phi'^{*}\left(\mathbf{x}^{\mathrm{T}}(k)\mathbf{w}_L(k)\right)e_L(k)\mathbf{x}^{*}(k)\\
&= \mathbf{w}(k) + \mu\sum_{t=1}^{L} e_t(k)\Phi'^{*}\left(\mathbf{x}^{\mathrm{T}}(k)\mathbf{w}_t(k)\right)\mathbf{x}^{*}(k)
\end{aligned} \tag{10.10}$$

The instantaneous data-reusing output error can be expressed as [109]

$$\begin{aligned}
e_t(k) &= d(k) - \Phi\left(\mathbf{x}^{\mathrm{T}}(k)\mathbf{w}_t(k)\right)\\
&= e_{t-1}(k) - \Big[\underbrace{\Phi\left(\mathbf{x}^{\mathrm{T}}(k)\mathbf{w}_t(k)\right) - \Phi\left(\mathbf{x}^{\mathrm{T}}(k)\mathbf{w}_{t-1}(k)\right)}_{\text{term depending on the nature of }\Phi}\Big].
\end{aligned} \tag{10.11}$$

From Equation (10.11), the performance of the data-reusing approach depends critically on whether the complex nonlinear activation function of a neuron is a contraction[6] or an expansion [190].

10.2.1 Convergence Analysis

To analyse the performance of the DR algorithm, it is important to establish the relationship between the *a priori* error $e(k)$ and *a posteriori* error $e_t(k)$. A detailed analysis for real valued nonlinear filters is provided in [63, 182, 184, 189, 190]. Premultiplying (10.8) by $\mathbf{x}^{\mathrm{T}}(k)$ and applying the nonlinear activation function Φ on either side, we have

$$\Phi\left(\mathbf{x}^{\mathrm{T}}(k)\mathbf{w}_{t+1}(k)\right) = \Phi\left[\mathbf{x}^{\mathrm{T}}(k)\mathbf{w}_t(k) + \mu e_t(k)\Phi'^{*}\left(\mathbf{x}^{\mathrm{T}}(k)\mathbf{w}_t(k)\right)\mathbf{x}^{T}(k)\mathbf{x}^{*}(k)\right] \tag{10.12}$$

[5]Symbol $(\cdot)'^{*}$ denotes $\left\{(\cdot)'\right\}^{*}$, for instance, $\Phi'^{*} = \left\{\Phi'\right\}^{*}$. Similarly $\Phi''^{*} = \left\{\Phi''\right\}^{*}$.

[6]For more detail, see Appendix P.

Subtract the teaching signal $d(k)$ from both sides of (10.8), to give the sequence[7]

$$|e_{t+1}(k)| > \left|1 - \mu(k)\Phi'^{*}\big(\mathbf{x}^{\mathrm{T}}(k)\mathbf{w}_t(k)\big)\mathbf{x}^{\mathrm{T}}(k)\mathbf{x}^{*}(k)\right| |e_t(k)| \tag{10.13}$$

Here it is assumed that the value of $\Phi'^{*}\big(\mathbf{x}^{\mathrm{T}}(k)\mathbf{w}_t(k)\big)$, $t = 1, 2, \ldots, L$ does not change significantly during successive iterations. Then, after L iterations of Equation (10.13), we have

$$|e_{L+1}(k)| > \left|\big[1 - \mu(k)\Phi'^{*}\big(\mathbf{x}^{\mathrm{T}}(k)\mathbf{w}(k)\big)\mathbf{x}^{\mathrm{T}}(k)\mathbf{x}^{*}(k)\big]^{L}\right| |e(k)| \tag{10.14}$$

which is the lower bound for the data reusing output error for a contractive nonlinear activation function. For the error in Equation (10.14) to be monotonically decreasing, the absolute value of the term $\big[1 - \mu(k)\Phi'^{*}\big(\mathbf{x}^{\mathrm{T}}(k)\mathbf{w}(k)\big)\mathbf{x}^{\mathrm{T}}(k)\mathbf{x}^{*}(k)\big]$ must be less than unity. In that case, the whole procedure is a fixed point iteration (see Appendix P). This gives the following constraint on the learning rate

$$0 < \mu(k) < \frac{1}{\left|\Phi'^{*}\big(\mathbf{x}^{\mathrm{T}}(k)\mathbf{w}(k)\big)\mathbf{x}^{\mathrm{T}}(k)\mathbf{x}^{*}(k)\right|} \tag{10.15}$$

Simulations. Figure 10.3 shows the averaged learning curves of the DRCNGD algorithm for a nonlinear FIR filter with a contractive activation function. The nonlinearity at the neuron was the logistic sigmoid function, given by

$$\Phi(\beta, x) = \frac{1}{1 + e^{-\beta x}}, \qquad x \in \mathbb{C}, \beta \in \mathbb{R}^{+} \tag{10.16}$$

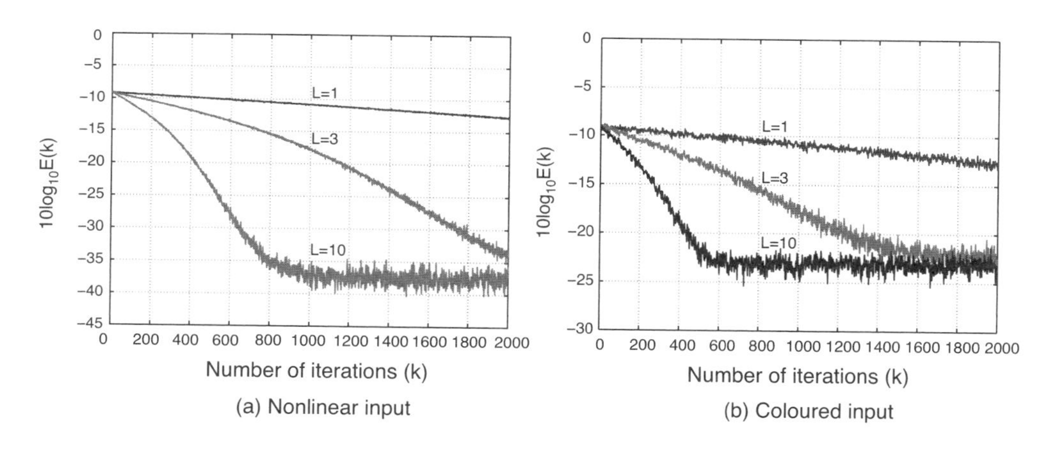

Figure 10.3 Performance of CNGD and data-reusing CNGD (DRCNGD) with $L = 1$, 3, and 10 for prediction of a coloured input (Equation 8.22) and nonlinear input (Equation 8.23)

[7] We use the absolute values since $\mathbb{C}$ is not an ordered field (see Appendix A).

The slope was $\beta = 1$ for contractive Φ, and the learning rate was chosen to be $\mu = 0.001$, in order to obtain a clear visualization of the performance of the algorithms. The data-reusing algorithm exhibited faster convergence than the standard algorithm ($L = 1$) for both types of input signals. The performance of this algorithm improves with order of the data-reusing iterations, and in the limit, for large L, it approaches the normalised version of the CNGD algorithm.

10.3 Data-reusing Algorithms for Complex RNNs

This analysis is based on the complex real time recurrent learning (CRTRL) algorithm for a recurrent perceptron, given in Section 7.3. The data-reusing weight update for this case can be written as [179, 190].

$$\mathbf{w}_{t+1}(k) = \mathbf{w}_t(k) + \mu e_t(k)\boldsymbol{\pi}_t^\star(k) \tag{10.17}$$

$$e_t(k) = d(k) - \Phi\big(\mathbf{I}^{\mathrm{T}}(k)\mathbf{w}_t(k)\big), \quad t = 1, \ldots, L \tag{10.18}$$

where the overall input to the network $\mathbf{I}(k)$ represents the concatenation of the feedback vector $\mathbf{y}(k)$, external input $\mathbf{x}(k)$, and bias input $(1 + j)$. The index t denotes the tth iteration of Equations (10.17) and (10.18), $\boldsymbol{\pi}_t^\star(k)$ is the vector of sensitivities for the kth recursion and tth iteration, and μ is the learning rate. The weight update iteration starts with $\mathbf{w}_1(k) = \mathbf{w}(k)$ (standard *a priori* update) and ends with $\mathbf{w}_{L+1}(k) = \mathbf{w}(k + 1)$.

From Equation (10.17), for $t = L$, we have the final data-reusing weight update

$$\begin{aligned}\mathbf{w}(k+1) = \mathbf{w}_{L+1}(k) &= \mathbf{w}_L(k) + \mu e_L(k)\boldsymbol{\pi}_L^\star(k)\\ &= \mathbf{w}_{L-1}(k) + \mu e_{L-1}(k)\boldsymbol{\pi}_{L-1}^\star(k) + \mu e_L(k)\boldsymbol{\pi}_L^\star(k)\\ &= \mathbf{w}(k) + \sum_{t=1}^{L} \mu e_t(k)\boldsymbol{\pi}_t^\star(k)\end{aligned} \tag{10.19}$$

The instantaneous error at the output neuron can be evaluated as

$$\begin{aligned}e_t(k) &= d(k) - \Phi\big(\mathbf{I}^{\mathrm{T}}(k)\mathbf{w}_t(k)\big)\\ &= \big[d(k) - \Phi\big(\mathbf{I}^{\mathrm{T}}(k)\mathbf{w}_{t-1}(k)\big)\big] - \big[\Phi\big(\mathbf{I}^{\mathrm{T}}(k)\mathbf{w}_t(k)\big) - \Phi(\mathbf{I}^{\mathrm{T}}(k)\mathbf{w}_{t-1}(k))\big]\\ &= e_{t-1}(k) - \Big[\underbrace{\Phi\big(\mathbf{I}^{\mathrm{T}}(k)\mathbf{w}_t(k)\big) - \Phi\big(\mathbf{I}^{\mathrm{T}}(k)\mathbf{w}_{t-1}(k)\big)}_{\text{term depending on the nature of } \Phi}\Big]\end{aligned} \tag{10.20}$$

The convergence of the DR iterations depends on the contractive/expansive notive of the terms in the square brackets.

Simulations. Figure 10.4 shows the learning curves of the DRCRTRL algorithm for a recurrent perceptron (nonlinear Infinite Impulse Response (IIR) filter) with a contractive activation function for both coloured (Equation 8.22) and nonlinear (Equation 8.23) inputs. The learning rate was $\mu = 0.001$ and the nonlinearity was the complex *tanh* with slope $\beta = 1$. The

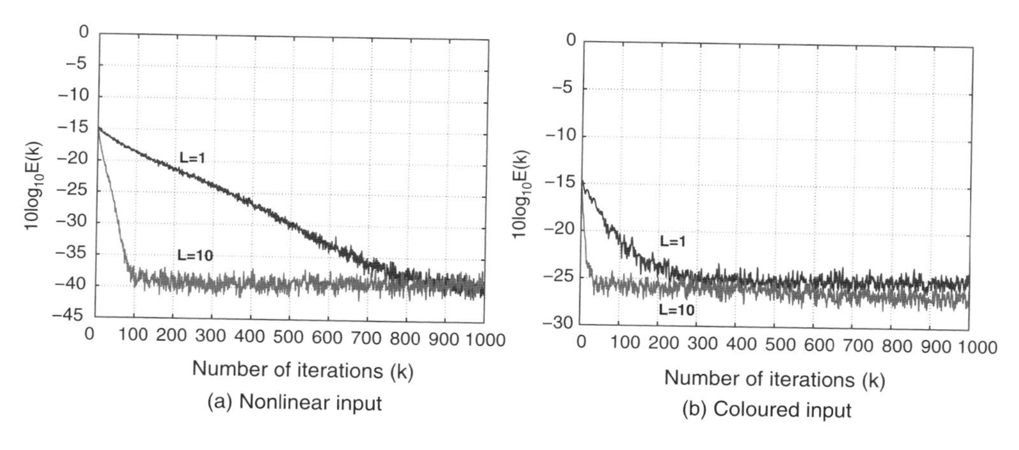

Figure 10.4 Performance of CRTRL and data-reusing CRTRL (DCRTRL) for prediction of coloured input (Equation 8.22) and nonlinear input (Equation 8.23) for a contractive nonlinear activation function

DRCRTRL algorithm converged faster than the standard CRTRL algorithm ($L = 1$) for both types of inputs. The performance of this algorithm improves with the order of data-reusing iterations and saturates for large L, approaching the normalised RTRL [187]. As desired, the advantages of the data reusing strategy were more pronounced for the 'difficult' nonlinear signal in Figure 10.4(a).

Figure 10.5 shows the performance of the data-reusing CRTRL algorithm for a recurrent perceptron with an expansive activation function (tanh with slope $\beta = 8$). The error curve does not converge and grows unbounded, the divergence is more emphasised with the order of data-reusing iteration.

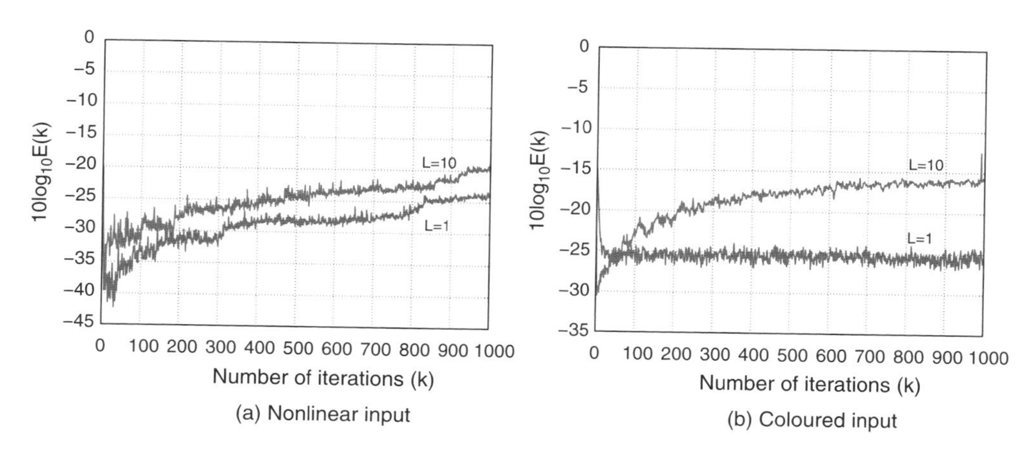

Figure 10.5 Performance of CRTRL and data-reusing CRTRL (DCRTRL) for prediction of coloured input (Equation 8.22) and nonlinear input (Equation 8.23) for an expansive nonlinear activation function

To summarise:

- Data reusing algorithms aim at refining the weight update by exploiting the relationship between the *a priori* and *a posteriori* error and by reusing the external input.
- This is achieved by combining the standard, recursive, mode of operation with the iterative, *a posteriori*, mode of operation, for more detail see Appendix G and Appendix M.
- There is a close relationship between the dynamics of learning and the properties of the nonlinearity of nonlinear adaptive filters in $\mathbb{C}$; this facilitates the use of fixed point theory in the analysis of DR algorithms.
- Due to the stochastic nature of the net input $net(k)$, the effect of the expansive nonlinearity $\Phi(\beta net(k))$ is more pronounced for DR algorithms, (see Figure 10.5(b)).
- Although the convergence of nonlinear filtering algorithms is well understood for strictly contractive and strictly expansive functions [148, 190, 256], in real world environments, there is a need to consider situations where the nature of nonlinearity Φ is randomly changing between contraction and expansion.
- For a rigorous analysis of nonlinear DR algorithms, we need to use random contraction mapping theorems [46, 164].

11

Complex Mappings and Möbius Transformations

It is often convenient to use matrix notation to represent complex processes, since this gives us the opportunity to use established tools from matrix theory and linear algebra for problem solving. Möbius transformations are amongst the most fundamental mappings in $\mathbb{C}$, and provide a formalism for dealing with a range of problems described in matrix notation – from brain mapping to relativity theory. This framework also helps to explain the mappings performed by complex valued nonlinear adaptive filters, and provides insight into their compositions (nesting), fixed points and invertibility. Our aim is to highlight links between the matrix description of complex processes, Möbius transformations, Riemann sphere, and mappings performed by complex neural networks. We also show that all-pass filters, the most versatile building blocks in signal processing, are Möbius transformations, and illustrate their use within fractional delay filters (FDF); these can be used to fine-tune digital filters by splitting the unit time delay between signal samples.

11.1 Matrix Representation of a Complex Number

A complex number $z = a + \jmath b$, can be expressed in an equivalent matrix notation as

$$a + \jmath b \rightarrow \begin{bmatrix} a & -b \\ b & a \end{bmatrix} = a \begin{bmatrix} 1 & 0 \\ 0 & 1 \end{bmatrix} + b \begin{bmatrix} 0 & -1 \\ 1 & 0 \end{bmatrix} \tag{11.1}$$

which implies

$$1 \leftrightarrow \begin{bmatrix} 1 & 0 \\ 0 & 1 \end{bmatrix} \quad \text{and} \quad \jmath \leftrightarrow \begin{bmatrix} 0 & -1 \\ 1 & 0 \end{bmatrix}$$

Complex Valued Nonlinear Adaptive Filters: Noncircularity, Widely Linear and Neural Models
Danilo P. Mandic and Vanessa Su Lee Goh

Thus for instance

$$|z|^2 = z\,z^* = \det(z) = \begin{vmatrix} a & -b \\ b & a \end{vmatrix} = a^2 + b^2$$

The analysis of hypercomplex processes can also benefit from their matrix representation. There are, for instance, two ways of representing a quaternion (see also Appendix C)

$$\vec{q} = q_0 + q_1\imath + q_2\jmath + q_3 k \ \in \mathbb{H}$$

in the matrix form (symbol $\mathbb{H}$ denotes the set of quaternions). For complex numbers

$$z = q_0 + \jmath q_1, \qquad w = q_2 + \jmath q_3$$

a quaternion can be represented by a 2×2 complex matrix

$$\vec{q} \leftrightarrow \begin{bmatrix} z & w \\ -w^* & z^* \end{bmatrix} = \begin{bmatrix} q_0 + \jmath q_1 & q_2 + \jmath q_3 \\ -q_2 + \jmath q_3 & q_0 - \jmath q_1 \end{bmatrix}$$

where the quaternion addition and multiplication correspond to matrix addition and matrix multiplication. Quaternions can be also represented as 4×4 real-valued matrices in the form of

$$\begin{bmatrix} q_0 & -q_1 & -q_2 & -q_3 \\ q_1 & q_0 & -q_3 & q_2 \\ q_2 & q_3 & q_0 & -q_1 \\ q_3 & -q_2 & q_1 & q_0 \end{bmatrix}$$

where, for instance, the conjugate of a quaternion corresponds to the transpose of the above matrix. A split-complex number in the matrix form can be expressed as

$$\begin{bmatrix} a & b \\ b & a \end{bmatrix}$$

In the same way, matrix representations can be extended to other hypercomplex numbers, such as *octonions*

$$a + b\imath_0 + c\imath_1 + d\imath_2 + e\imath_3 + f\imath_4 + g\imath_5 + h\imath_6 \ \in \mathbb{O}$$

since each triplet of bases $(\imath_k, \imath_l, \imath_m),\ k, l, m = 0, \ldots, 6,\ k \neq l \neq m$ behaves like a quaternion $(\imath, \jmath, k)$.

The matrix representation of complex numbers is very straightforward and elegant and allows us to use the established results from matrix theory in order to solve problems in the complex domain. Thus, for instance, a *sum* of two complex numbers $z_1 = a_1 + \jmath b_1$ and $z_2 = a_2 + \jmath b_2$, can be represented as[1]

$$\begin{bmatrix} a_1 & -b_1 \\ b_1 & a_1 \end{bmatrix} + \begin{bmatrix} a_2 & -b_2 \\ b_2 & a_2 \end{bmatrix} = \begin{bmatrix} a_1 + a_2 & -(b_1 + b_2) \\ b_1 + b_2 & a_1 + a_2 \end{bmatrix} \leftrightarrow z_1 + z_2 \tag{11.2}$$

Similarly, the *product* of complex numbers $z_1 z_2$ can be represented in the matrix form as

$$\begin{bmatrix} a_1 & -b_1 \\ b_1 & a_1 \end{bmatrix} \times \begin{bmatrix} a_2 & -b_2 \\ b_2 & a_2 \end{bmatrix} = \begin{bmatrix} a_1 a_2 - b_1 b_2 & -(b_1 a_2 + a_1 b_2) \\ b_1 a_2 + a_1 b_2 & a_1 a_2 - b_1 b_2 \end{bmatrix} \leftrightarrow z_1 \times z_2 \tag{11.3}$$

whereas *rotation*[2] of a complex quantity by an angle θ can be represented as

$$z_1 = z e^{\jmath\theta} \leftrightarrow \begin{bmatrix} a & -b \\ b & a \end{bmatrix} \times \begin{bmatrix} \cos\theta & -\sin\theta \\ \sin\theta & \cos\theta \end{bmatrix} = \begin{bmatrix} a\cos\theta - b\sin\theta & -(b\cos\theta + a\sin\theta) \\ b\cos\theta + a\sin\theta & a\cos\theta - b\sin\theta \end{bmatrix} \tag{11.4}$$

and is visualised in Figure 11.1.

From Equations (11.1) and (11.4) the matrix representation[3] of $\jmath$ and z^* are rotations, and complex multiplication corresponds to the simultaneous amplification and rotation (*amplitwist* [218]). Finally, for a nonzero complex number z, its inverse $z^{-1} = 1/z$ can be expressed in the

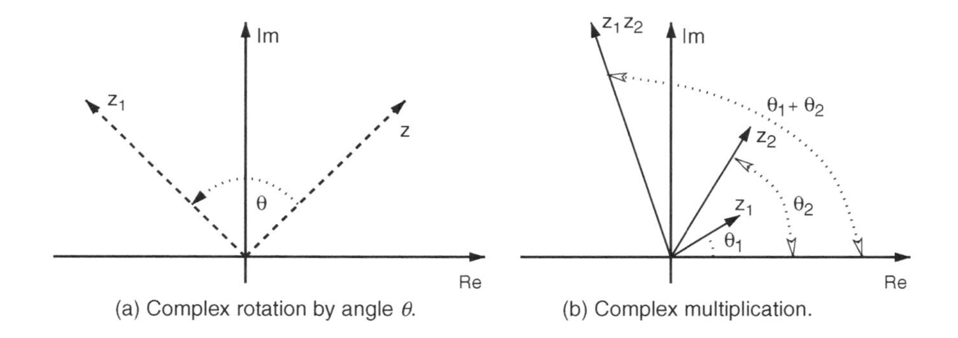

(a) Complex rotation by angle θ.
(b) Complex multiplication.

Figure 11.1 Complex rotation and multiplication

[1] The matrix representation of complex numbers also facilitates the use of software tools, such as Matlab.

[2] In matrix theory, a rotation matrix $\mathbf{A}$ is a real square matrix for which $\mathbf{A}^T\mathbf{A} = \mathbf{I}$, and $\det(\mathbf{A}) = 1$. For instance, matrix

$$\mathbf{A} = \begin{bmatrix} 0 & -1 \\ 1 & 0 \end{bmatrix}$$

corresponds to planar rotation by $\pi/2$.

[3] Operator $\jmath$ represents a rotation by $\pi/2$.

matrix form as

$$\begin{bmatrix} \dfrac{a}{a^2+b^2} & \dfrac{b}{a^2+b^2} \\ \dfrac{-b}{a^2+b^2} & \dfrac{a}{a^2+b^2} \end{bmatrix} \leftrightarrow z^{-1}$$

Matrix notation can be also used in conjunction with augmented complex statistics (see Chapter 12), to represent *widely linear* models. The *standard* $\rightarrow$ *augmented* complex variable transformation can be expressed as

$$z \rightarrow z^a \quad \leftrightarrow \quad \begin{bmatrix} z \\ z^* \end{bmatrix} = \begin{bmatrix} 1 & \jmath \\ 1 & -\jmath \end{bmatrix} \begin{bmatrix} x \\ y \end{bmatrix} \tag{11.5}$$

whereas in the case of complex valued signals, we have[4]

$$\mathbf{z} \rightarrow \mathbf{z}^a \quad \leftrightarrow \quad \begin{bmatrix} \mathbf{z} \\ \mathbf{z}^* \end{bmatrix} = \begin{bmatrix} \mathbf{I} & \jmath\mathbf{I} \\ \mathbf{I} & -\jmath\mathbf{I} \end{bmatrix} \begin{bmatrix} \mathbf{x} \\ \mathbf{y} \end{bmatrix} \tag{11.6}$$

11.2 The Möbius Transformation

We next show that some aspects of complex valued nonlinear adaptive filters can be further formalised within the framework of Möbius transformations.[5]

Definition 1 *(Möbius mapping [16]). Let $a, b, c,$ and d denote four complex constants with the restriction that $ad \neq bc$. The function*

$$w = f(z) = \frac{az+b}{cz+d} \tag{11.7}$$

is called a Möbius transformation, bilinear transformation, or linear fractional transformation.[6]

The condition $ad \neq bc$ is necessary, since for complex variables z_1 and z_2,

$$f(z_1) - f(z_2) = \frac{(ad-bc)(z_1-z_2)}{(cz_1+d)(cz_2+d)}$$

[4]To make this transform unitary, we may scale by $1/\sqrt{2}$, to give

$$\left(\frac{1}{\sqrt{2}}\begin{bmatrix} \mathbf{I} & \jmath\mathbf{I} \\ \mathbf{I} & -\jmath\mathbf{I} \end{bmatrix}\right) \times \left(\frac{1}{\sqrt{2}}\begin{bmatrix} \mathbf{I} & \jmath\mathbf{I} \\ \mathbf{I} & -\jmath\mathbf{I} \end{bmatrix}^H\right) = \mathbf{I}.$$

[5]Augustus Ferdinand Möbius was a German mathematician who held a chair in theoretical astronomy at Leipzig. He is best known for his discovery of the Möbius strip, a one-sided surface formed by giving a rectangular strip a half-twist and then joining the ends together. Interestingly, the Möbius strip has found a commercial application as a conveyor belt, allowing for the surface of the belt to last longer.

[6]Möbius transformations should not be confused with Möbius transforms (in number theory) or Möbius functions (multiplicative functions in number theory and combinatorics).

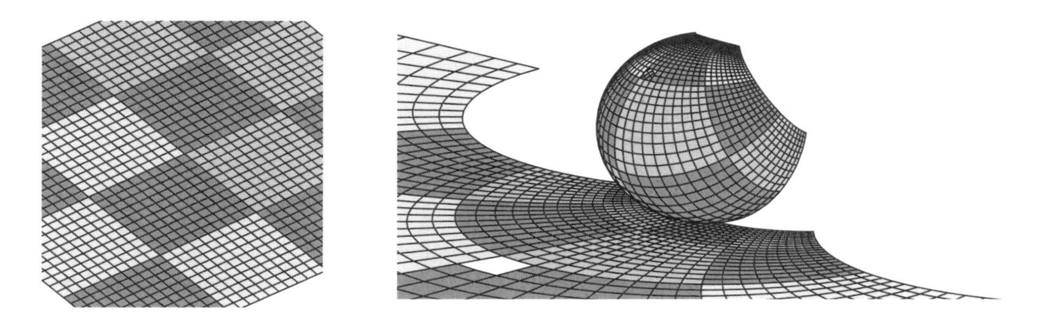

Figure 11.2 Geometric interpretation of the mapping performed by the bilinear transformation (Equation 11.8). Left: the complex plane; right: the image of the complex plane and the mapping onto the Riemann sphere (for clarity scales are not matched)

is constant for $ad - bc = 0$. Möbius transformations are also called *homographic* transformations, *bilinear* transformations, or *fractional linear* transformations.

The Möbius transformation has the following properties:

- It is analytic everywhere except for the pole at $z = -d/c$;
- The mapping is one-to-one and onto a half plane, and vice versa;
- The inverse of a Möbius transformation is also a Möbius transformation;
- The Möbius transformation does not determine the coefficients a,b,c,d uniquely, that is, if $\varphi \in \mathbb{C} \setminus \{0\}$, then coefficients φa, φb, φc, φd correspond to the same transformation;
- Every Möbius transformation (except $f(z) = z$) has one or two fixed points z^*, that is, points[7] for which $f(z^*) = z^*$ [16].

One classical example of Möbius transformations is the so called *bilinear transform*, used in the design of digital filters, given by

$$s \quad \rightarrow \quad \frac{2}{T}\frac{1 - z^{-1}}{1 + z^{-1}} \tag{11.8}$$

where T is the sampling period, and s and z are respectively the complex variable associated with the Laplace and $\mathcal{Z}$ transform. This is a conformal mapping which converts a transfer function of an analogue filter to a transfer function of a linear, shift-invariant digital filter. The $\jmath\omega$ axis[8], $\mathtt{Re}[s]= 0$, in the s-plane is mapped onto the unit circle $|z| = 1$, whereas the left-hand side half-plane in the s domain, $\mathtt{Re}[s]< 0$, is mapped onto the inside of the unit circle in the z plane ($|z| < 1$). Figure 11.2 illustrates the mapping of a plane performed by the

[7]These can be calculated to be

$$z^*_{1,2} = \frac{(a-d) \pm \sqrt{(a-d)^2 + 4bc}}{2c}$$

For $c = 0$, we have only one fixed point $z^* = -b/(a - d)$.

[8]We use both $\mathtt{Re}[\cdot]$ and $\Re\{\cdot\}$ to denote the real part of a complex quantity, and $\mathtt{Im}[\cdot]$ and $\Im\{\cdot\}$ to denote the imaginary part of a complex quantity.

bilinear transform, and the one-to-one correspondence between this mapping and the Riemann sphere.[9]

11.3 Activation Functions and Möbius Transformations

Observe that

(i) The map $g : \mathbb{C} \rightarrow \mathbb{C}$, with $g(z) = e^z$ is holomorphic on $\mathbb{C}$;
(ii) The sigmoidal nonlinear activation function $f(z)$ of a neuron is holomorphic and conformal.

Thus, for instance, by matching the coefficients associated with the equal powers of z in the complex tanh function to those in a general form of the Möbius transformation (Equation 11.7), we have

$$\frac{1 - e^{-\beta net}}{1 + e^{-\beta net}} = \frac{az + b}{cz + d} \tag{11.9}$$

and for $z = e^{-\beta net}$, the complex tanh is a Möbius transformation, with [178]

$$a = -1, \quad b = 1, \quad c = 1, \quad d = 1 \tag{11.10}$$

Since the condition $ad - bc \neq 0$ is also satisfied, the hyperbolic tangent activation function is a *Möbius transformation and holomorphic*. So too is the logistic function

$$\sigma(z) = \frac{1}{1 + e^{-\beta z}} \tag{11.11}$$

for which

$$a = 0, b = 1, c = 1, d = 1, \quad and \quad ad - bc \neq 0$$

Following upon this result, it was further established in [65] that

The sigmoidal transformation $f(z)$ performed by a neuron in a neural network on a complex signal $z = \alpha + \jmath\beta$ is a Möbius transformation.

Examples of the complex logistic and tanh mappings are shown in Figure 11.3, whereas Figure 11.4 illustrates the correspondence between the Möbius mappings from Figure 11.3 and the Riemann sphere.

[9] Another conformal mapping from $\mathbb{C}$ to $\mathbb{C}$ is the Schwarz–Christoffel transformation, which maps the upper half plane $\Im\{s\} > 0$ onto a simply connected domain D in the z-plane (polygon). For instance,

$$z = f(s) = B + A \int \left(\frac{1}{\sqrt{s+1}\sqrt{s}\sqrt{s-1}} \right) \mathrm{d}s$$

will map the upper half plane from the s domain onto a square in the z domain [54, 206].

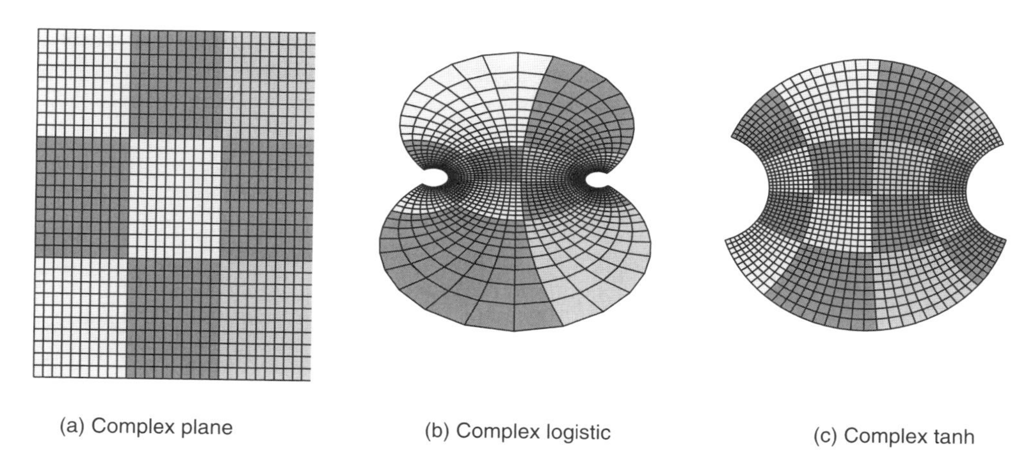

Figure 11.3 Geometric view of the mappings performed by the complex logistic and tanh functions (for clarity scales are not matched)

We can consider a Möbius transformation as a composition of a sequence of simpler transformations

- *translation* $f(z) = z + d/c$
- *inversion* $g(z) = 1/z$
- *dilation and rotation* $h(z) = \frac{-(ad-bc)}{c^2}z$
- *translation* $k(z) = z + ac$

that is, the Möbius transformation can be expressed as

$$\frac{az+b}{cz+d} = k(z) \circ h(z) \circ g(z) \circ f(z)$$

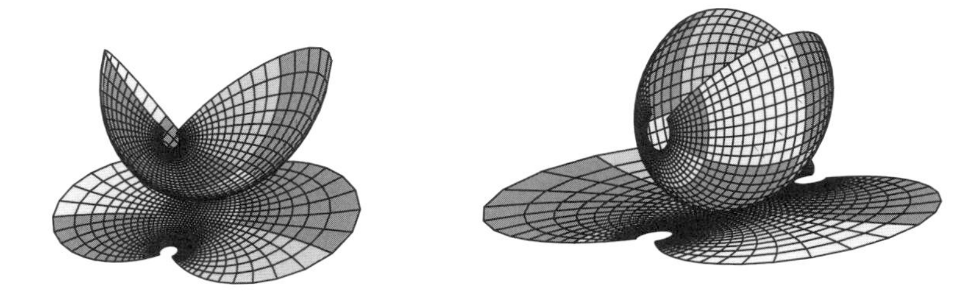

(a) Logistic function as a Möbius mapping (b) Function tanh as a Möbius mapping

Figure 11.4 Geometric view of the mapping performed by the complex logistic (Equation 11.11) and tanh functions (Equation 11.9), and the corresponding mappings onto the Riemann sphere (for clarity scales are not matched)

Möbius transformations may be also be seen as stereographic projections from a plane to a sphere, followed by a rotation and inverse stereographic projection (see Figures 2.2 and Figure 11.4).

We can now associate the matrix

$$A = \begin{bmatrix} a & b \\ c & d \end{bmatrix} \tag{11.12}$$

with the Möbius transformation. Since a Möbius transformation remains unchanged if all the coefficients (a, b, c, d) from Equation (11.7) are multiplied by the same nonzero constant, we can normalise A to yield $\det(A) = ad - bc = 1$. The condition $ad - bc \neq 0$ from Equation (11.7) is equivalent to a nonzero determinant of this matrix (nonsingular matrix). The matrix A then becomes unique up to the sign, and this allows us to express the Möbius transformation from Equation (11.7) as

$$f(z) \quad \leftrightarrow \quad A\,z \tag{11.13}$$

If F and G are matrices associated with Möbius transformations f and g, then the composition (nesting) $f \circ g$ can be expressed as the matrix product $F\,G$, that is

$$f(g(x)) \quad = \quad f \circ g \quad \leftrightarrow \quad F\,G \tag{11.14}$$

To analyse an n–dimensional nested nonlinear system, as examplified by a feedforward or recurrent neural network with hidden neurons, we can use the notion of a *modular group* [16]. In order to show that the set of all Möbius transformations forms a modular group under composition, we must show the existence of the identity and inverse transformations.

The identity transformation

$$f(z) = z = \frac{1z + 0}{0z + 1}$$

is described by the identity matrix I, which is the neutral element of the group. The inverse matrix A^{-1} is associated with the Möbius inverse

$$f^{-1}(z) = \frac{dz - b}{-cz + a} \tag{11.15}$$

Now, we can state that [178, 218]

> *The set of all Möbius transformations of the form $f(z) = (az + b)/(cz + d)$, where a, b, c, d are integers and $ad - bc = 1$ forms a* modular group *under composition, and is denoted by* Γ.

It can be proved [16] that the modular group[10] Γ is generated by a composition of two simple transformations:

- translation $Tz = z + 1$;
- inversion $Sz = -\frac{1}{z}$.

[10]The modular Möbius group is also an automorphism of the Riemann sphere, and under certain conditions it is a Lie group.

In the matrix form, it can be shown that:

The modular group Γ *is generated by the two matrices*

$$T = \begin{bmatrix} 1 & 1 \\ 0 & 1 \end{bmatrix} \quad \textit{and} \quad S = \begin{bmatrix} 0 & -1 \\ 1 & 0 \end{bmatrix} \tag{11.16}$$

and every $A \in \Gamma$ *can be expressed in the form*

$$A = T^{n_1} S T^{n_2} S \cdots S T^{n_k} \tag{11.17}$$

where n_i, $i = 1, \ldots, k$ *are integers.*

Thus, for instance,

$$A = \begin{bmatrix} 4 & 9 \\ 11 & 25 \end{bmatrix} \quad \leftrightarrow \quad ST^{-3}ST^{-4}ST^{2}$$

however, in general, this solution is not unique.

The modular group is central to the theory of fractals and iterated function systems (see Appendix P).

Example 1. *Show that the matrix form of the input–output transfer function of two cascaded nonlinear elements (neurons) belongs to a modular group* Γ.

Solution. For cascaded nonlinear elements (neurons), their transfer functions are *nested*[11] and not *multiplied*. From Equation (11.14) we have a composition of the two nonlinear mappings

$$h_1(z) = \frac{a_1 z + b_1}{c_1 z + d_1} \quad \text{and} \quad h_2(z) = \frac{a_2 z + b_2}{c_2 z + d_2},$$

that is

$$h_1 \circ h_2 = \frac{a_1 \frac{a_2 z + b_2}{c_2 z + d_2} + b_1}{c_1 \frac{a_2 z + b_2}{c_2 z + d_2} + d_1} = \frac{(a_1 a_2 + b_1 c_2) z + a_1 b_2 + b_1 d_2}{(a_2 c_1 + c_2 d_1) z + b_2 c_1 + d_1 d_2} \tag{11.18}$$

If the Möbius mappings performed by h_1 and h_2 are respectively described by matrices

$$H_1 = \begin{bmatrix} a_1 & b_1 \\ c_1 & d_1 \end{bmatrix} \text{ and } H_2 = \begin{bmatrix} a_2 & b_2 \\ c_2 & d_2 \end{bmatrix}$$

[11]The nesting process is explained in Chapter 3.

we have

$$h_1(z) \circ h_2(z) \Leftrightarrow (H_1 \times H_2)\, z = \left(\begin{bmatrix} a_1 & b_1 \\ c_1 & d_1 \end{bmatrix} \times \begin{bmatrix} a_2 & b_2 \\ c_2 & d_2 \end{bmatrix} \right) z$$
$$= \begin{bmatrix} a_1a_2 + b_1c_2 & a_1b_2 + b_1d_2 \\ a_2c_1 + c_2d_1 & b_2c_1 + d_1d_2 \end{bmatrix} z = H\,z \tag{11.19}$$

which belongs to the modular group Γ of compositions of Möbius transformations. □

This proves that the global I/O relationship in a neural network may be formalised within the framework of Möbius transformations, for more detail see [16, 65, 178].

11.4 All-pass Systems as Möbius Transformations

All-pass filters, described by the transfer function

$$H(z) = \frac{z^{-1} - p^*}{1 - pz^{-1}}, \quad z, p \in \mathbb{C}, \quad |p| < 1 \tag{11.20}$$

are a fundamental building block in signal processing. Their applications include those in multimedia signal processing [273], seismic processing [225], array processing [138], and also in the modelling of nonuniform time delays in digital filters [32]. A detailed overview of all–pass filters can be found in [249].

Matching the coefficients associated with the powers of z in the transfer function of an all-pass filter (Equation 11.20) with those in a general form of the Möbius transformation (Equation 11.7), we have

$$\frac{z^{-1} - p^*}{1 - pz^{-1}} = \frac{-p^*z + 1}{z - p} = \frac{az + b}{cz + d} \tag{11.21}$$

Therefore, the first-order all-pass filter is a Möbius transformation given by

$$A = \begin{bmatrix} -p^* & 1 \\ 1 & -p \end{bmatrix}, \qquad ad - bc \neq 0, \quad |p| \neq 1 \tag{11.22}$$

Since for first-order all-pass sections the pole p is always real, fixed points of this mapping are at $z_1 = \pm 1$. This corresponds to $\omega = 0$ and $\omega = \pi$ in frequency, that is, to the minimum and maximum of the group delay [178].

To model cascaded all-pass systems, we can define a finite *Blaschke product* as [22, 201, 206]

$$f(z) = \lambda \prod_{j=1}^{n} \left(\frac{z - a_j}{1 - a_j^* z} \right) \tag{11.23}$$

where $|a_j| < 1$, $j = 1, \ldots, n$ and $|\lambda| = 1$. This is a composition of Möbius transformations,[12] and as such it has a fixed point (see Section 11.2). For example, a second-order all-pass system, given by

$$H(z) = \frac{-p_1 z + 1}{z - p_1} \frac{-p_2 z + 1}{z - p_2} \quad \Leftrightarrow \begin{bmatrix} -p_1 & 1 \\ 1 & -p_1 \end{bmatrix} \times \begin{bmatrix} -p_2 & 1 \\ 1 & -p_2 \end{bmatrix} \tag{11.24}$$

is such a Blaschke product.

11.5 Fractional Delay Filters

In numerous applications (audio, music, time delay estimation), it is not only the sampling frequency, but also the actual sampling instants that are of crucial importance. Fractional delay filters (FDF) provide a very useful building block that can be used for fine–tuning of sampling instants [162]. Typical examples include:

- *digital audio*, where, e.g. for sampling rate conversion between 44.1 kHz and 48 kHz, at every time instant an FDF can be used to compute output signal samples at a different delay value;
- *communications*, where continuous-time pulse sequences arrive with different propagation delays, but should be sampled exactly at the middle of each pulse, that is, the sampling frequency and sampling instants must be synchronised;
- *synthetic musical instruments*, where propagation delays in musical resonators (tubes, guitar body) can make instrument sound out of tune; discretisation of differential equations describing acoustic vibrations should be based on changing the sampling instants online.

The transfer function of a tap delay line is given by

$$G(z) = \sum_{m=0}^{M} \theta_m z^{-m} \tag{11.25}$$

where θ_m are filter coefficients and filter memory is of length M; for the modelling of long channels (e.g. strong resonances) a significant amount of memory is required. An ideal fractional delay element should be a digital version of a continuous time delay line; all-pass filters are particularly well suited to such approximation since their magnitude response is unity at all frequencies.

Using the first-order all-pass filter from Equation (11.20), the transfer function of a fractional delay filter becomes a weighted sum of transfer functions of all-pass filters, given by

$$G(z) = \sum_{m=0}^{M} \theta_m \left(\frac{z^{-1} - p^*}{1 - pz^{-1}} \right)^m \tag{11.26}$$

[12] In fact, both the numerator and denominator term in Equation (11.23) are Klein terms $f_1(z) = z + a \,:\, M_1 = \{\{1, a\}, \{0, 1\}\}$ and $f_2(z) = 1/(az + 1) \,:\, M_2 = \{\{0, 1\}, \{a, 1\}\}$.

Essentially, FDFs are interpolators which approximate signal values in between the sample points k and $(k+1)$ by a linear combination of sample values on either side of the desired 'fractional' sampling instant. The memory of an FDF (Equation 11.26) depends on the value of pole p – the closer $|p|$ to the unit circle the longer the tail of the impulse response. The transfer function (Equation 11.26) can also be expressed in a 'nested' form as

$$\tilde{\theta}_0 + \left(\tilde{\theta}_1 + \left(\tilde{\theta}_2 \times \begin{bmatrix} -p^* & 1 \\ 1 & -p \end{bmatrix} + \cdots + \tilde{\theta}_M \times \begin{bmatrix} -p^* & 1 \\ 1 & -p \end{bmatrix}\right) \cdots \right) \times \begin{bmatrix} -p^* & 1 \\ 1 & -p \end{bmatrix} \tag{11.27}$$

thus providing a link between Möbius transformations and fractional delay filters. Design of FDFs requires careful selection of the underlying all-pass building blocks so as to form an orthonormal basis.[13] Examples include Legendre [234] and Laguerre [205, 299] filters which have real poles, and Kautz [146, 228] filters for which the poles are complex.

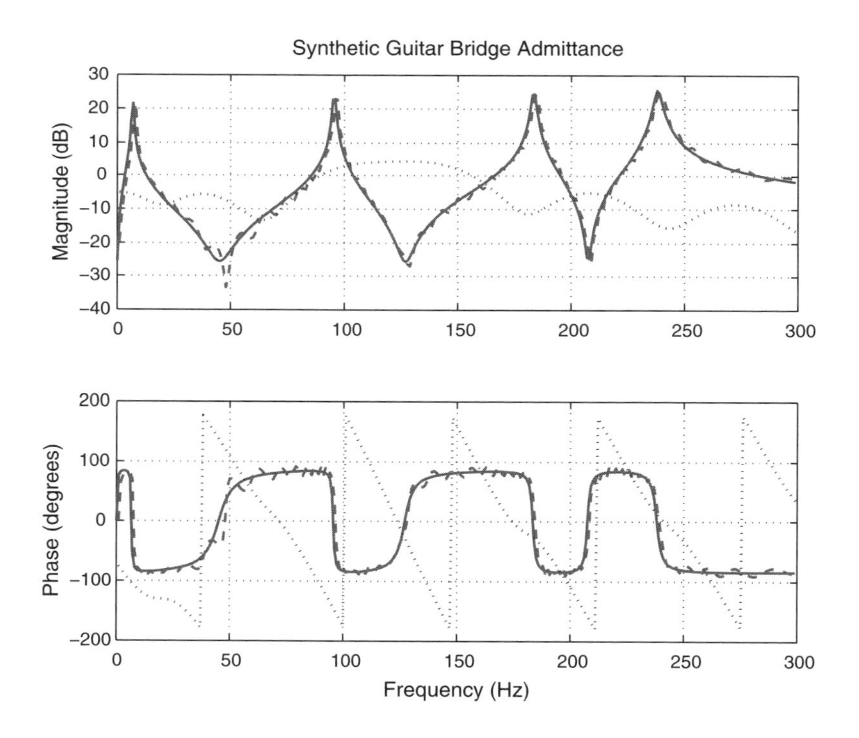

Figure 11.5 The magnitude and phase response of the resonance of a classical guitar (solid line) and its approximation using an FIR filter of order 256 (dotted line) and a Kautz filter with nine taps (dashed line)

[13]To ensure real valued output, complex poles must exist in conjugate pairs.

The Kautz filter approximates any transfer function by a linear combination of all-pass components

$$H(z) = \sum_{m=0}^{M/2} \left(\theta_{2m} K_{2m}(z) + \theta_{2m+1} K_{2m+1}(z)\right) \tag{11.28}$$

where K_{2m} and K_{2m+1} are respectively the transfer functions of the even and odd all-pass taps, given by

$$K_{2m}(z) = \kappa_0(z, \zeta) A^m(z)$$
$$K_{2m+1}(z) = \kappa_1(z, \zeta) A^m(z)$$

and

$$\kappa_0(z, \zeta_0) = |1 + \zeta_0| \sqrt{\frac{1 - \zeta_0 \zeta_0^*}{2}} \frac{z^{-1} - 1}{(1 - \zeta_0 z^{-1})(1 - \zeta_0^* z^{-1})},$$
$$\kappa_1(z, \zeta_1) = |1 - \zeta_1| \sqrt{\frac{1 - \zeta_1 \zeta_1^*}{2}} \frac{z^{-1} + 1}{(1 - \zeta_1 z^{-1})(1 - \zeta_1^* z^{-1})},$$
$$A(z, \zeta_i) = \frac{(1 - \zeta_i z^{-1})}{(z^{-1} - \zeta_i^*)} \frac{(z^{-1} - \zeta_i)}{(1 - \zeta_i^* z^{-1})}$$

where the poles of the all-pass blocks are denoted by ζ_i, $i = 0, \ldots, M$. For $\zeta \in \mathbb{R}$, Kautz filters degenerate into Laguerre filters, while for $\zeta = 0$ they become standard FIR filters.

The modelling capabilities of Kautz filters are illustrated in Figure 11.5, where approximations of the frequency response of the resonance of a classical guitar are calculated using an FIR filter of order 256 and a Kautz filter of order 9. A relatively short Kautz filter was able to provide a very good approximation, whereas a long FIR filter failed to deliver.

12

Augmented Complex Statistics

The emergence of new applications of complex statistical signal processing has highlighted problems related to the standard theory of complex stochastic processes; these include:

- Although the complex Gaussian model is commonly used and is well understood [101, 110, 113, 211, 316], not many theoretical results are available for the generality of complex random variables and signals;
- In general, a complex random vector $\mathbf{z}$ and its conjugate $\mathbf{z}^*$ are correlated, hence the covariance matrix

$$\mathcal{C} = \text{cov}(\mathbf{z}) = E\left[\mathbf{z}\mathbf{z}^{\mathrm{H}}\right] \tag{12.1}$$

  does not completely describe the second order statistics of $\mathbf{z}$, and another quantity

$$\mathcal{P} = \text{pcov}(\mathbf{z}) = E\left[\mathbf{z}\mathbf{z}^{\mathrm{T}}\right] \tag{12.2}$$

  called the *pseudocovariance* [219, 239] or *complementary covariance* [268], needs to be taken into account;
- The probability density function of Gaussian complex random variables has a form similar to that for real Gaussian variables only for *proper*, or *second order circular*, random processes $\mathbf{z}$ for which the pseudocovariance $E\left[\mathbf{z}\mathbf{z}^{\mathrm{T}}\right]$ vanishes. However, general complex random processes are *improper*, that is, they are correlated with their complex conjugates,[1] and $E\left[\mathbf{z}\mathbf{z}^{\mathrm{T}}\right] \neq 0$;
- The notion of a *proper* complex process is closely related to the notion of a *circular* complex process, however, 'properness' is a second order statistical property and circularity is a property of the probability density function. Properness therefore does not reveal anything about the actual multivariate signal distribution.

[1] For illustration, consider a random variable $z = z_r + \jmath z_i$. Then $zz^* = z_r^2 + z_i^2$ and $zz^T = z_r^2 - z_i^2 + 2\jmath z_r z_i$. Upon applying the statistical expectation operator, $E\left[zz^*\right] > 0$ (unless $z = 0 + \jmath 0$), whereas $E\left[zz^{\mathrm{T}}\right] = E[z_r^2] - E[z_i^2] + 2\jmath E[z_r z_i]$ vanishes only if z_r and z_i are uncorrelated and with the same variance.

Complex Valued Nonlinear Adaptive Filters: Noncircularity, Widely Linear and Neural Models
Danilo P. Mandic and Vanessa Su Lee Goh

Therefore, to access the information contained in the pseudocovariance, second order statistical modelling in $\mathbb{C}$ should examine joint statistical properties of $\mathbf{z}$ and $\mathbf{z}^*$, that is, it should be based on the 'augmented' random vector $\mathbf{z}^a = \left[\mathbf{z}^{\mathrm{T}}, \mathbf{z}^{\mathrm{H}}\right]^{\mathrm{T}}$ and the associated *augmented complex statistics*. Standard statistical signal processing approaches, however, implicitly assume second order circularity of the complex signals [5, 17, 110, 113, 120], resulting in suboptimal solutions.

This chapter reviews the theory of complex random variables and signals with a particular emphasis on complex circularity and augmented complex statistics, giving new perspective to the notions of white noise and autoregressive modelling in $\mathbb{C}$. This is then combined with estimation theory in Chapter 13, in order to introduce *widely linear* mean square estimation (WLMSE) and the augmented CLMS (ACLMS), a widely linear extension of the standard CLMS.

12.1 Complex Random Variables (CRV)

Complex random variables (CRV) arise as special cases of bivariate real random variables, and despite the ubiquity of complex random processes, their analysis is often parameterised in terms of the distribution of the real and imaginary components[2] [224], rather than performed directly in $\mathbb{C}$. When dealing with complex random variables, we can therefore either:

- consider a complex random variable $Z = X + \jmath Y$ as a two-dimensional 'composite' real random variable (RRV) $(X, Y) \in \mathbb{R}^2$; this way complex numbers are nothing else but pairs of real numbers and the theory of complex variables loses most of its appeal;
- develop tools for statistical analysis directly in $\mathbb{C}$.

Standard second-order statistical approaches for the analysis of complex signals are straightforward extensions from the corresponding approaches in $\mathbb{R}$. Thus, for instance, for a random variable Z and random vector $\mathbf{z}$, we have the following correspondence

$$\begin{aligned} & \qquad\qquad\qquad\qquad \mathbb{R} \qquad\quad \mathbb{C} \\ c_Z &= \mathrm{cov}(Z) = E\left[Z^2\right] \longrightarrow E\left[ZZ^*\right] \qquad (12.3) \\ \mathcal{C} &= \mathrm{cov}(\mathbf{z}) = E\left[\mathbf{z}\mathbf{z}^T\right] \longrightarrow E\left[\mathbf{z}\mathbf{z}^H\right] \qquad (12.4) \end{aligned}$$

Consequently, the concepts of mean square error, linear regression, and Wiener filtering can be readily extended from $\mathbb{R}$ to $\mathbb{C}$.

There are, however, several difficulties with this approach:

[2]This is because, technically speaking, a vector in $\mathbb{C}$ can be represented by a vector in $\mathbb{R}^2$, however, much of the power, beauty, and simplicity of complex representation would be lost [268].

- For $Z = X + \jmath Y,\ X, Y \in \mathbb{R}$, the cumulative distribution F and probability density function (pdf) f

$$F_X(x) = P(X \leq x) \qquad F_Y(y) = P(Y \leq y)$$
$$f_X(x) = \frac{\mathrm{d}}{\mathrm{d}s} F_X(s)_{|s=x} \qquad f_Y(y) = \frac{\mathrm{d}}{\mathrm{d}s} F_Y(s)_{|s=y} \tag{12.5}$$

  specify the probability distribution for Z, however, the expressions (12.5) *do not* apply directly to complex random variables, since the field of complex numbers is not an ordered set, and the relation $\leq$ is meaningless in $\mathbb{C}$ (see Appendix A);
- Complex variable Z and its conjugate Z^* are related by a deterministic transformation, and we cannot assign a probability density function to Z and Z^* independently.[3]

12.1.1 Complex Circularity

Circularity is intimately related to rotation in the geometric sense; a random variable Z is said to be *circular* if its statistical properties are '*invariant under a rotation*'. For a complex random variable Z, rotation by angle ϕ is achieved by multiplication by $e^{\jmath\phi}$, giving $Z_\phi = Ze^{\jmath\phi}$ (see Figure 11.1).

A *circular complex random variable* (CCRV) can be therefore defined as [53]

> *A complex random variable Z is called circular if for any angle ϕ both Z and $Ze^{\jmath\phi}$, that is its rotation by angle ϕ, have the same probability distribution.*

This means that the statistics of second order circular signals are *invariant to phase transformations*.[4]

Design of Complex Circular Random Variables

For convenience, we shall describe the generation process of a circular random variable $Z = \rho\cos(\theta) + \jmath\rho\sin(\theta)$ in the polar coordinate system. The circular symmetry of the probability density function is then described by $f_Z(\rho, \theta) = f_Z(\rho, \theta - \phi)$, and Z can be generated as follows [11]:

1. Generate a real valued random variable ρ with an arbitrary pdf $f(\rho)$. This determines the properties of the complex random variable Z;
2. Generate another real valued random variable θ, which is uniformly distributed on $[0, 2\pi]$ and independent of ρ;
3. Construct[5] $Z = X + \jmath Y$ as

$$X = \rho\cos(\theta), \qquad Y = \rho\sin(\theta) \tag{12.6}$$

[3] Although for mathematical tractability it is sometimes assumed that Z and Z^* are not algebraically linked [11].
[4] The covariance $c_Z = cov(Z) = E[ZZ^*] = E[Ze^{\jmath\phi}Z^*e^{-\jmath\phi}]$ is always invariant under rotation, whereas pseudocovariance $\mathrm{pcov}(Z)$ is invariant under rotation for circular signals.
[5] This way we effectively combine two independent real random processes, hence the pseudocovariance vanishes and the signal in hand is circular.

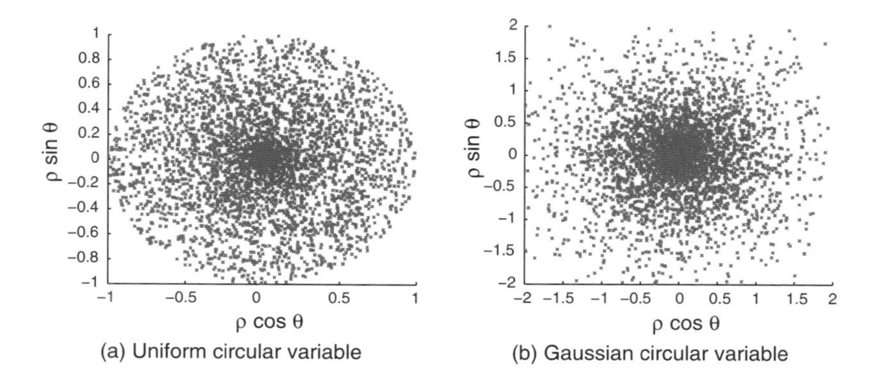

Figure 12.1 Realisations of 2000 samples of circular complex random variables

Figure 12.1 illustrates shapes of the probability distributions for the so generated uniform and Gaussian complex circular variables.

12.1.2 The Multivariate Complex Normal Distribution

Following the approach from [31], we shall now introduce a multivariate Generalised Complex Normal Distribution (GCND), a first step towards building statistical signal processing algorithms which account for the noncircularity of a signal.

The standard multivariate complex normal distribution (CND) [316], given by

$$f(\mathbf{z}) = \frac{1}{\pi^N \det \mathbf{Z}} e^{-\mathbf{z}^H \mathbf{Z}^{-1} \mathbf{z}}, \quad \mathbf{z} = [z_1, \ldots, z_N]^{\mathrm{T}}, \quad z_i = x_i + \jmath y_i, \; x_i, y_i \in \mathbb{R}, \; i = 1, \ldots, N \tag{12.7}$$

has been widely used in complex valued signal processing. Apart from the standard requirement of positive semidefiniteness, the covariance matrix of this distribution

$$\mathbf{Z} = E\left[(\mathbf{z} - E[\mathbf{z}])(\mathbf{z} - E[\mathbf{z}])^{\mathrm{H}}\right] \overset{\text{for } E[\mathbf{z}]=\mathbf{0}}{\Longleftrightarrow} E\left[\mathbf{z}\mathbf{z}^{\mathrm{H}}\right] \tag{12.8}$$

satisfies a number of other restrictions,[6] that is, (12.7) describes effectively a real multivariate distribution of the *composite real vector*

$$\mathbf{w} = \left[x_1, y_1, \ldots, x_N, y_N\right]^{\mathrm{T}}, \quad \mathbf{w} \in \mathbb{R}^{2N \times 1} \tag{12.9}$$

[6]This is because the complex normal distribution was introduced as a straightforward extension of the real valued one for the analysis of $I - Q$ signals, which are by their nature circular (see Figure 2.8b).

for which the covariance matrix (Equation 12.8) satisfies the conditions (see also Equation (12.33)

$$E[x_k x_l] = E[y_k y_l] \quad \Rightarrow \quad \mathrm{var}(x_k) = \mathrm{var}(y_k) \tag{12.10}$$

$$E[x_k y_l] = -E[x_l y_k] \quad \Rightarrow \quad \mathrm{cov}(x_k, y_k) = 0 \qquad \forall x_k, y_l, \ k, l = 1, \ldots, N \tag{12.11}$$

The first condition (Equation 12.10) implies that the variance of x_k is equal to the variance of y_k, whereas the second condition that the covariance $\mathrm{cov}(x_k, y_k) = E[x_k y_k] = 0$ implies[7]

$$E[z_k z_l] = 0 \tag{12.12}$$

This significantly restricts the freedom in the choice of allowable covariance matrices of the real and imaginary parts of z_n, and therefore the probability density function (Equation 12.7) is suitable only for the modelling of complex *second-order circular* random variables (CCRV), for which the conditions (12.10) and (12.11) are satisfied.

Generalised complex normal distribution. To obtain a mathematical expression for the normal distribution for general complex stochastic signals, we start from a complex stochastic variable

$$z_k = x_k + \jmath y_k, \quad x_k \in \mathcal{N}(0, \sigma_x^2), \quad y_k \in \mathcal{N}(0, \sigma_y^2), \quad n = 1, \ldots, N \tag{12.13}$$

and a composite real random vector $\mathbf{w} \in \mathbb{R}^{2N \times 1}$, given in Equation (12.9), for which the covariance matrix is given by

$$\mathbf{W} = cov(\mathbf{w}) = E\left[\mathbf{w}\mathbf{w}^{\mathrm{T}}\right] \tag{12.14}$$

and for which the real multivariate normal distribution can be expressed as

$$f(\mathbf{w}) = \frac{1}{(2\pi)^N \sqrt{\det(\mathbf{W})}} \, e^{-\frac{1}{2}\mathbf{w}^T \mathbf{W}^{-1} \mathbf{w}} \tag{12.15}$$

As already shown in Equation (11.5), the mapping between the composite real variable $w_k = (x_k, y_k) \in \mathbb{R}^2$ and the complex variable z_k and its complex conjugate z_k^* can be expressed as (to make $\mathbb{J}$ a unitary matrix, we may scale by $1/\sqrt{2}$)

$$\begin{bmatrix} z_k \\ z_k^* \end{bmatrix} = \mathbb{J} \begin{bmatrix} x_k \\ y_k \end{bmatrix} = \begin{bmatrix} 1 & \jmath \\ 1 & -\jmath \end{bmatrix} \begin{bmatrix} x_k \\ y_k \end{bmatrix}$$

[7]Since

$$E[z_k z_l] = E[(x_k + \jmath y_k)(x_l + \jmath y_l)] = E[\underbrace{x_k x_l - y_k y_l}_{=0 \text{ by } (12.10)}] + \jmath E[\underbrace{x_k y_l + x_l y_k}_{=0 \text{ by } (12.11)}] \quad .$$

For convenience, the 'augmented' complex vector $\mathbf{v} \in \mathbb{C}^{2N\times 1}$ can be introduced as

$$\mathbf{v} = \left[z_1, z_1^*, \ldots, z_N, z_N^*\right]^{\mathrm{T}}$$
$$\mathbf{v} = \mathbf{A}\mathbf{w} \tag{12.16}$$

where matrix $\mathbf{A} = \mathrm{diag}(\mathbb{J}, \ldots, \mathbb{J}) \in \mathbb{C}^{2N\times 2N}$ is block diagonal and transforms the composite real vector $\mathbf{w}$ into the augmented complex vector $\mathbf{v}$, for which the covariance matrix

$$\mathbf{V} = \mathrm{cov}(\mathbf{v}) = E[\mathbf{v}\mathbf{v}^H] = \mathbf{A}\mathbf{W}\mathbf{A}^H \tag{12.17}$$

To obtain a generalised multivariate complex normal distribution, we need to replace the terms $\mathbf{w}$, $\det(\mathbf{W})$, and $\mathbf{w}^{\mathrm{T}}\mathbf{W}^{-1}\mathbf{w}$ in (12.15) by the corresponding expressions in terms of $\mathbf{v}$, that is [31]

$$\mathbf{w} = \mathbf{A}^{-1}\mathbf{v} = \frac{1}{2}\mathbf{A}^{\mathrm{H}}\mathbf{v}$$
$$\det(\mathbf{W}) = \left(\frac{1}{2}\right)^{2N} \det(\mathbf{V})$$
$$\mathbf{w}^{\mathrm{T}}\mathbf{W}^{-1}\mathbf{w} = \mathbf{v}^{\mathrm{H}}\mathbf{V}^{-1}\mathbf{v} \tag{12.18}$$

The multivariate *generalised complex normal distribution* (GCND) can now be expressed as [31]

$$f(\mathbf{v}) = \frac{1}{\pi^N\sqrt{\det(\mathbf{V})}}\, e^{-\frac{1}{2}\mathbf{v}^{\mathrm{H}}\mathbf{V}^{-1}\mathbf{v}} \tag{12.19}$$

and has been derived without any restriction[8] on the covariance matrix $\mathbf{W}$ of the composite real vector $\mathbf{w}$ (except nonsingularity).

The standard multivariate complex normal distribution (CND) (Equation 12.7) can be obtained from the generalised complex multivariate normal distribution as a special case, by applying restrictions (12.10) and (12.11). So too can the univariate complex normal distribution (for $N = 1$, and $z = x + \jmath y$), that is

$$f(z) = \frac{1}{\pi\sigma_z^2}\, e^{-\frac{zz^*}{\sigma_z^2}} = \frac{1}{\pi\sigma_z^2}\, e^{-\frac{|z|^2}{\sigma_z^2}} \tag{12.20}$$

Augmented complex observation. The augmented complex observation is given in Equation (12.16), however, in this work we most often use a slightly rearranged augmented complex

[8]The generalised complex multivariate normal distribution assumes that the complex variable and its complex conjugate have a joint probability density, however, the complex conjugation operator is a deterministic mapping from Z to Z^*. Also, due to $\mathbb{C}$ not being an ordered set (only lexicographic ordering is readily available, see Appendix A), the problem of general complex cumulative and probability density functions still remains widely open. One recent attempt to introduce density functions which can be interpreted directly in Z can be found in [224].

vector $\mathbf{z}^a$, given by[9]

$$\mathbf{z}^a = \left[z_1, \ldots, z_N, z_1^*, \ldots, z_N^*\right]^T = \left[\mathbf{z}^T, \mathbf{z}^H\right]^{\mathrm{T}} \tag{12.21}$$

To convert the expressions derived for the augmented vector $\mathbf{v}$ (Equation 12.16) to the corresponding expressions for $\mathbf{z}^a$, we may employ a permutation matrix[10] $\mathbf{P} \in \mathbb{R}^{2N \times 2N}$, to give

$$\begin{aligned} \mathbf{z}^a &= \mathbf{P}\mathbf{v} \\ \det(\mathbf{V}) &= \det(\mathbf{Z}^a) = (\det(\mathbf{Z}))^2 \end{aligned} \tag{12.22}$$

where $\mathbf{Z}^a = cov(\mathbf{z}^a)$; the permutation matrix $\mathbf{P}$ provides therefore a deterministic transformation from $\mathbf{v}$ to $\mathbf{z}^a$, and all the expressions derived for $\mathbf{v}$ hold also for $\mathbf{z}^a$.

Generally, characterisation of complex processes in terms of signal distribution is difficult (since $\mathbb{C}$ is not an ordered field, see also Appendix A), and in practical applications we need to resort to the description in terms of the moments, cumulants and characteristic functions, as shown in Section 12.2.

12.1.3 Moments of Complex Random Variables (CRV)

To provide a full statistical description (first-, second-, and higher-order moments) of CRVs, consider the characteristic function[11] (for more detail, see [11, 12, 113, 243])

$$\Phi_{Z,Z^*}(w, w^*) = E\left[\exp\left(\jmath \frac{Z^* w + Z w^*}{2}\right)\right] \tag{12.23}$$

where $w = u + \jmath v$ and $w^* = u - \jmath v$ are complex variables and $E[\cdot]$ is the ensemble mean. Similarly to the real case, to obtain the statistical moments of Z, we can apply Taylor series expansion (TSE) to the characteristic function (12.23), to yield

$$E\left[\exp\left(\jmath \frac{Z^* w + Z w^*}{2}\right)\right] = \sum_{n=1}^{\infty} \frac{1}{n!} \frac{\jmath^n}{2^n} \sum_{p=0}^{n} b(n, p) w^{n-p} w^{*p} E\left[Z^{*n-p} Z^p\right] \tag{12.24}$$

where $b(n, p)$ are binomial coefficients. From Equation (12.24), for a given order n, there are $(n + 1)$ different moments[12] $E[Z^{*n-p} Z^p]$. For example, for $n = 2$, we have three different moments: $E[Z^2]$, $E[ZZ^*]$, and $E[Z^{*2}]$; for real random variables, all these moments are equal.

---

[9] Both $\mathbf{z}$ and $\mathbf{z}^*$ are column vectors.

[10] All the elements of this matrix are either zero or unity, and there is precisely one nonzero element in every row and column of $\mathbf{P}$. In addition, $\mathbf{P}$ is an orthogonal matrix, that is, $\mathbf{P}^{\mathrm{T}} = \mathbf{P}^{-1}$ and $|\det(\mathbf{P})| = 1$.

[11] This is a 2D FFT of a 2D probability density function.

[12] Practically speaking, there are only $m = (\text{floor}(n/2) + 1)$ correlation functions required for a complete nth-order description, since e.g. $E[Z^{*2}]$ is a deterministic transformation of $E[Z^2]$.

Moments of an arbitrary order n can be derived from the characteristic function Φ_{Z,Z^*} as[13]

$$E\left[Z^{*n-p}Z^p\right] = \frac{2^n}{\jmath^n}\frac{\partial^n \Phi_{Z,Z^*}(0,0)}{\partial w^{n-p}\partial w^{*p}}. \tag{12.25}$$

The corresponding $(n+1)$ *cumulants* of order n can be derived from the so called 'second' characteristic function

$$\Psi_{Z,Z^*}(w, w^*) = \log\left(\Phi_{Z,Z^*}(w, w^*)\right) \tag{12.26}$$

and are given by (see also Footnote 12)

$$\mathrm{Cum}\big[\underbrace{Z^*,\ldots,Z^*}_{n-p},\underbrace{Z,\ldots,Z}_{p}\big] = \mathrm{Cum}\left[Z^{*n-p}, Z^p\right] = \frac{2^n}{\jmath^n}\frac{\partial^n \Psi_{Z,Z^*}(0,0)}{\partial w^{n-p}\partial w^{*p}} \tag{12.27}$$

Cumulants, then:

a) provide a measure of independence of complex random variables;
b) for Gaussian complex random variables, cumulants of order $n \geq 2$ vanish;
c) joint cumulants of two statistically independent complex random variables are zero;
d) property *c*) implies that for two statistically independent random variables $S, T \in \mathbb{C}$

$$\mathrm{Cum}\,[S+T] = \mathrm{Cum}\,[S] + \mathrm{Cum}\,[T] \tag{12.28}$$

that is, cumulants of the sum of two independent random variables are equal to the sum of cumulants of those variables.

Moments and cumulants in the multidimensional case using tensorial notation are addressed in [11].

12.2 Complex Circular Random Variables

The notion of 'circularity' has been commonly used to characterise Gaussian complex random variables (second-order circularity) [101, 211], however, for rigorous statistical description of circularity, we also need to involve general signal distributions and higher-order statistics (HOS). The property of circularity can be expressed equivalently in terms of the *probability density function* (pdf), *characteristic function*, and *cumulants*, as follows:

- *Circularity in terms of the probability density function.* A complex random variable Z is circular if its *pdf* is a function of only the product zz^*, that is[14]

$$p_{Z,Z^*}(z, z^*) = p_{Z_\phi,Z_\phi^*}(z_\phi, z_\phi^*) \tag{12.29}$$

[13]It is assumed that w and w^* are independent variables, a common and convenient assumption in the literature.
[14]This also implies that the pdf of a circular complex random variable is function of only the modulus of z.

and for Gaussian CCRVs we obtain the classical result [101] (see also Equation 12.20)

$$p_{Z,Z^*}(z, z^*) = \frac{1}{\pi\sigma_z^2} e^{-zz^*/\sigma_z^2} \tag{12.30}$$

where σ_z^2 denotes the variance of the CRV and $Z_\phi = Ze^{\jmath\phi}$ (see also Section 12.1.1).

- *Circularity in terms of the characteristic function.* A complex random variable Z is circular if its (first or second) characteristic function depends only on the product ww^*, that is[15]

$$\Phi_{Z_\phi, Z_\phi^*}(w, w^*) = \Phi_{Z,Z^*}(we^{-\jmath\phi}, w^* e^{\jmath\phi}) \tag{12.31}$$

- *Circularity in terms of the cumulants.* A complex random variable Z is circular if the only nonzero moments[16] and cumulants are those that have the same power in Z and Z^*, that is

$$\text{Cum}[Z^2] = 0, \quad \text{Cum}[Z^4] = 0, \quad \ldots \quad \text{Cum}[Z^{2p}] = 0, \quad \ldots \tag{12.32}$$

Gaussian circularity. From Equation (12.32), for a zero mean *circular* Gaussian random variable $Z = X + \jmath Y$, we have

$$E[Z^2] = E[X^2] - E[Y^2] + \jmath E[XY] = 0 \tag{12.33}$$

and to test for circularity, it is therefore sufficient to verify that for $X, Y \in \mathcal{N}(0, \sigma^2)$

$$E[X^2] = E[Y^2] \quad \text{and} \quad E[XY] = 0 \tag{12.34}$$

that is, X and Y *have the same power and are not correlated.* This has already been shown in Equations (12.10) and (12.11), starting from the standard complex multivariate normal distribution.

12.3 Complex Signals

A natural extension of the theory of complex random variables is to the analysis of complex signals. Mathematical tools for the analysis of complex stationary signals include multicorrelations and multispectra, whereas the analysis of nonstationary signals is typically based on higher-order time–frequency distributions. A low-rank approximation of improper complex random vectors is provided in [266], and its applications in Wiener filtering are given in [267]. In [270] a generalised likelihood ratio test for impropriety of complex signals is proposed. A comprehensive introduction to higher-order signal analysis tools, together

[15] Equations (12.29) and (12.31) apply for any circular distribution. The dependence on the product zz^* is more obvious for distributions in their expanded form, like the Gaussian in Equation (12.30).

[16] Partial derivatives in Equation (12.27) are nonzero only when they are of the same order in w and w^*.

with numerous examples, is provided in [12]. An account of second-order circularity for a generalised single-sideband modulator is provided in [268], whereas some applications in the areas of telecommunications and information theory are outlined in [219].

12.3.1 Wide Sense Stationarity, Multicorrelations, and Multispectra

For a complex signal $z(t)$, the multicorrelation of order $(p+q)$ is defined as (for more detail see [12, 113, 243])

$$C_{z,p+q,p}(\mathbf{t}) = \mathrm{Cum}[z(t_0), \ldots, z(t_{p-1}), z^*(t_p), \ldots, z^*(t_{p+q-1})] \tag{12.35}$$

where index $\mathbf{t} = \{t_0, \ldots, t_{p+q-1}\}$, q refers to the conjugated and p to the nonconjugated terms. Statistical properties of multicorrelations are in direct relation to the corresponding properties of cumulants, that is

- multicorrelations are multilinear;
- for Gaussian signals multicorrelations are equal to zero for $(p+q) > 2$.

Thus, if the random variables $z(t_0), \ldots, z(t_{p+q-1})$ are statistically independent, that is signal z is *white in the strict sense*, all the multicorrelations vanish, except those with the same index t_i in both z and z^*.

Complex stationarity. If $\mathbf{z}_n = [z(t_1), \ldots, z(t_n)]^{\mathrm{T}} \in \mathbb{C}^n$ is an n-dimensional induced vector of the complex signal $z(t)$, then [12]

- We say that signal $z(t)$ is *stationary* if the statistics of all the induced random vectors $\mathbf{z}_n$ are invariant under the time shift operation.
- Signal $z(t)$ is said to be *stationary of order k* if the pdf of $\mathbf{z}_n$ is invariant under the time shift operation for all $n \leq k$.

Alternatively, the property of stationarity can be expressed in terms of the multicorrelation of order $(p+q)$ as

$$C_{z,p+q,p}(\mathbf{t}+\boldsymbol{\tau}) = C_{z,p+q,p}(\boldsymbol{\tau}) \qquad \forall p, q \text{ such that } p+q \leq n \text{ and } \forall \boldsymbol{\tau} \tag{12.36}$$

In other words, for stationary complex signals, the multicorrelation is no longer $(p+q)$-dimensional but $(p+q-1)$-dimensional, and is a function of only time lag $\boldsymbol{\tau} = (\tau_1, \ldots, \tau_{p+q-1})$, that is

$$C_{z,p+q,p}(\boldsymbol{\tau}) = \mathrm{Cum}[z(t), z(t+\tau_1), \ldots, z(t+\tau_{p-1}), z^*(t-\tau_p), \ldots, z^*(t-\tau_{p+q-1})] \tag{12.37}$$

The multispectrum of order $(p+q)$ is the Fourier transform (FT) of the corresponding multicorrelation, that is

$$S_{z,p+q,p}(\boldsymbol{\nu}) = \int C_{z,p+q,p}(\boldsymbol{\tau}) \exp(-\jmath 2\pi \boldsymbol{\tau}^{\mathrm{T}} \boldsymbol{\nu}) d\boldsymbol{\tau} \tag{12.38}$$

Extensions of these results to cross-multicorrelations and cross-multispectra and to various other classes of multidimensional signals are straightforward and can be found in [12].

12.3.2 *Strict Circularity and Higher-order Statistics*

We shall now extend the definition of circularity for complex random variables from Section 12.2 to that for general complex valued signals.

- We say that a signal $z(t)$ is circular of order n if the induced vectors $\mathbf{z}_m = [z(t_1), \ldots, z(t_m)]^{\mathrm{T}} \in \mathbb{C}^m$ of order $m \leq n$ are circular.
- For a circular complex random vector of order n, moments of the order lower or equal to n vanish when the number of their conjugated terms is different from that of the nonconjugated terms, that is

$$\forall_{p,q} \quad C_{z,p+q,p}(\boldsymbol{\tau}) = 0 \quad \text{for} \quad p+q \leq n \quad \text{and} \quad p \neq q. \tag{12.39}$$

Alternatively, in the frequency domain, the condition of complex circularity can be expressed as:

- A complex signal is *strictly circular*, that is, circular for all orders n, if the only nonzero multispectra are $S_{z,2,0}(\boldsymbol{\nu})$, $S_{z,2,1}(\boldsymbol{\nu})$ and $S_{z,2,2}(\boldsymbol{\nu})$. If the signal is also analytic, the only nonzero multispectrum is $S_{z,2,1}(\boldsymbol{\nu})$.

Thus, for instance, for band-limited signals (see Chapter 2), the only nonzero multispectra are of the type $S_{z,2p,p}(\boldsymbol{\nu})$.

12.4 Second-order Characterisation of Complex Signals

Analysis of signals in terms of their second-order structure or cumulant properties, rather than in terms of density functions is very convenient, as this helps to avoid having to assign a density to a complex quantity. As has already been shown, a unique feature of second-order statistical analysis in $\mathbb{C}$ is that, in order to use the full statistical information, we need to examine properties of both the covariance and pseudocovariance function, that is, to consider so-called augmented statistics.

The analysis presented is a natural extension of the analysis of random variables from Section 12.2, and reveals the intimate relationship between second-order stationarity, correlation structure, and circularity of complex random signals.

12.4.1 *Augmented Statistics of Complex Signals*

For a discrete time zero mean complex signal $z(k)$, its second-order statistics are usually described by the covariance function, defined by[17]

$$c_z(k_1, k_2) = E[z(k_1)z^*(k_2)] \tag{12.40}$$

[17]Note that Equations (12.40) and (12.41) are 2D functions evaluated for two time indexes k_1 and k_2 using standard (scalar) signal representation, unlike (12.49) which uses a vector signal representation.

Following on from the analysis of complex random variables, to entirely describe the second-order statistics of $z(k)$, we need to introduce another function called *pseudocovariance* [219] or *complementary covariance* [268], given by

$$p_z(k_1, k_2) = E[z(k_1)z(k_2)] \tag{12.41}$$

In various instances the pseudocovariance $p_z(k_1, k_2)$ is equal to zero and can be disregarded.[18] However, in general, the pseudocovariance (Equation 12.41) is nonzero, and to describe the second-order statistics of complex signals completely, it is necessary to consider both their 'proper' (12.40) and 'improper' (12.41) characteristics,[19] that is, to 'augment' the standard complex statistics.

We shall now define the properties of statistical stationarity and orthogonality of general complex signals in light of augmented complex statistics.

Orthogonality of random vectors. Two zero mean complex random vectors $\mathbf{z}_1$ and $\mathbf{z}_2$ are *uncorrelated* (or orthogonal if they are zero mean), if[20]

$$\text{cov}(\mathbf{z}_z, \mathbf{z}_2) = \text{pcov}(\mathbf{z}_1, \mathbf{z}_2) = \mathbf{0} \tag{12.42}$$

The joint Gaussianity and uncorrelatedness in $\mathbb{C}$ therefore imply statistical independence.

Stationarity of complex signals. Complex stationarity, defined in the usual way in terms of the shift invariance of the statistics of induced random vectors, has already been introduced in Section 12.3.1 however, for rigour, statistical stationarity in $\mathbb{C}$ must be considered both in terms of the covariance and pseudocovariance functions, as follows:

- A complex signal $z(k)$ is said to be *wide sense stationary* (WSS) if its mean is constant and its covariance $c_z(k_1, k_2)$ (Equation 12.40) is only a function of the lag $\tau = k_1 - k_2$. Wide sense stationarity does not imply any condition on the pseudocovariance $p_z(k_1, k_2)$ (12.41).
- A complex signal $z(k)$ is said to be *second-order (SO) stationary* if it is wide sense stationary and if its pseudocovariance $p_z(k_1, k_2)$ (Equation 12.41) depends on only the time lag τ.

Whereas for real valued signals, the concepts of wide sense stationarity and second order stationarity are equivalent, *for complex signals, wide sense stationarity does not imply second order stationarity*.

Since for the generality of complex signals, the covariance and pseudocovariance functions are not independent, the augmented covariance matrix exhibits a special matrix structure, and to satisfy the restriction of non-negative definiteness, relationships between the covariance and pseudocovariance matrices need to be established in both the Fourier and time domain.

[18]This is the case with the analytic signal of any stationary process, and more generally, for any circular signal, such as those in communications [237].

[19]The covariance and pseudocovariance are not arbitrary functions, e.g. c_z is non-negative definite and p_z is symmetric [239].

[20]One may argue that uncorellatedness should be defined only in terms of cov, not pcov. In that case, in $\mathbb{C}$ the joint Gaussianity and uncorrelatedness is not sufficient for statistical independence between $\mathbf{z}_1$ and $\mathbf{z}_2$.

Augmented Complex Statistics in the Fourier Domain

It is often convenient to consider second-order statistical properties of a signal in the Fourier domain, since this provides insight into the properties of power spectra. As has already been seen, for stationary signals the pseudocorrelation function $p_z(\tau)$ is symmetric, hence its Fourier transform

$$R_z(\nu) = \mathcal{F}\{p_z(\tau)\}$$

is also symmetric, that is

$$R_z(\nu) = R_z(-\nu) \tag{12.43}$$

whereas the Fourier transform of the covariance function

$$\Gamma_z(\nu) = \mathcal{F}\{c_z(\tau)\} \tag{12.44}$$

represents the power spectrum of $z(k)$.

Following the approach from [236, 239], to illustrate the properties of augmented complex signals in the Fourier domain, consider a 2×1 dimensional augmented complex random vector

$$\mathbf{z}^a(k) = [z(k), z^*(k)]^{\mathrm{T}} \tag{12.45}$$

for which the covariance function is the 2×2 dimensional matrix $E[\mathbf{z}^a(k)\mathbf{z}^{a\mathrm{H}}(k-\tau)]$. The Fourier transform

$$\Gamma_{\mathbf{z}}(\nu) = \mathcal{F}\Big\{E[\mathbf{z}^a(k)\mathbf{z}^{a\mathrm{H}}(k-\tau)]\Big\} = \begin{bmatrix} \Gamma_z(\nu) & R_z(\nu) \\ R_z^*(\nu) & \Gamma_z(-\nu) \end{bmatrix} \tag{12.46}$$

is called the *spectral matrix* of $\mathbf{z}^a(k)$ or *spectral covariance matrix*, and contains all the information necessary for the second-order description of $\mathbf{z}(k)$. The spectral covariance matrix is non-negative definite (NND) [236], hence its diagonal elements and the determinant $\det(\Gamma_{\mathbf{z}}(\nu))$ are non-negative. From (12.46), for $R_z(\nu)$ to be the spectral pseudocovariance function of a complex signal $z(k)$ with power spectrum $\Gamma_z(\nu)$, due to the symmetric nature of $R_z(\nu)$, it has to satisfy [239]

$$|R_z(\nu)|^2 \leq \Gamma_z(\nu)\Gamma_z(-\nu) \tag{12.47}$$

Based on Equation (12.47), the necessary condition for nonstationary complex signals becomes

$$|R_z(\nu_1, \nu_2)|^2 \leq \Gamma_z(\nu_1, -\nu_1)\Gamma_z(\nu_2, -\nu_2). \tag{12.48}$$

Time domain counterparts of relations (12.47) and (12.48) are given below.

Augmented Complex Statistics in the Time Domain

It has already been shown that for a zero mean complex vector $\mathbf{z}(k) = \mathbf{x}(k) + \jmath\mathbf{y}(k) \in \mathbb{C}^n$, its covariance $\mathcal{C}$ and pseudocovariance $\mathcal{P}$ matrices are given by

$$\mathcal{C} = E[\mathbf{z}(k)\mathbf{z}^{\mathrm{H}}(k)] \quad \text{and} \quad \mathcal{P} = E[\mathbf{z}(k)\mathbf{z}^{\mathrm{T}}(k)] \tag{12.49}$$

Similarly to the analysis of augmented complex signals, we can consider the augmented complex vector[21] $\mathbf{z}^a(k) = \left[\mathbf{z}^{\mathrm{T}}(k), \mathbf{z}^{\mathrm{H}}(k)\right]^{\mathrm{T}}$ (see also Equation 12.21), for which the covariance and pseudocovariance matrix can be combined into an *augmented covariance matrix*, denoted by $\mathcal{C}_a$ and given by [268]

$$\mathcal{C}_a = \operatorname{cov}\left(\mathbf{z}^a(k)\right) = E\left[\mathbf{z}^a(k)\mathbf{z}^{aH}(k)\right] = \begin{bmatrix} \mathcal{C} & \mathcal{P} \\ \mathcal{P}^* & \mathcal{C}^* \end{bmatrix} \tag{12.50}$$

The time domain counterpart of condition (12.47) is therefore that the Schur complement

$$\mathcal{C}^* - \mathcal{P}^{\mathrm{H}}\mathcal{C}^{-1}\mathcal{P} \tag{12.51}$$

is non-negative definite [238].

The augmented covariance matrix $\mathcal{C}_a$ provides a complete second order statistical description of a general complex valued signal, and belongs to a matrix algebra[22]

$$\mathcal{W} = \left\{ \begin{bmatrix} h_1(t_1, t_2) & h_2(t_1, t_2) \\ h_2^*(t_1, t_2) & h_1^*(t_1, t_2) \end{bmatrix} \right\} \tag{12.52}$$

where $h_1, h_2 : [0, T]^2 \rightarrow \mathbb{C}$. The matrix algebra $\mathcal{W}$ is closed under addition, multiplication, and inversion. It is also closed under multiplication with a real scalar, but not with a complex scalar.

12.4.2 Second-order Complex Circularity

We have already seen that the covariance functions for the random vector $\mathbf{z}(k)$ and its rotated version $\mathbf{z}(k)e^{\jmath\phi}$ are equal for all ϕ, however, the pseudocovariance functions for $\mathbf{z}(k)$ and $\mathbf{z}(k)e^{\jmath\phi}$ are equal only if the pseudocovariance vanishes. Circularity, in terms of the properties of induced vectors and multispectra, has already been addressed in Section 12.3.2, however, to define second-order circularity we need to consider properties of both the covariance and pseudocovariance functions:

- A complex valued random vector $\mathbf{z}(k)$ is called *second-order circular*, *proper* [219], or *circularly symmetric*, if its pseudocovariance matrix $\mathcal{P}$ vanishes, that is, the second-order statistics are completely characterised by the covariance matrix $\mathcal{C}$. Otherwise, vector $\mathbf{z}(k)$ is called *noncircular* or *improper*;
- For proper random vectors, the augmented covariance matrix $\mathcal{C}_a$ (Equation 12.50) is diagonal, or equivalently, the spectral covariance matrix $\Gamma_{\mathbf{z}}(\nu)$ (Equation 12.46) is diagonal. Complex valued random vectors $\mathbf{z}_1(k)$ and $\mathbf{z}_2(k)$ are called *jointly proper* (jointly second-order circular), if the composite random vector having $\mathbf{z}_1(k)$ and $\mathbf{z}_2(k)$ as subvectors is proper [219].

---

[21] This is typically the case in adaptive filtering applications where the vector $\mathbf{z}(k)$ is the 'tap input vector', that is the segment of signal $z(k)$ in the filter memory.

[22] For more detail see [268, 271].

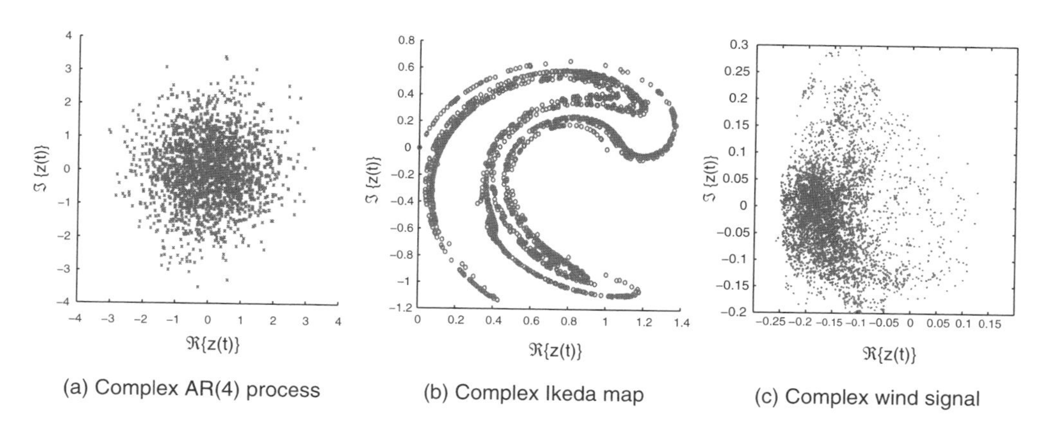

Figure 12.2 Geometric view of circularity via 'real–imaginary' scatter plots: (a) circular AR(4) signal; (b) noncircular chaotic signal; (c) noncircular wind signal

It is important to notice that

- complex Gaussianity is preserved by linear transformations;
- circularity is also preserved by linear transformations;
- in general, complex Gaussianity and orthogonality of two zero mean random signals do not imply statistical independence, unlike for real valued Gaussian signals.

Complex Circularity: Some Examples

To further illustrate the complex circularity, Figure 12.2 presents the real-imaginary ($\Re - \Im$) scatter diagrams for three signals: (a) complex linear autoregressive AR(4) process driven by complex white Gaussian noise (complex circular); (b) the first two coordinates of the Ikeda map (noncircular signal); (c) complex wind signal (intermittent and noncircular for the

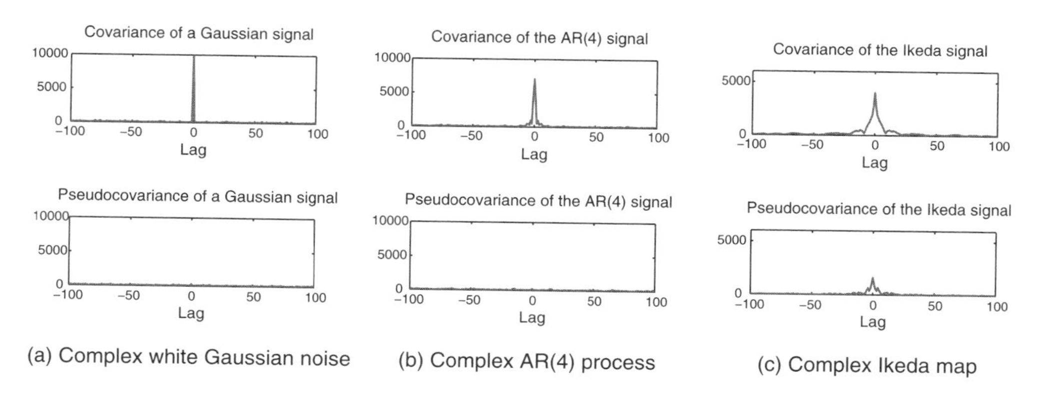

Figure 12.3 Circularity via 'covariance–pseudocovariance' plots: (a) circular complex white Gaussian noise (b) circular complex AR(4) process; (b) noncircular complex Ikeda map

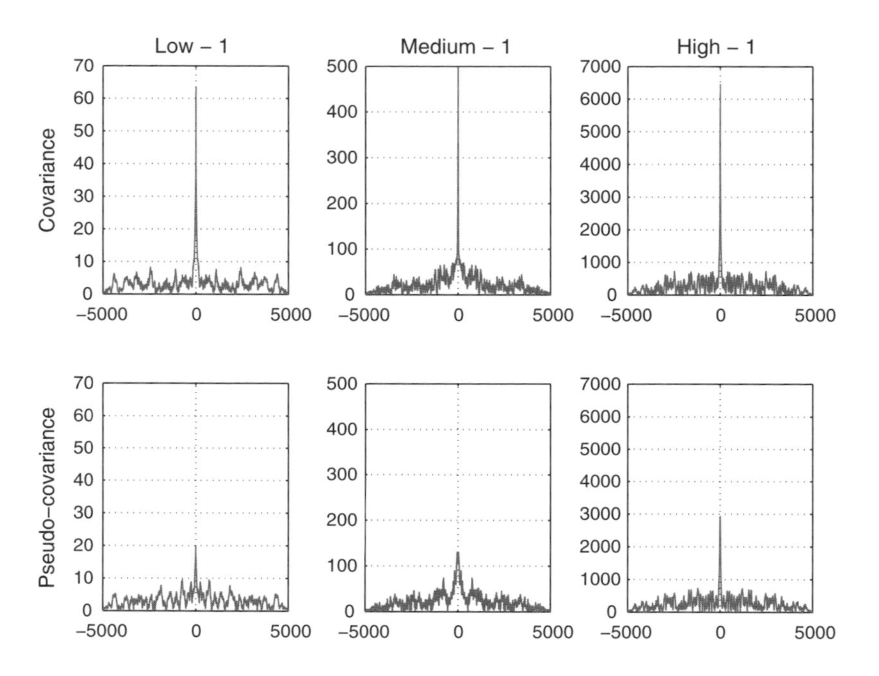

(a) Covariance and pseudocovariance functions for raw complex wind data

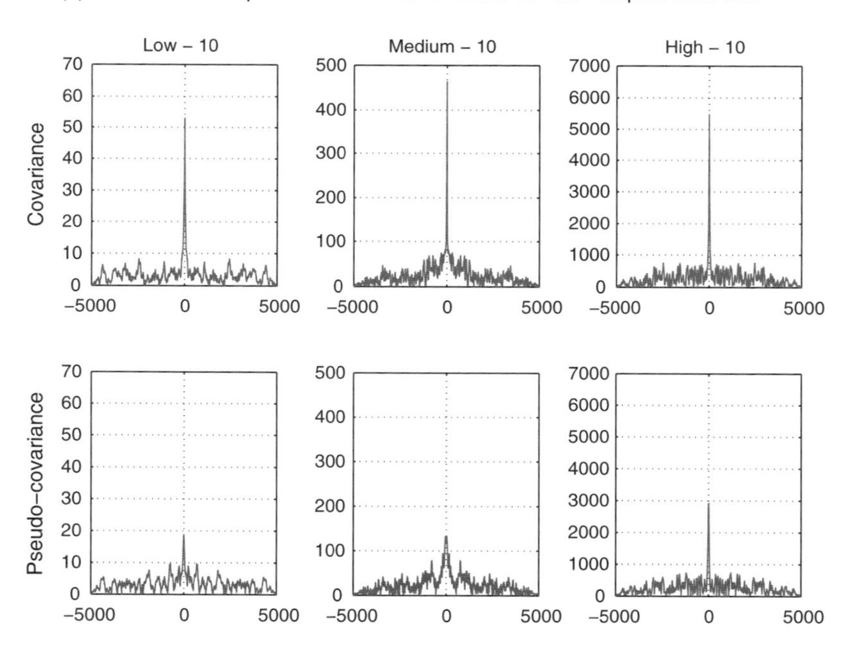

(b) Covariance and pseudocovariance functions for wind data averaged over 10 samples

Figure 12.4 Second-order statistical properties of complex wind data for different wind regimes ('low', 'medium', 'high'): (a) illustrates the covariance and pseudocovariance for the original data, whereas (b) shows statistical properties for data averaged over 10 samples

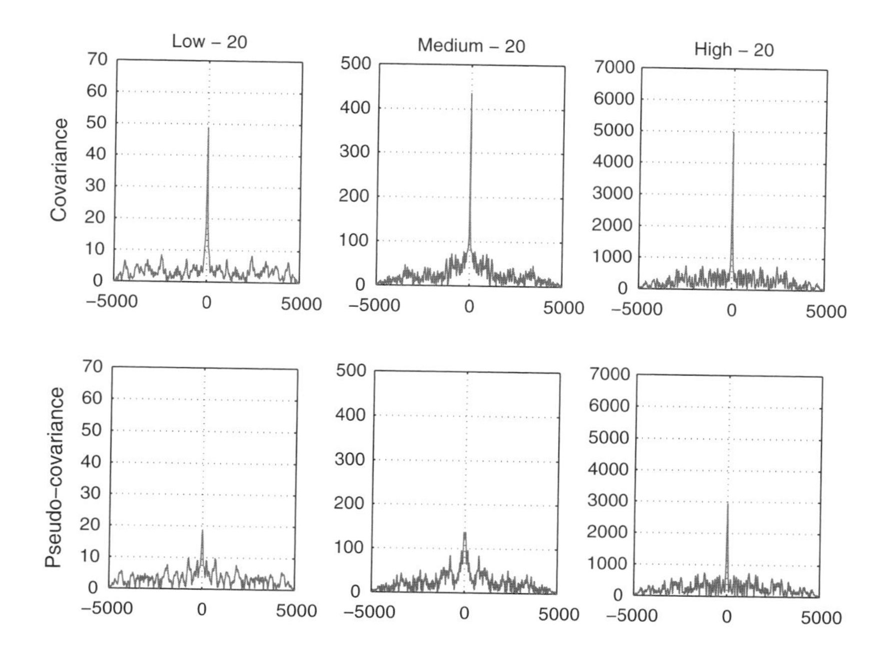

(a) Covariance and pseudocovariance functions for wind data averaged over 20 samples

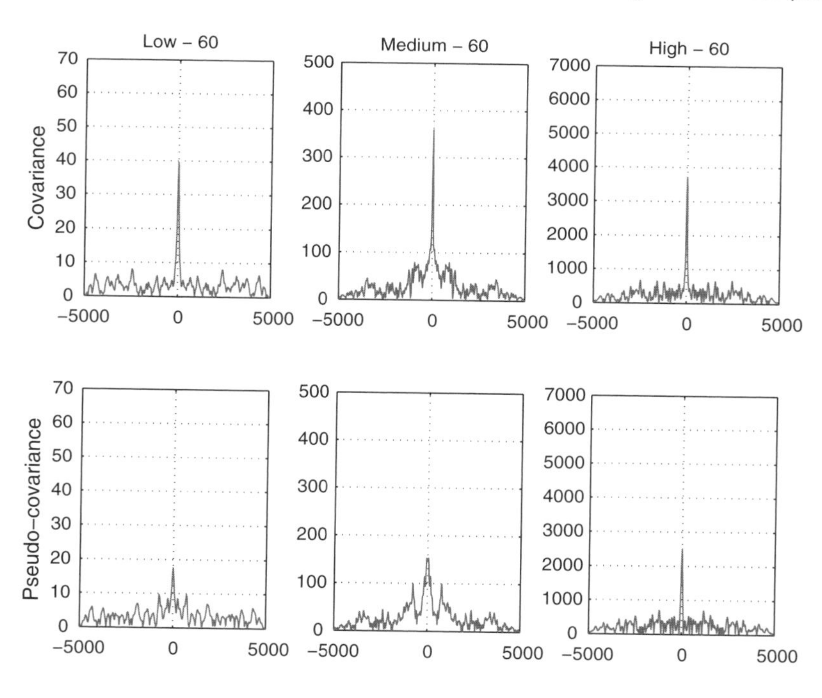

(b) Covariance and pseudocovariance functions for wind data averaged over 60 samples

Figure 12.5 Second-order statistical properties of complex wind data for different wind regimes ('low', 'medium', 'high'): (a) illustrates the covariance and pseudocovariance for data averaged over 20 samples, whereas (b) shows statistical properties for data averaged over 60 samples

segment analysed). From the shapes of these signal distributions, the autoregressive signal clearly exhibits a circularly symmetric distribution, which is not the case for the Ikeda and wind signals.

The covariance and pseudocovariance of the second-order circular complex white Gaussian noise and the AR(4) process from Figure 12.2(a) (see also Equation 13.44), together with those for the noncircular Ikeda map (see Equation 13.46), are illustrated in Figure 12.3. From Figure 12.3(a), the covariance of the complex WGN is nonzero only for lag $m = 0$, whereas the pseudocovariance vanishes. Since the AR(4) process was generated by passing circular white Gaussian noise though a stable autoregressive linear filter, the covariance function of such a process is nonzero, whereas the pseudocovariance function vanishes, as shown in Figure 12.3(b). This conforms with the analysis of second-order circular processes and illustrates that circularity is preserved under linear transformations; the small values of the pseudocovariance in the bottom panel of Figure 12.3(b) and Figure 12.3(a) are due to the artifacts caused by the finite lengths of the AR(4) process and the driving white Gaussian noise. On the other hand, the complex Ikeda map is strongly noncircular, hence both the covariance and pseudocovariance functions should be taken into account when modelling such a process. Indeed, Figure 12.3(c) shows that the pseudocovariance function for the complex Ikeda signal does not vanish.

To further illustrate the second-order circularity properties for a 'complex by convenience of representation' wind dataset (see also Figure 13.5a), Figure 12.4 shows the covariance and pseudocovariance functions for three wind regimes with different dynamics ('low', 'medium', and 'high') for the original data and data averaged over 10 samples; Figure 12.5 shows the covariance and pseudocovariance functions for data averaged over 20 and 60 samples. Observe that the larger the dynamical changes in the data, the more emphasised the pseudocovariance.

Properties, related to their second-order circularity, of some complex valued signals commonly encountered in practical applications are listed below (for further reading see [269]).

- Signal $z(t) \in \mathbb{C}^n$ is second-order circular if its pseudocovariance $p_z(k_1, k_2) = 0$; as a consequence, *real valued signals cannot be circular*.
- The equivalent lowpass signal (see Figure 2.4(b)) corresponding to a wide sense stationary real signal is always proper.
- Since thermal noise is usually assumed to be wide sense stationary, the equivalent lowpass (complex valued) representation of such noise is proper.
- In general, nonstationary complex random signals are improper, and therefore analytic signals constructed from nonstationary real signals are also improper.
- Most signals made complex by convenience of representation are improper (radar, sonar, beamforming), together with some artificial signals, such as in the case of Binary Phase Shift Keying (BPSK) modulation.

13

Widely Linear Estimation and Augmented CLMS (ACLMS)

It has been shown in Chapter 12 that the full-second order statistical description of a general complex valued process can be obtained only by using the augmented complex statistics, that is, by considering both the covariance and pseudocovariance functions. It is therefore natural to ask how much we can gain in terms of the performance of statistical signal processing algorithms by doing so. To that end, this chapter addresses linear estimation for both circular and noncircular (proper and improper) complex signals; this is achieved based on a finite impulse response (FIR) system model and for both the second-order regression modelling with fixed coefficients (autoregressive modelling) and for linear adaptive filters for which the filter coefficients are adaptive. Based mainly on the work by Picinbono [239, 240] and Schreier and Scharf [268], Sections 13.1 – 13.3 show that for general complex signals (noncircular), the optimal linear model is the 'widely linear' (WL) model, which is linear both in z and z^*. Next, based on the widely linear model, for adaptive filtering of general complex signals, the augmented complex least mean square (ACLMS) algorithm is derived, and by comparing the performances of ACLMS and CLMS, we highlight how much is lost by treating improper signals in the conventional way.

13.1 Minimum Mean Square Error (MMSE) Estimation in $\mathbb{C}$

The estimation of one signal from another is at the very core of statistical signal processing, and is illustrated in Figure 13.1, where $\{z(k)\}$ is the input signal, $\mathbf{z}(k) = [z(k-1), \ldots, z(k-N)]^T$ is the regressor vector in the filter memory, $d(k)$ is the teaching signal, $e(k)$ is the instantaneous output error, $y(k)$ is the filter output[1], and $\mathbf{h} = [h_1, \ldots, h_N]^T \in \mathbb{C}^{N\times 1}$ is the vector of filter coefficients.

[1]For prediction applications $d(k) = z(k)$ and $y(k) = \hat{z}_L(k)$, where the subscript 'L' refers to the standard linear estimator.

Complex Valued Nonlinear Adaptive Filters: Noncircularity, Widely Linear and Neural Models
Danilo P. Mandic and Vanessa Su Lee Goh

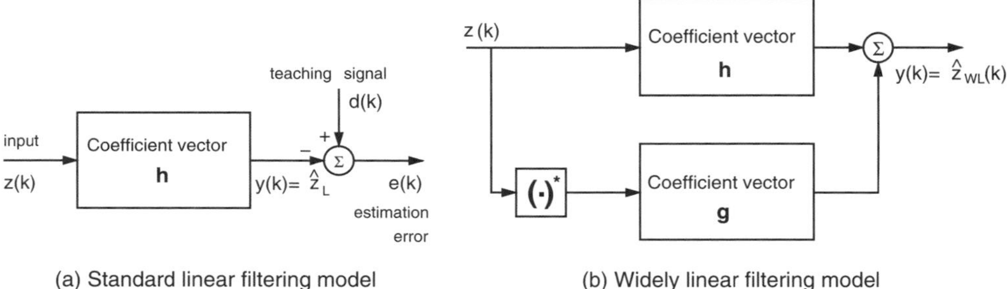

(a) Standard linear filtering model (b) Widely linear filtering model

Figure 13.1 Complex linear estimators

A solution that minimises the mean squared (MS) error is linear regression [110, 113]

$$\hat{z}_{\mathrm{L}}(k) = E[y(k)|\mathbf{z}(k)] \quad \Leftrightarrow \quad y(k) = \mathbf{h}^{\mathrm{T}}\mathbf{z}(k) \tag{13.1}$$

which estimates a scalar random variable y (real or complex) from an observation vector $\mathbf{z}(k)$, and is linear in $\mathbf{z}$ if both y and $\mathbf{z}$ are zero mean, and jointly normal. To perform this linear regression, we need to decide on the order N of the system model (typically a finite impulse response (FIR) system), and also on how to measure best fit of the data (error criterion). The estimator chooses those values of $\mathbf{h}$ which make $z(k)$ closest to $d(k)$, where closeness is measured by an error criterion, which should be reasonably realistic for the task in hand and it should be analytically tractable. Depending on the character of estimation, the commonly used error criteria are:

- *Deterministic error criterion*, given by

$$J = \min_{\mathbf{h}} \sum_k |e(k)|^p = \sum_k \big(d(k) - y(k)\big)^p \tag{13.2}$$

which for $p = 2$ is known as the *Least Squares (LS) problem*, and its solution is known as the Yule–Walker solution (the basis for autoregressive (AR) modelling in $\mathbb{C}$), given by [33]

$$\mathbf{h} = \mathbf{R}^{-1}\mathbf{r} \tag{13.3}$$

where $\mathbf{R}$ is the input correlation matrix and $\mathbf{r} = E[z(k)\mathbf{z}^*(k)]$.
- *Stochastic error criterion*, given by

$$J = \min_{\mathbf{h}} E\{|e(k)|^p\} \tag{13.4}$$

For $p = 2$ this optimisation problem is known as the *Wiener filtering problem*, for which the solution is given by Wiener–Hopf equations[2]

$$\mathbf{h} = \mathbf{R}_{\mathbf{z},\mathbf{z}}^{-1}\mathbf{r}_{d,\mathbf{z}} \tag{13.5}$$

where $\mathbf{R}_{\mathbf{z},\mathbf{z}}$ is the tap input correlation matrix and $\mathbf{r}_{d,\mathbf{z}}$ is the cross-correlation vector between the teaching signal and the tap input.

From (13.1) and (13.5), the output of the linear estimator is given by

$$y(k) = \hat{z}_L(k) = \left(\mathbf{R}_{\mathbf{z},\mathbf{z}}^{-1}\mathbf{r}_{d,\mathbf{z}}\right)^{\mathrm{T}}\mathbf{z}(k) \tag{13.6}$$

A stochastic gradient based iterative solution to this problem, which bypasses the requirement of piece-wise stationarity of the signal is the complex least mean square (CLMS) algorithm [307].

13.1.1 Widely Linear Modelling in $\mathbb{C}$

As shown in Chapter 12, for complete second-order statistical description of general complex signals we need to consider the statistics of the augmented input vector (in the prediction setting)

$$\mathbf{z}^a(k) = \left[z(k-1), \ldots, z(k-N), z^*(k-1), \ldots, z^*(k-N)\right]^{\mathrm{T}} = \left[\mathbf{z}^{\mathrm{T}}(k), \mathbf{z}^{\mathrm{H}}(k)\right]^{\mathrm{T}} \tag{13.7}$$

Thus, a linear estimator in $\mathbb{C}$ should be linear in both $\mathbf{z}$ and $\mathbf{z}^*$, that is

$$\hat{z}_{WL}(k) = y(k) = \mathbf{h}^{\mathrm{T}}\mathbf{z}(k) + \mathbf{g}^{\mathrm{T}}\mathbf{z}^*(k) = \mathbf{q}^{\mathrm{T}}\mathbf{z}^a(k) \tag{13.8}$$

where $\mathbf{h}$ and $\mathbf{g}$ are complex vectors of filter coefficients and $\mathbf{q} = \left[\mathbf{h}^{\mathrm{T}}, \mathbf{g}^{\mathrm{T}}\right]^{\mathrm{T}}$. Statistical moments of random variable $y(k)$ in Equation (13.8) are defined by the corresponding moments of the augmented input $\mathbf{z}^a(k)$, and the signal model (Equation 13.8) is referred to as a *wide sense linear* or *widely linear* (WL) estimator [239, 240, 271], depicted in Figure 13.1(b).

From Equations (13.6) and (13.8), the optimum widely linear coefficient vector is given by

$$\mathbf{q} = \left[\mathbf{h}^{\mathrm{T}}, \mathbf{g}^{\mathrm{T}}\right]^{\mathrm{T}} = \mathcal{C}_a^{-1}\mathbf{r}_{d,\mathbf{z}^a} \tag{13.9}$$

where $\mathcal{C}_a$ is the augmented covariance matrix given in Equation (12.50) and $\mathbf{r}_{d,\mathbf{z}^a}$ is the cross-correlation vector between the augmented input $\mathbf{z}^a(k)$ and the teaching signal $d(k)$. The widely linear MMSE solution can now be expressed as

$$y(k) = \hat{z}_{WL}(k) = \mathbf{q}^{\mathrm{T}}\mathbf{z}^a(k) = \left(\mathcal{C}_a^{-1}\mathbf{r}_{d,\mathbf{z}^a}\right)^{\mathrm{T}}\mathbf{z}^a(k) = \left(\begin{bmatrix} \mathcal{C} & \mathcal{P} \\ \mathcal{P}^* & \mathcal{C}^* \end{bmatrix}^{-1}\begin{bmatrix} \mathcal{C}_{d,\mathbf{z}} \\ \mathcal{P}_{d,\mathbf{z}} \end{bmatrix}\right)^{\mathrm{T}}\begin{bmatrix} \mathbf{z}(k) \\ \mathbf{z}^*(k) \end{bmatrix} \tag{13.10}$$

[2] To indicate the block nature of the solution, a piece-wise stationary segment of the data is considered.

The widely linear signal model therefore utilises information from both the covariance $\mathcal{C}$ and pseudocovariance $\mathcal{P}$ matrices (given in Equation 12.49) and from ordinary crosscorrelations $\mathcal{C}_{d,\mathbf{z}}$ and $\mathcal{P}_{d,\mathbf{z}}$, and as such it is suitable for the estimation of general complex signals. In the special case when the complementary statistics vanish, that is $\mathcal{P} = 0$ and $\mathcal{P}_{d,\mathbf{z}} = 0$, the widely linear estimator (Equation 13.8) degenerates into the standard linear estimator (Equation 13.1).

Practical applications of wide sense linear filters are only emerging, this is due to the fact that almost all existing applications assume circularity (explicitly or implicitly), however, this assumption cannot be generally accepted. A stochastic gradient based iterative solution to the widely linear estimation problem, which caters for nonstationary signals is called the augmented CLMS (ACLMS) and is introduced in Section 13.4.

13.2 Complex White Noise

Central to the autoregressive modelling and prediction is the concept of white noise;[3] a wide sense stationary signal $z(k)$ is said to be white if its power spectrum $\Gamma(\nu)$ is constant, or equivalently, if its covariance function c_z is a Dirac delta function, that is (see also Equation 12.44)

$$c_z(m) = c_z\delta(m) \quad \Leftrightarrow \quad \Gamma_z(\nu) = \text{const}$$

The concept of white noise is inherited from the analysis of real random variables, however, the fact that the power spectrum of a wide sense stationary white noise is constant does not imply any constraint on the pseudocovariance[4] p_z or spectral pseudocovariance $R_z(\nu) = \mathcal{F}\{p_z\}$ (see also Equation 12.43).

It has been shown in Section 12.4.1 that spectral covariance $\Gamma_z(\nu)$ and spectral pseudocovariance $R_z(\nu)$ of a second-order stationary complex signal z need to satisfy

$$\begin{aligned} \Gamma_z(\nu) &\geq 0 \\ R_z(\nu) &= R_z(-\nu) \\ |R_z(\nu)|^2 &\leq \Gamma_z(\nu)\Gamma_z(-\nu) \end{aligned} \tag{13.11}$$

In other words, for a *second-order stationary white noise* signal we have [239]

$$\begin{aligned} c_z(m) &= c_z\delta(m) \\ |R_z(\nu)|^2 &\leq \Gamma_z(\nu)\Gamma_z(-\nu) = c_z^2 \end{aligned} \tag{13.12}$$

It is often implicitly assumed that the spectral pseudocovariance function $R_z(\nu)$ of a second-order white signal vanishes, however, this would only mean that such white signal is *circular* [239].

- A *second-order circular white noise* signal is characterised by a constant power spectrum and vanishing spectral pseudocovariance function, that is, $\Gamma_z(\nu) = \text{const}$ and $R_z(\nu) = 0$. The real and imaginary parts of circular white noise are white and uncorrelated.

[3] Whiteness, in terms of multicorrelations, has already been introduced in Section 12.3.1.

[4] It can even be nonstationary.

Since the concept of whiteness is intimately related with the correlation structure of a signal (only instantaneous relationships are allowed, that is, there is no memory in the system), whiteness can also be defined in the time domain. One special case of a second-order white signal, which is a direct extension of the real valued white noise, is called *doubly white* noise.

- A second-order white signal is called *doubly white*, if

$$\begin{aligned} c_z(m) &= c_z \delta(m) \\ p_z(m) &= p_z \delta(m) \end{aligned} \tag{13.13}$$

where the only condition on the pseudocovariance function is $|p_z| \leq c_z$. The spectral covariance and pseudocovariance functions of doubly white noise are then given by [239]

$$\Gamma_w(\nu) = c_w \qquad \text{and} \qquad R_w(\nu) = p_w \tag{13.14}$$

13.3 Autoregressive Modelling in $\mathbb{C}$

The task of autoregressive (AR) modelling is, given a set of data, to find a regression of order p which approximates the given dataset. The standard autoregressive model in $\mathbb{C}$ takes the same form as the AR model for real valued signals, that is

$$z(k) = h_1 z(k-1) + \cdots + h_p z(k-p) + w(k) = \mathbf{h}^{\mathrm{T}} \mathbf{z}(k), \quad \mathbf{h} \in \mathbb{C}^{N \times 1} \tag{13.15}$$

where $\{z(k)\}$ is the random process to be modelled and $\{w(k)\}$ is white Gaussian noise (also called the driving noise), $\mathbf{h} = \left[h_1, \ldots, h_p\right]^{\mathrm{T}}$ and $\mathbf{z} = \left[z(k-1), \ldots, z(k-p)\right]^{\mathrm{T}}$. If the driving noise is assumed to be doubly white, we need to find the coefficient vector $\mathbf{h}$ and the covariance and pseudocovariance of the noise which provide best fit to the data in the minimum mean square error sense. Equivalently, in terms of the transfer function $H(\nu)$, we have (for more detail see [239])

$$\begin{aligned} |H(\nu)|^2 &= \Gamma(\nu) \\ p_w H(\nu) H(-\nu) &= R(\nu) \quad \Leftrightarrow \quad |R(\nu)|^2 = |p_w|^2 \Gamma(\nu) \Gamma(-\nu) \end{aligned} \tag{13.16}$$

In autoregressive modelling, it is usually assumed that the driving noise has zero mean and unit variance, and we can assume $c_w = 1$, which then implies[5] $|p_w| \leq 1$. Then, linear autoregressive modelling (based on the deterministic error criterion, Equation 13.2) has a solution only if both Equations (13.11) and (13.16) are satisfied. This happens, for instance, when $R(\nu) = 0$ and $p_w = 0$, that is, when the driving noise is white and second-order circular, which is the usual assumption in standard statistical signal processing literature [110, 113]. In this case, the solution has the same form as in the real case [33], that is

$$\mathbf{h} = \mathbf{R}^{-1} \mathbf{r} \tag{13.17}$$

where $\mathbf{R} = E\left[\mathbf{z}(k)\mathbf{z}^{\mathrm{H}}(k)\right]$ and $\mathbf{r} = E\left[z(k)\mathbf{z}^*(k)\right]$.

[5]The pseudocovariance p_w can even be a complex quantity, hence the modulus operator.

Thus, a general complex signal cannot be modelled by a linear filter driven by doubly white noise.

13.3.1 Widely Linear Autoregressive Modelling in $\mathbb{C}$

The widely linear autoregressive (WLAR) model in $\mathbb{C}$ is linear in both $\mathbf{z}$ and $\mathbf{z}^*$, that is

$$\begin{aligned} z(k) &= \sum_{i=1}^{p} h_i z(k-i) + \sum_{i=1}^{p} g_i z^*(k-i) + h_0 w(k) + g_0 w^*(k) \\ &= \mathbf{h}^{\mathrm{T}}\mathbf{z}(k) + \mathbf{g}^{\mathrm{T}}\mathbf{z}^*(k) + [h_0, g_0] w^a(k) \end{aligned} \tag{13.18}$$

and has more degrees of freedom and hence potentially improved performance over the standard linear model. The gain in performance, however, depends on the degree of circularity of the signal at hand. Autoregressive modelling is intimately related to prediction, that is (since $E[w(k)] = 0$)

$$\hat{z}_{\mathrm{WL}}(k) = E\left[\mathbf{h}^{\mathrm{T}}\mathbf{z}(k) + \mathbf{g}^{\mathrm{T}}\mathbf{z}^*(k) + [h_0, g_0]w^a(k)\right] = \mathbf{h}^{\mathrm{T}}\mathbf{z}(k) + \mathbf{g}^{\mathrm{T}}\mathbf{z}^*(k) = \mathbf{q}^{\mathrm{T}}\mathbf{z}^a(k) \tag{13.19}$$

When the driving noise w is circular, the widely linear model has no advantage over the standard linear model, whereas for noncircular signals we expect improvement in the performance proportional to the degree of noncircularity within the signal.

13.3.2 Quantifying Benefits of Widely Linear Estimation

The goal of widely linear estimation is to find coefficient vectors $\mathbf{h}$ and $\mathbf{g}$ that minimise the mean squared error $E[|d(k) - y(k)|^2]$ of the regression

$$\hat{y} = \mathbf{h}^{\mathrm{T}}\mathbf{z} + \mathbf{g}^{\mathrm{T}}\mathbf{z}^* \tag{13.20}$$

Following the approach from [240], to find the solution, apply the principle of orthogonality to obtain[6]

$$E[\hat{y}^*\mathbf{z}] = E[y^*\mathbf{z}], \quad \text{and} \quad E[\hat{y}^*\mathbf{z}^*] = E[y^*\mathbf{z}^*] \tag{13.21}$$

and replace $\hat{y}$ in (13.21) with its widely linear estimate (13.20), to yield[7]

$$\mathcal{C}\mathbf{h} + \mathcal{P}\mathbf{g} = \mathbf{u} \tag{13.22}$$

$$\mathcal{P}^*\mathbf{h} + \mathcal{C}^*\mathbf{g} = \mathbf{v}^* \tag{13.23}$$

where $\mathcal{C}$ and $\mathcal{P}$ are defined in (12.49), $\mathbf{u} = E[y^*\mathbf{z}]$, and $\mathbf{v} = E[y\mathbf{z}]$.

---

[6] We have $(y - \hat{y}) \perp \mathbf{z}$ and $(y - \hat{y}) \perp \mathbf{z}^*$, and as a consequence the orthogonality can be expressed in terms of expectations, as given in Equation (13.21).

[7] For convenience of the derivation, in Equations (13.22) and (13.28) the expression $z(k) = \mathbf{h}^{\mathrm{T}}\mathbf{z}(k) + \mathbf{g}^{\mathrm{T}}\mathbf{z}^*(k)$ is replaced by $z(k) = \mathbf{h}^{\mathrm{H}}\mathbf{z}(k) + \mathbf{g}^{\mathrm{H}}\mathbf{z}^*(k)$. This is a deterministic transformation and does not affect the generality of the results.

From Equations (13.22) and (13.23), the coefficient vectors that minimise the MSE of the widely linear model (Equation 13.8) are given by

$$\mathbf{h} = [\mathcal{C} - \mathcal{P}\mathcal{C}^{-1^*}\mathcal{P}^*]^{-1}[\mathbf{u} - \mathcal{P}\mathcal{C}^{-1^*}\mathbf{v}^*] \tag{13.24}$$

$$\mathbf{g} = [\mathcal{C}^* - \mathcal{P}^*\mathcal{C}^{-1}\mathcal{P}]^{-1}[\mathbf{v}^* - \mathcal{P}^*\mathcal{C}^{-1}\mathbf{u}] \tag{13.25}$$

and the corresponding widely linear mean square error (WLMSE) e^2_{WL} is given by

$$e^2_{\mathrm{WL}} = E[|y|^2] - (\mathbf{h}^{\mathrm{H}}\mathbf{u} + \mathbf{g}^{\mathrm{H}}\mathbf{v}^*) \tag{13.26}$$

whereas the mean square error (LMSE) e^2_{L} obtained with standard linear estimation is given by

$$e^2_{\mathrm{L}} = E[|y|^2] - \mathbf{u}^{\mathrm{H}}\mathcal{C}^{-1}\mathbf{u} \tag{13.27}$$

The advantage of widely linear estimation over standard linear estimation can be illustrated by comparing the corresponding mean square estimation errors $\delta e^2 = e^2_{\mathrm{L}} - e^2_{\mathrm{WL}}$, that is

$$\delta e^2 = [\mathbf{v}^* - \mathcal{P}^*\mathcal{C}^{-1}\mathbf{u}]^{\mathrm{H}}[\mathcal{C}^* - \mathcal{P}^*\mathcal{C}^{-1}\mathcal{P}]^{-1}[\mathbf{v}^* - \mathcal{P}^*\mathcal{C}^{-1}\mathbf{u}] \tag{13.28}$$

Due to the positive definiteness of the term $[\mathcal{C}^* - \mathcal{P}^*\mathcal{C}^{-1}\mathcal{P}]$ from Equation (13.28)

δe^2 is always non-negative;

$\delta e^2 = 0$ only when $[\mathbf{v}^* - \mathcal{P}^*\mathcal{C}^{-1}\mathbf{u}] = \mathbf{0}$.

that is, *widely linear estimation outperforms standard linear estimation for general complex signals; the two models produce identical results for circular signals.*

Exploitation of widely linear modelling promises several benefits, including:

- identical performance for circular signals and improved performance for noncircular signals;
- in blind source separation we may be able to deal with more sources than observations;
- improved signal recovery in communications modulation schemes (BPSK, GMSK);
- different and more realistic bounds on minimum variance unbiased (MVU) estimation;
- improved 'direction of arrival' estimation in augmented array signal processing;
- the analysis of augmented signal processing algorithms benefits from special matrix structures which do not exist in standard complex valued signal processing.

13.4 The Augmented Complex LMS (ACLMS) Algorithm

We now consider the extent to which widely linear mean square estimation has advantages over standard linear mean square estimation in the context of linear adaptive prediction. To answer this question, consider a widely linear adaptive prediction model for which the tap input $\mathbf{z}(k)$ to a finite impulse response filter of length N at the time instant k is given by

$$\mathbf{z}(k) = [z(k-1), z(k-2), \ldots, z(k-N)]^{\mathrm{T}} \tag{13.29}$$

Within widely linear regression, the augmented tap input delay vector $\mathbf{z}^a(k) = [\mathbf{z}^T(k), \mathbf{z}^H(k)]^T$ is 'widely linearly' combined with the adjustable filter weights $\mathbf{h}(k)$ and $\mathbf{g}(k)$ to form the output[8]

$$y(k) = \sum_{n=1}^{N} \left[h_n(k)z(k-n) + g_n(k)z^*(k-n)\right] \quad \Longleftrightarrow \quad y(k) = \mathbf{h}^T(k)\mathbf{z}(k) + \mathbf{g}^T(k)\mathbf{z}^*(k) \tag{13.30}$$

where $\mathbf{h}(k)$ and $\mathbf{g}(k)$ are the $N \times 1$ column vectors comprising the filter weights at time instant k, and $y(k)$ is the estimate of the desired signal $d(k)$.

For adaptive filtering applications,[9] similarly to the derivation of the standard complex least mean square (CLMS) algorithms, we need to minimise the cost function [113, 307]

$$J(k) = \tfrac{1}{2}|e(k)|^2 = \tfrac{1}{2}\left[e_r^2(k) + e_i^2(k)\right], \quad \text{with} \quad e(k) = d(k) - y(k) \tag{13.31}$$

where $e_r(k)$ and $e_i(k)$ are the respectively the real and imaginary part of the instantaneous output error $e(k)$. For simplicity, consider a generic weight update in the form[10]

$$\Delta w_n(k) = -\mu \nabla_{w_n} J(k) = -\mu \frac{\partial J(k)}{\partial w_n(k)} = -\mu \left(\frac{\partial J(k)}{\partial w_n^r(k)} + \jmath \frac{\partial J(k)}{\partial w_n^i(k)} \right) \tag{13.32}$$

where $w_n(k) = w_n^r(k) + \jmath w_n^i(k)$ is a complex weight and μ is the learning rate, a small positive constant. The real and imaginary parts of the gradient $\nabla_{w_n} J(k)$ can be expressed respectively as

$$\frac{\partial J(k)}{\partial w_n^r(k)} = e_r(k)\frac{\partial e_r(k)}{\partial w_n^r(k)} + e_i(k)\frac{\partial e_i(k)}{\partial w_n^r(k)} = -e_r(k)\frac{\partial y_r(k)}{\partial w_n^r(k)} - e_i(k)\frac{\partial y_i(k)}{\partial w_n^r(k)} \tag{13.33}$$

$$\frac{\partial J(k)}{\partial w_n^i(k)} = e_r(k)\frac{\partial e_r(k)}{\partial w_n^i(k)} + e_i(k)\frac{\partial e_i(k)}{\partial w_n^i(k)} = -e_r(k)\frac{\partial y_r(k)}{\partial w_n^i(k)} - e_i(k)\frac{\partial y_i(k)}{\partial w_n^i(k)}. \tag{13.34}$$

Similarly to Equations (13.33) and (13.34), the error gradients with respect to the elements of the weight vectors $\mathbf{h}(k)$ and $\mathbf{g}(k)$ of the widely linear variant of CLMS can be calculated as

[8] For consistent notation, we follow the original derivation of the complex LMS from [307].

[9] The widely linear LMS (WLLMS) and widely linear blind LMS (WLBLMS) algorithms for multiple access interference suppression in DS-CDMA communications were derived in [262].

[10] We here provide a step by step derivation of ACLMS. The $\mathbb{CR}$ calculus (see Chapter 5) will be used in Chapter 15 to simplify the derivations for feedback and nonlinear architectures.

$$\frac{\partial J(k)}{\partial h_n^r(k)} = -e_r(k)\frac{\partial y_r(k)}{\partial h_n^r(k)} - e_i(k)\frac{\partial y_i(k)}{\partial h_n^r(k)} = -e_r(k)z_r(k-n) - e_i(k)z_i(k-n) \quad (13.35)$$

$$\frac{\partial J(k)}{\partial h_n^i(k)} = -e_r(k)\frac{\partial y_r(k)}{\partial h_n^i(k)} - e_i(k)\frac{\partial y_i(k)}{\partial h_n^i(k)} = e_r(k)z_i(k-n) - e_i(k)z_r(k-n) \quad (13.36)$$

$$\frac{\partial J(k)}{\partial g_n^r(k)} = -e_r(k)\frac{\partial y_r(k)}{\partial g_n^r(k)} - e_i(k)\frac{\partial y_i(k)}{\partial g_n^r(k)} = -e_r(k)z_r(k-n) + e_i(k)z_i(k-n) \quad (13.37)$$

$$\frac{\partial J(k)}{\partial g_n^i(k)} = -e_r(k)\frac{\partial y_r(k)}{\partial g_n^i(k)} - e_i(k)\frac{\partial y_i(k)}{\partial g_n^i(k)} = -e_r(k)z_i(k-n) - e_i(k)z_r(k-n) \quad (13.38)$$

giving the updates

$$\begin{aligned}\Delta h_n(k) &= -\mu\frac{\partial J(k)}{\partial h_n(k)} = -\mu\left(\frac{\partial J(k)}{\partial h_n^r(k)} + \jmath\frac{\partial J(k)}{\partial h_n^i(k)}\right)\\ &= \mu\Big[\big(e_r(k)z_r(k-n) + e_i(k)z_i(k-n)\big) + \jmath\big(e_i(k)z_r(k-n) - e_r(k)z_i(k-n)\big)\Big]\\ &= \mu e(k)z^*(k) \quad (13.39)\\ \Delta g_n(k) &= -\mu\frac{\partial J(k)}{\partial g_n(k)} = -\mu\left(\frac{\partial J(k)}{\partial g_n^r(k)} + \jmath\frac{\partial J(k)}{\partial g_n^i(k)}\right)\\ &= \mu\Big[\big(e_r(k)z_r(k-n) - e_i(k)z_i(k-n)\big) + \jmath\big(e_r(k)z_i(k-n) + e_i(k)z_r(k-n)\big)\Big]\\ &= \mu e(k)z(k) \quad (13.40)\end{aligned}$$

These weight updates can we written in vector form as

$$\mathbf{h}(k+1) = \mathbf{h}(k) + \mu e(k)\mathbf{z}^*(k) \quad (13.41)$$

$$\mathbf{g}(k+1) = \mathbf{g}(k) + \mu e(k)\mathbf{z}(k) \quad (13.42)$$

To further simplify the notation, we can introduce an augmented weight vector $\mathbf{w}^a(k) = \left[\mathbf{h}^{\mathrm{T}}(k), \mathbf{g}^{\mathrm{T}}(k)\right]^{\mathrm{T}}$, and rewrite the ACLMS in its compact form as [132, 194]

$$\mathbf{w}^a(k+1) = \mathbf{w}^a(k) + \mu e(k)\mathbf{z}^{a*}(k) \quad (13.43)$$

where the 'augmented' instantaneous error[11] is $e(k) = d(k) - \mathbf{z}^{a\mathrm{T}}(k)\mathbf{w}^a(k)$. This completes the derivation of the augmented CLMS (ACLMS) algorithm, a widely linear extension of standard CLMS.

The ACLMS algorithm has the same generic form as the standard CLMS, it is simple to implement, yet it takes into account the full available second-order statistics of complex valued inputs (noncircularity).

[11]The output error itself is not augmented, but it is calculated based on the linear combination of the augmented input and weight vectors.

13.5 Adaptive Prediction Based on ACLMS

Simulations were performed for a 4-tap (4 taps of **h** and 4 taps of **g**) FIR filter trained with ACLMS and the performances were compared to those of standard CLMS for a range of both synthetic and real world data, denoted by **DS1** – **DS4**. The synthetic benchmark signals were a linear circular complex $AR(4)$ process and two noncircular chaotic series,[12] whereas wind was used as a real world dataset.

DS1. *Linear AR(4) process* ('AR4'), given by [180]

$$y(k) = 1.79y(k-1) - 1.85y(k-2) + 1.27y(k-3) - 0.41y(k-4) + n(k) \tag{13.44}$$

where $n(k)$ is a complex, white Gaussian noise with variance $\sigma^2 = 1$.

DS2. *Wind* ('wind'), containing wind speed and direction data averaged over one minute.[13]

DS3. *Lorenz Attractor* ('lorenz'), is a nonlinear, three-dimensional, deterministic system given by the coupled differential equations [173]

$$\frac{\mathrm{d}x}{\mathrm{d}t} = \sigma(y-x), \quad \frac{\mathrm{d}y}{\mathrm{d}t} = x(\rho - z) - y, \quad \frac{\mathrm{d}z}{\mathrm{d}t} = xy - \beta z \tag{13.45}$$

where (typically) $\sigma = 10$, $\beta = 8/3$, $\rho = 28$.

DS4. *Ikeda Map* ('ikeda'), described by [104]

$$\begin{aligned} x(k+1) &= 1 + u\,(x(k)\cos[t(k)] - y(k)\sin[t(k)]) \\ y(k+1) &= u\,(x(k)\sin[t(k)] + y(k)\cos[t(k)]) \end{aligned} \tag{13.46}$$

where u is a parameter (typically $u = 0.8$) and

$$t(k) = 0.4 - \frac{6}{1 + x^2(k) + y^2(k)}. \tag{13.47}$$

Both batch[14] and online[15] learning scenarios were considered, and the standard prediction gain $R_p = 10 \log \frac{\sigma_y^2}{\sigma_e^2}$ was used as a quantitative measure of performance.

Batch learning scenario. Learning curves for the batch learning scenario are shown in the left-hand part of Figure 13.2. The dotted lines correspond to the learning curves of the CLMS algorithm, whereas the solid lines correspond to those of the ACLMS algorithm. For 'AR4'

[12] The two chaotic time series are generated by coupled difference equations, and were made complex by 'convenience of representation', that is, by taking the x and y components from Equations (13.45) and (13.46) and building a complex signal $z(k) = x(k) + \jmath y(k)$.

[13] The data used are from AWOS (Automated Weather Observing System) sensors obtained from the Iowa Department of Transportation. The Washington (AWG) station was chosen, and the dataset analysed corresponds to the wind speed and direction observed in January 2004. This dataset is publicly available from http://mesonet.agron.iastate.edu/request/awos/1min.php.

[14] For 1000 epochs with $\mu = 0.001$ and for 1000 data samples; for more detail on batch learning see Appendix G.

[15] With $\mu = 0.01$ and for 1000 samples of **DS1** – **DS4**.

signal (strictly circular) and 'wind' signal (almost circular for the given averaging interval and data length), there was almost no difference in performances of CLMS and ACLMS. A completely different situation occurred for the strongly noncircular 'lorenz' (see also Figure 12.2) and 'ikeda' signals. After the training, the prediction gains for the ACLMS algorithm were respectively about 3.36 (for 'lorenz') and 2.24 (for 'ikeda') times bigger than those of the corresponding CLMS algorithm. These results are perfectly in line with the background theory, that is, for noncircular signals, the minimum mean square error solution is based on augmented complex statistics.

Online learning scenario. Learning curves for adaptive one step ahead prediction are shown in Figure 13.2 (right), where the solid line corresponds to the real part of the original 'lorenz' signal, the dotted line represents the prediction based on CLMS, and the dashed line corresponds to the ACLMS based prediction. For the same filter setting, ACLMS was able to track the desired signal more accurately than CLMS.

Figure 13.2 also shows that the more noncircular the signal in question, the greater the performance advantage of ACLMS over CLMS, which conforms with the analysis in Section 13.3.2.

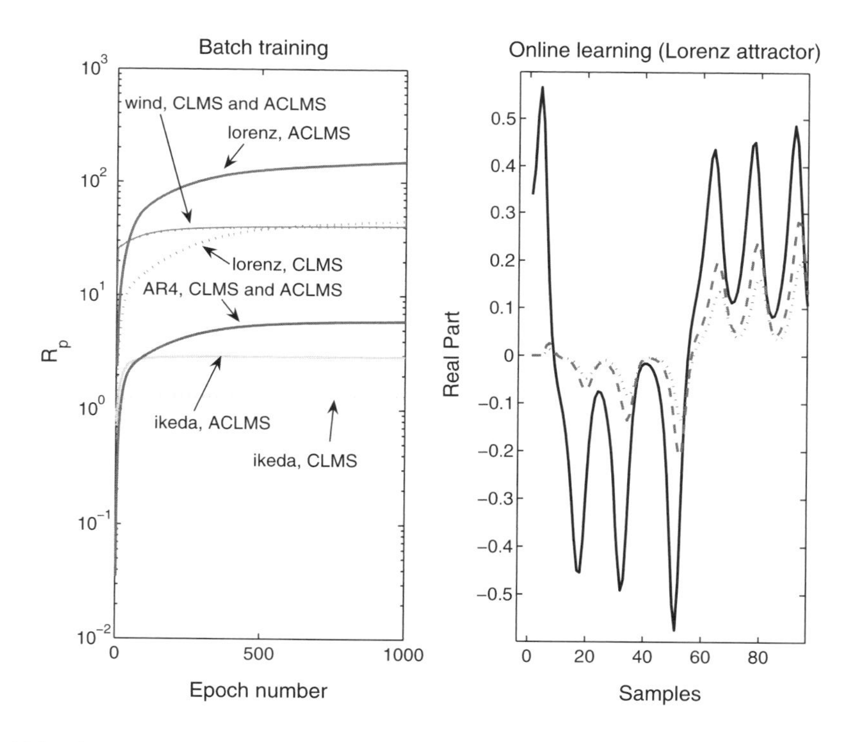

Figure 13.2 Comparison of the ACLMS and standard CLMS. Left: Prediction gains R_p for signals **DS1 – DS4**. Right: Tracking performance for the Lorenz signal; solid line represents the original signal, dotted line the CLMS based prediction, and dashed line the ACLMS based prediction

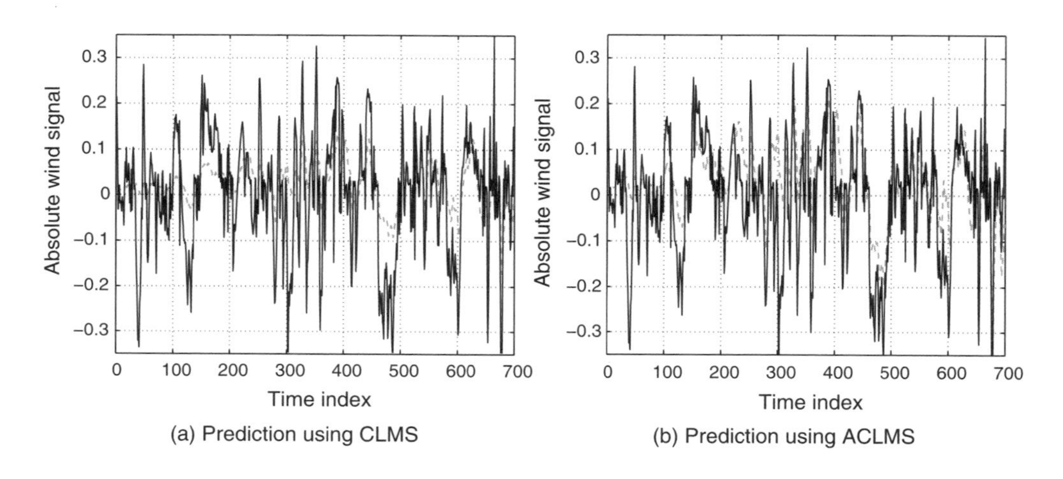

Figure 13.3 The original signal (solid line) and one step ahead prediction (dashed line)

13.5.1 Wind Forecasting Using Augmented Statistics

In the first experiment, adaptive one step ahead prediction of the original wind signal[16] was performed for a $N = 10$ tap FIR filter trained with CLMS and ACLMS. The time waveforms of the original and predicted signal are shown in Figure 13.3, indicating that the ACLMS was better suited to the statistics of the wind signal considered. A segment from Figure 13.3 is enlarged in Figure 13.4, showing the ACLMS being able to track the changes in wind dynamics more accurately than CLMS.

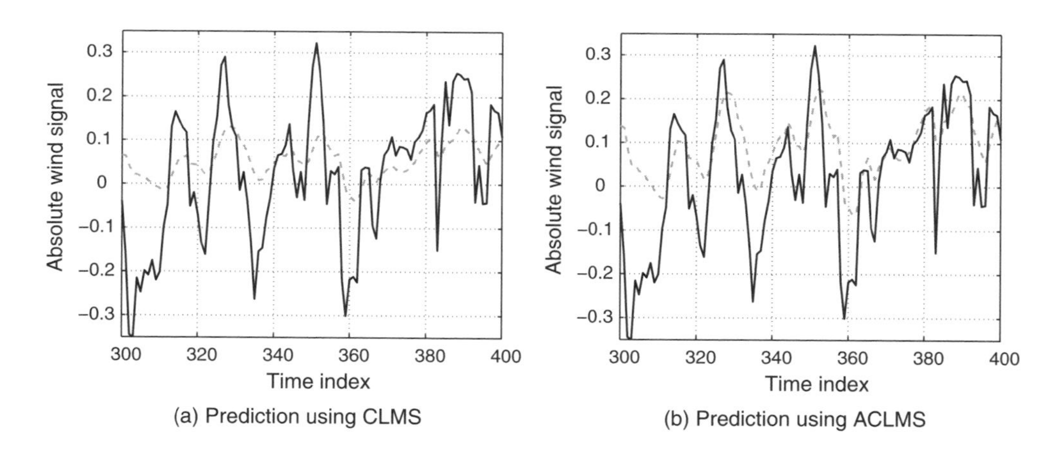

Figure 13.4 The original signal (solid line) and one step ahead prediction (dashed line)

[16]IOWA wind data averaged over 3 hours (see Footnote 13), which facilitates Gaussianity and widely linear modelling.

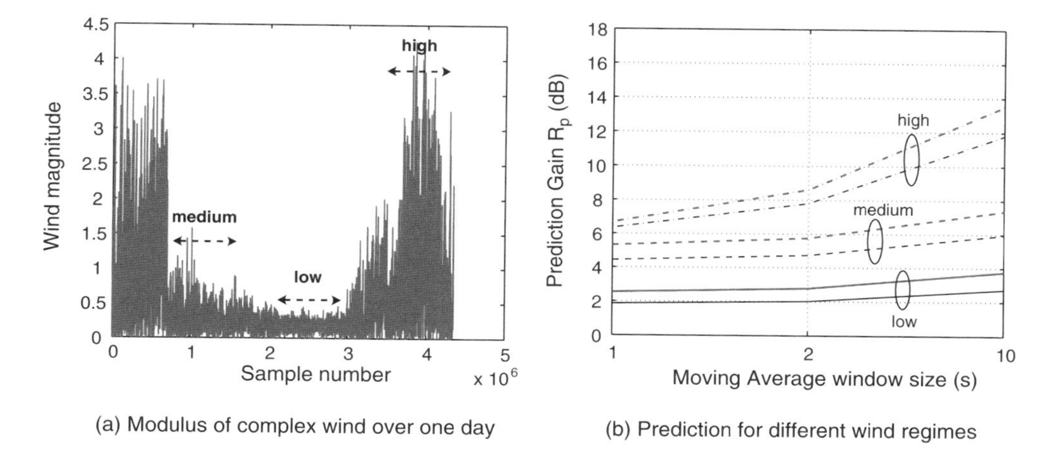

Figure 13.5 Performance of CLMS and ACLMS for different wind regimes. Thick lines correspond to ACLMS and thin lines to CLMS

In the second experiment, tracking performances of CLMS and ACLMS were investigated over a long period of time. Figure 13.5 shows the modulus of complex wind measurements recorded over one day at a sampling frequency of 50 Hz, and the performances of CLMS and ACLMS for the wind regimes denoted (according to the wind dynamics) by 'low', 'medium' and 'high'. The prediction was performed on the raw data, and also on the data averaged over 2 and 10 seconds, and in all the cases, due to the noncircular nature of the considered wind data, the widely linear ACLMS outperformed standard CLMS.

To summarise:

- For nonlinear and noncircular signals (chaotic Lorenz and Ikeda maps), and signals with a large variation in the dynamics ('high' wind from Figure 13.5), the augmented statistics based modelling exhibited significant advantages over standard modelling for both the batch and online learning paradigms;
- For signals with relatively mild dynamics (linear 'AR4', heavily averaged wind from Figure 13.3, and the 'medium' and 'low' regions from Figure 13.5), the widely linear model outperformed the standard complex model; the improvement in the performance, however, varied depending on the degree of circularity within the signal;
- In practical applications, the pseudocovariance matrix is estimated from short segments of data in the filter memory and in the presence of noise; such estimate will be nonzero even for circular sources, and widely linear models are a natural choice.

It is therefore natural to ask whether it is possible to design a rigorous statistical testing framework which would reveal the second-order circularity properties of the 'complex by convenience' class of signals, a subject of Chapter 18.

14

Duality Between Complex Valued and Real Valued Filters

The duality between the fields of real numbers and complex numbers is usually addressed in terms of the relation between a complex number $z = x + \jmath y$ and a number $(x, y) \in \mathbb{R}^2$, which can be expressed as

$$z = \begin{bmatrix} 1 & 0 \\ 0 & \jmath \end{bmatrix} \begin{bmatrix} x \\ y \end{bmatrix} \tag{14.1}$$

Similarly, the mapping between a 'composite' real variable $w = (x, y) \in \mathbb{R}^2$ and the complex variable z and its complex conjugate z^* can be expressed as

$$\begin{bmatrix} z \\ z^* \end{bmatrix} = \mathbb{J} \begin{bmatrix} x \\ y \end{bmatrix} = \begin{bmatrix} 1 & \jmath \\ 1 & -\jmath \end{bmatrix} \begin{bmatrix} x \\ y \end{bmatrix} \quad \Leftrightarrow \quad \begin{bmatrix} x \\ y \end{bmatrix} = \frac{1}{2} \begin{bmatrix} 1 & 1 \\ -\jmath & \jmath \end{bmatrix} \begin{bmatrix} z \\ z^* \end{bmatrix} \tag{14.2}$$

These mappings can help to establish relations between the statistics in $\mathbb{R}^2$ and $\mathbb{C}$. A deeper insight into this problem is given in Chapter 12, where some unique properties of complex probability distributions (such as complex circularity) are highlighted, whereas the benefits of signal processing in the complex domain are summarised in Chapter 2. Chapter 13 introduces learning algorithms for widely linear adaptive filters which are suitable for processing noncircular signals; one such example is the augmented complex least mean square (ACLMS) algorithm.

Since the mappings (Equations 14.1 and 14.2) establish a one to one correspondence between the points in $\mathbb{C}$ and $\mathbb{R}^2$, it is natural to consider the duality between learning algorithms for the corresponding adaptive filters in this context. In this chapter, an insight into the duality between adaptive filters in $\mathbb{R}^2$ and $\mathbb{C}$ is provided based on linear finite impulse response (FIR) filters trained by dual channel real least mean square (DCRLMS), complex least mean square (CLMS), and ACLMS algorithms. This is supported by simulation results for both circular and noncircular benchmark and real world data.

Complex Valued Nonlinear Adaptive Filters: Noncircularity, Widely Linear and Neural Models
Danilo P. Mandic and Vanessa Su Lee Goh

14.1 A Dual Channel Real Valued Adaptive Filter

Consider a dual channel real adaptive filter (DCRAF) in the prediction setting, shown in Figure 14.1. The operation of such a filter is described by (for more detail see [23])

$$\begin{aligned}\hat{x}(k) &= \mathbf{a}^{\mathrm{T}}(k)\mathbf{x}(k) + \mathbf{b}^{\mathrm{T}}(k)\mathbf{y}(k)\\ \hat{y}(k) &= \mathbf{c}^{\mathrm{T}}(k)\mathbf{x}(k) + \mathbf{d}^{\mathrm{T}}(k)\mathbf{y}(k)\end{aligned} \tag{14.3}$$

where $\mathbf{a}, \mathbf{b}, \mathbf{c}, \mathbf{d} \in \mathbb{R}^{L\times 1}$ are column vectors of filter coefficients, L denotes the filter length, $\hat{x}(k)$ and $\hat{y}(k)$ are the predictions of channels $x(k)$ and $y(k)$ and

$$\begin{aligned}\mathbf{x}(k) &= [x(k-1), \ldots, x(k-L)]^{\mathrm{T}}\\ \mathbf{y}(k) &= \left[y(k-1), \ldots, y(k-L)\right]^{\mathrm{T}}\end{aligned}$$

are the past samples from the x and y channel contained in the filter memory. In the prediction setting, the teaching signals for the x and y channel are respectively $d_x(k) = x(k)$ and $d_y(k) = y(k)$.

The output errors at the x and y channel of the DCRAF are defined as

$$\begin{aligned}e_x(k) &= d_x(k) - \hat{x}(k) = x(k) - \mathbf{a}^{\mathrm{T}}(k)\mathbf{x}(k) - \mathbf{b}^{\mathrm{T}}(k)\mathbf{y}(k)\\ e_y(k) &= d_y(k) - \hat{y}(k) = y(k) - \mathbf{c}^{\mathrm{T}}(k)\mathbf{x}(k) - \mathbf{d}^{\mathrm{T}}(k)\mathbf{y}(k)\end{aligned} \tag{14.4}$$

whereas the cost function used for the calculation of filter coefficient updates is given by

$$J = J(\mathbf{a}, \mathbf{b}, \mathbf{c}, \mathbf{d}) = \frac{1}{2}\left(e_x^2(k) + e_y^2(k)\right) \tag{14.5}$$

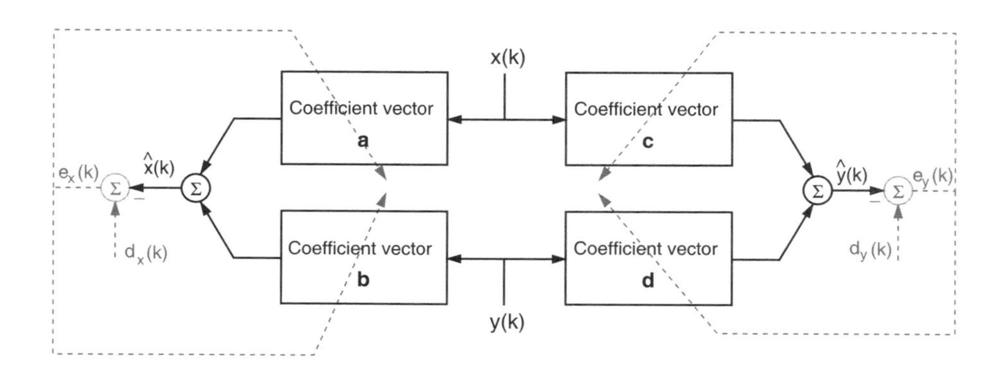

Figure 14.1 A dual channel real valued adaptive filter (DCRAF). Solid lines illustrate the signal flow graph; broken lines illustrate signal flow in the filter coefficient update. The filter outputs are generated based on a linear combination of the x and y channel

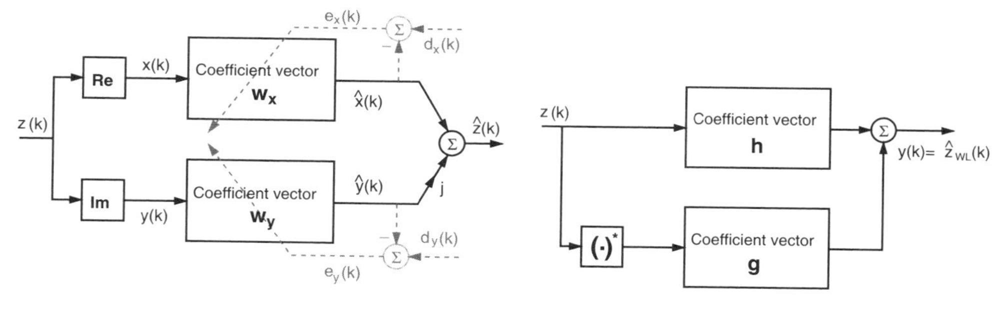

(a) Dual univariate real adaptive filter

(b) Widely linear complex adaptive filter

Figure 14.2 A dual univariate real valued adaptive filter (DUAF) and a widely linear complex adaptive filter (WLCAF). Solid lines illustrate the signal flow graph; broken lines illustrate signal flow in the filter coefficient update

Stochastic gradient updates for the coefficient vectors from (14.3) now become[1]

$$\mathbf{a}(k+1) = \mathbf{a}(k) - \mu_a \nabla J_{\mathbf{a}|\mathbf{a}=\mathbf{a}(k)} \quad \mathbf{b}(k+1) = \mathbf{b}(k) - \mu_b \nabla J_{\mathbf{b}|\mathbf{b}=\mathbf{b}(k)}$$
$$\mathbf{c}(k+1) = \mathbf{c}(k) - \mu_c \nabla J_{\mathbf{c}|\mathbf{c}=\mathbf{c}(k)} \quad \mathbf{d}(k+1) = \mathbf{d}(k) - \mu_d \nabla J_{\mathbf{d}|\mathbf{d}=\mathbf{d}(k)} \tag{14.6}$$

Since neither $\mathbf{x}(k)$ nor $\mathbf{y}(k)$ are generated through the filter, the past samples from channels x and y contained in the filter memory do not depend on the filter coefficients (similar to the equation error model for feedback filters [190]). With this in mind, coefficient updates of the dual channel real LMS (DCRLMS) algorithm are calculated similarly to the standard LMS and are given by[2]

$$\begin{aligned}
\mathbf{a}(k+1) &= \mathbf{a}(k) + \mu e_x(k)\mathbf{x}(k) \\
\mathbf{b}(k+1) &= \mathbf{b}(k) + \mu e_x(k)\mathbf{y}(k) \\
\mathbf{c}(k+1) &= \mathbf{c}(k) + \mu e_y(k)\mathbf{x}(k) \\
\mathbf{d}(k+1) &= \mathbf{d}(k) + \mu e_y(k)\mathbf{y}(k)
\end{aligned} \tag{14.7}$$

A comparison with a rather simplistic dual univariate approach, shown in Figure 14.2(a) (see also Chapter 6) and described by

$$\begin{aligned}
\hat{x}(k) &= \mathbf{w}_x^{\mathrm{T}}(k)\mathbf{x}(k) & \hat{y}(k) &= \mathbf{w}_y^{\mathrm{T}}(k)\mathbf{y}(k) \\
e_x(k) &= d_x(k) - \hat{x}(k) & e_y(k) &= d_y(k) - \hat{y}(k)
\end{aligned}$$
$$\begin{aligned}
\mathbf{w}_x(k+1) &= \mathbf{w}_x(k) + \mu e_x(k)\mathbf{x}(k) \\
\mathbf{w}_y(k+1) &= \mathbf{w}_y(k) + \mu e_y(k)\mathbf{y}(k)
\end{aligned} \tag{14.8}$$

---

[1] Stepsizes $\mu_a, \mu_b, \mu_c, \mu_d$ correspond to the updates of the coefficient vectors $\mathbf{a}(k), \mathbf{b}(k), \mathbf{c}(k), \mathbf{d}(k)$.

[2] For convenience, $\mu_a = \mu_b = \mu_c = \mu_d = \mu$. Alternatively, we can have different learning rates associated with errors $e_x(k)$ and $e_y(k)$, that is $\mu_x = \mu_a = \mu_b$ and $\mu_y = \mu_c = \mu_d$.

shows that, similarly to complex adaptive filters,[3] the dual channel real adaptive filter from Figure 14.1 combines the information from both the x and y channels in order to produce filter output, whereas the dual univariate adaptive filter processes the real and imaginary channel independently.

The dual univariate LMS algorithm from (14.8) therefore performs adequately only if the real and imaginary parts of a complex process are not correlated (see Chapter 12).

14.2 Duality Between Real and Complex Valued Filters

To compare the operation of adaptive filters in $\mathbb{R}^2$ and $\mathbb{C}$, we shall first rewrite the mathematical description of the standard complex valued adaptive filter and widely linear complex valued adaptive filter so as to have a form similar to that of DCRAF from Equation (14.3). This is then followed by a comparison of the corresponding learning algorithms.

14.2.1 Operation of Standard Complex Adaptive Filters

The output of a complex adaptive FIR filter is given by

$$\hat{z}(k) = \mathbf{h}^{\mathrm{T}}(k)\mathbf{z}(k), \quad \mathbf{z}(k) = [z(k-1), \ldots, z(k-L)]^{\mathrm{T}} = \mathbf{x}(k) + \jmath\mathbf{y}(k) \tag{14.9}$$

or in an expanded form as

$$\hat{x}(k) + \jmath\hat{y}(k) = \left(\mathbf{h}_r^{\mathrm{T}}(k) + \jmath\mathbf{h}_i^{\mathrm{T}}(k)\right)\left(\mathbf{x}(k) + \jmath\mathbf{y}(k)\right)$$

or

$$\begin{aligned} \hat{x}(k) &= \mathbf{h}_r^{\mathrm{T}}(k)\mathbf{x}(k) - \mathbf{h}_i^{\mathrm{T}}(k)\mathbf{y}(k) \\ \hat{y}(k) &= \mathbf{h}_i^{\mathrm{T}}(k)\mathbf{x}(k) + \mathbf{h}_r^{\mathrm{T}}(k)\mathbf{y}(k) \end{aligned} \tag{14.10}$$

where $\mathbf{h} \in \mathbb{C}^{L\times 1}$ is a column vector of filter coefficients. A comparison with the DCRAF from Figure 14.1 and Equation (14.3) shows that the outputs of the two filters are equivalent for

$$\begin{aligned} \mathbf{a}(k) &= \mathbf{h}_r(k) \qquad & \mathbf{b}(k) &= -\mathbf{h}_i(k) \\ \mathbf{c}(k) &= \mathbf{h}_i(k) \qquad & \mathbf{d}(k) &= \mathbf{h}_r(k) \end{aligned} \tag{14.11}$$

that is, for fixed coefficient vectors, the standard complex valued filter can be considered a constrained version of the dual channel real filter.

In order to compare the dynamics of the corresponding weight updates, we shall cast the standard complex least mean square algorithm, given by [307]

$$\mathbf{h}(k+1) = \mathbf{h}(k) + \mu e(k)\mathbf{z}^*(k) \tag{14.12}$$

[3]The widely linear adaptive filter is shown in Figure 14.2(b), it simplifies into a standard complex valued filter for $\mathbf{g} = \mathbf{0}$, see also Chapter 12.

into the same form as the DCRLMS (Equation 14.7), that is (for convenience denote $e(k) = e_r(k) + \jmath e_i(k) = e_x(k) + \jmath e_y(k)$)

$$\mathbf{h}_r(k+1) = \mathbf{h}_r(k) + \mu\left[e_x(k)\mathbf{x}(k) + e_y(k)\mathbf{y}(k)\right]$$
$$\mathbf{h}_i(k+1) = \mathbf{h}_i(k) + \mu\left[e_y(k)\mathbf{x}(k) - e_x(k)\mathbf{y}(k)\right] \tag{14.13}$$

A comparison between Equations (14.13) and (14.7) shows that, unlike the DCRLMS, the real and imaginary parts of the filter coefficient vector ($\mathbf{h}_r(k)$ and $\mathbf{h}_i(k)$) are updated based on both the errors from the x and y channels and the tap inputs $\mathbf{x}(k)$ and $\mathbf{y}(k)$. The complex least mean square and the dual channel real least mean square are therefore, in general, totally different [322]. However, for a doubly white circular input, the complex correlation matrix is equal to the sum of correlation matrices of the real and imaginary part of the complex input. Hence, its eigenvalues are twice the eigenvalues of the correlation matrix of the real (or imaginary) part and the CLMS and DCRLMS converge to the same solution, with the CLMS converging twice faster.

14.2.2 Operation of Widely Linear Complex Filters

Widely linear estimators are introduced with the aim to utilise complete second-order statistical information within the signal in hand (see Chapter 13). To achieve this, a widely linear complex adaptive filter (WLCAF) from Figure 14.2(b) should be linear in both $\mathbf{z}$ and $\mathbf{z}^*$, that is

$$\hat{z}(k) = \mathbf{h}^{\mathrm{T}}(k)\mathbf{z}(k) + \mathbf{g}^{\mathrm{T}}(k)\mathbf{z}^*(k) = \mathbf{q}^{\mathrm{T}}(k)\mathbf{z}^a(k) \tag{14.14}$$

where the so-called augmented input vector is given by

$$\mathbf{z}^a(k) = \left[z(k-1), \ldots, z(k-L), z^*(k-1), \ldots, z^*(k-L)\right]^{\mathrm{T}} = \left[\mathbf{z}^{\mathrm{T}}(k), \mathbf{z}^{\mathrm{H}}(k)\right]^{\mathrm{T}} \tag{14.15}$$

whereas $\mathbf{h}(k) \in \mathbb{C}^{L\times 1}$ and $\mathbf{g}(k) \in \mathbb{C}^{L\times 1}$ are vectors of filter coefficients, and $\mathbf{q}(k) = \left[\mathbf{h}^T(k), \mathbf{g}^T(k)\right]^{\mathrm{T}}$.

To compare the outputs of the widely linear complex filter and the dual channel real adaptive filter, rewrite Equation (13.8) to have the same form as Equation (14.3), to give

$$\hat{x}(k) = \big(\underbrace{\mathbf{h}_r(k) + \mathbf{g}_r(k)}_{\mathbf{a}(k)}\big)^{\mathrm{T}}\mathbf{x}(k) + \big(\underbrace{\mathbf{g}_i(k) - \mathbf{h}_i(k)}_{\mathbf{b}(k)}\big)^{\mathrm{T}}\mathbf{y}(k)$$
$$\hat{y}(k) = \big(\underbrace{\mathbf{h}_i(k) + \mathbf{g}_i(k)}_{\mathbf{c}(k)}\big)^{\mathrm{T}}\mathbf{x}(k) + \big(\underbrace{\mathbf{h}_r(k) - \mathbf{g}_r(k)}_{\mathbf{d}(k)}\big)^{\mathrm{T}}\mathbf{y}(k) \tag{14.16}$$

that is, the conditions for the outputs of the corresponding filters to be identical.

The weight updates for the ACLMS algorithm (see Chapter 13) are given by

$$\mathbf{h}(k+1) = \mathbf{h}(k) + \mu_h e(k)\mathbf{z}^*(k)$$
$$\mathbf{g}(k+1) = \mathbf{g}(k) + \mu_g e(k)\mathbf{z}(k) \tag{14.17}$$

To compare the updates of the augmented complex least mean square and the dual channel real least mean square algorithms, rewrite the output error of the widely linear filter $e(k) = z(k) - \hat{z}(k) = e_r(k) + \jmath e_i(k) = e_x(k) + \jmath e_y(k)$ as

$$\begin{aligned} e_x(k) &= e_r(k) = x(k) - \hat{x}(k) \\ e_y(k) &= e_i(k) = y(k) - \hat{y}(k) \end{aligned} \tag{14.18}$$

For $\mu_h = \mu_g = \mu$, the ACLMS update can be cast into the same form as the updates for DCRLMS (Equation 14.7), to give

$$\begin{aligned} \mathbf{h}_r(k+1) &= \mathbf{h}_r(k) + \mu\left[e_x(k)\mathbf{x}(k) + e_y(k)\mathbf{y}(k)\right] \\ \mathbf{h}_i(k+1) &= \mathbf{h}_i(k) + \mu\left[e_y(k)\mathbf{x}(k) - e_x(k)\mathbf{y}(k)\right] \\ \mathbf{g}_r(k+1) &= \mathbf{g}_r(k) + \mu\left[e_x(k)\mathbf{x}(k) - e_y(k)\mathbf{y}(k)\right] \\ \mathbf{g}_i(k+1) &= \mathbf{g}_i(k) + \mu\left[e_y(k)\mathbf{x}(k) + e_x(k)\mathbf{y}(k)\right] \end{aligned} \tag{14.19}$$

From Equation (14.16) and (14.7), the corresponding weight updates

$$\begin{aligned} \Delta\mathbf{a}(k) &= \frac{1}{2}\big(\Delta\mathbf{h}_r(k) + \Delta\mathbf{g}_r(k)\big) \qquad \Delta\mathbf{b}(k) = \frac{1}{2}\big(\Delta\mathbf{g}_i(k) - \Delta\mathbf{h}_i(k)\big) \\ \Delta\mathbf{c}(k) &= \frac{1}{2}\big(\Delta\mathbf{h}_i(k) + \Delta\mathbf{g}_i(k)\big) \qquad \Delta\mathbf{d}(k) = \frac{1}{2}\big(\Delta\mathbf{h}_r(k) - \Delta\mathbf{g}_r(k)\big) \end{aligned} \tag{14.20}$$

The dual channel real adaptive filter and the widely linear complex adaptive filter, trained with the corresponding learning algorithms DCRLMS and ACLMS, are therefore equivalent when the stepsize of the DCRLMS is twice the stepsize of ACLMS. From Equations (5.20) and (5.26–5.29), for the correlation matrix of the bivariate real input

$$\mathrm{R}_{\omega\omega} - \lambda^{\omega}\mathrm{I} = \frac{1}{4}\mathrm{A}^H[\mathcal{C}^a - 2\lambda^{\omega}\mathrm{I}]\mathrm{A} = 0$$

that is, the eigenvalues of the augmented complex covariance matrix are twice the eigenvalues of the bivariate real correlation matrix. This is then reflected in the modes of convergence and stability bounds, and the ACLMS converges twice as fast as the DCRLMS for the same learning rate.

14.3 Simulations

To verify the performance of the Dual Univariate LMS (DULMS), Dual Channel Real LMS (DCRLMS), Complex LMS (CLMS) and Augmented CLMS (ACLMS), in all the experiments, the order of the adaptive FIR filter was chosen to be $L = 4$, and the learning rate was set to $\mu = 0.05$. The measurement used to assess the performance was the standard prediction gain (6.80). Simulations were undertaken by averaging independent trials on prediction of:

- Stochastic complex valued coloured AR(4) process, given in Equation (13.44), driven by
  - circular white Gaussian noise (CWGN) characterised by a constant power spectrum and vanishing spectral pseudocovariance function, that is, $\Gamma_z(\nu) = const$ and $R_z(\nu) = 0$, generated as in Section 12.1.1;
  - doubly white noise (DWN), given in Equation (13.13);

Table 14.1 Comparison of prediction gains R_p [dB] for the various classes of signals

Algorithm	AR4 (DWN)	AR4 (CWGN)	Ikeda	Lorenz	Wind-medium
DULMS	5.8518	5.5666	−0.3910	18.2365	6.5284
DCRLMS	5.8423	5.5588	3.9733	21.1833	7.1528
CLMS	6.6380	6.2664	2.4278	20.7674	7.4499
ACLMS	6.6096	6.2465	4.0330	23.2565	7.9914

- Synthetic benchmark Ikeda process, given in Equation (13.46), and synthetic Lorenz chaotic series (x and y dimension), given in Equation (13.45), with the initial values $x(0) = 5$, $y(0) = 5$ and $z(0) = 20$;
- Single trial two-dimensional real world wind signal described in Figure 13.5.

Table 14.1 summarises the prediction gains for DCRLMS, DULMS, CLMS and ACLMS, and the above classes of signals. As expected, the dual univariate least mean square (DULMS) algorithm had performance similar to those of other algorithms under consideration only for complex valued AR(4) signals driven by a doubly white and circular Gaussian noise. The complex least mean square algorithm had similar performance to that of the dual channel real least mean square for most of the test signals, whereas the augmented complex least mean square algorithm had the best performance. However, the performances of DCRLMS and ACLMS were identical for $\mu_{\mathrm{DCRLMS}} = 2\mu_{\mathrm{ACLMS}}$.

Figure 14.3 further illustrates the relationship between the prediction gain and the learning rates associated with the x and y channel within the DCRLMS algorithm. Parameter $\mu_x = \mu_a = \mu_b$ was the learning rate used in the updates of coefficient vectors $\mathbf{a}(k)$ and $\mathbf{b}(k)$, whereas $\mu_y = \mu_c = \mu_d$ was the learning rate used in the updates of coefficient vectors $\mathbf{c}(k)$ and $\mathbf{d}(k)$ (see also Footnote 2).

Figure 14.4 illustrates the relationship between the prediction gain and the learning rates μ_h and μ_g associated with the updates of coefficient vectors $\mathbf{h}(k)$ and $\mathbf{g}(k)$ within the ACLMS algorithm (Equation 14.17).

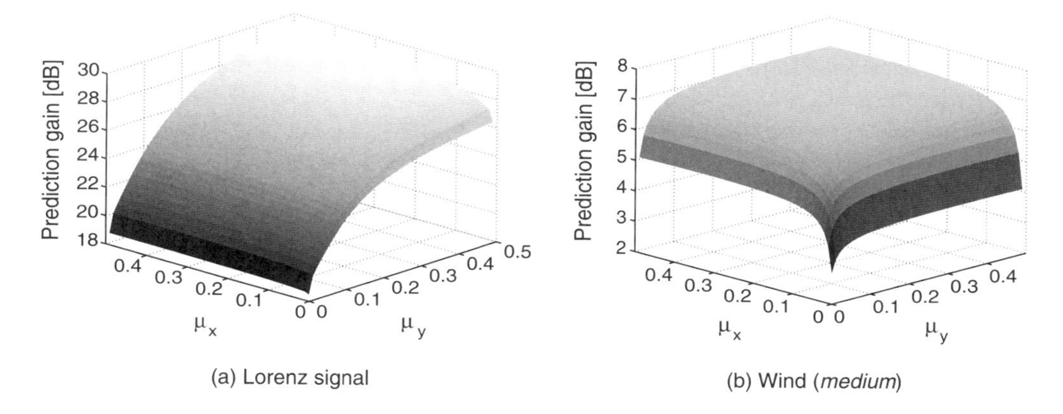

Figure 14.3 Relationship between the prediction gain and learning rates $\mu_x = \mu_a = \mu_b$ and $\mu_y = \mu_c = \mu_d$ within DCRLMS (14.7) for prediction of noncircular Lorenz and wind signals

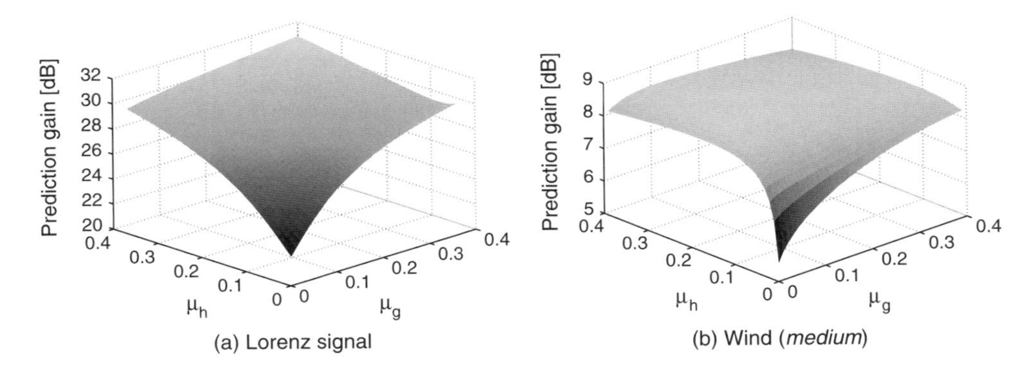

Figure 14.4 Relationship between the prediction gain and learning rates μ_h and μ_g of the updates of coefficient vectors **h** and **g** within ACLMS (14.17) for prediction of noncircular Lorenz and wind signals

A comparison between Figures 14.3(a) and 14.4(a) shows that for DCRLMS, variations in the learning rate μ_x had little effect on the performance, whereas for ACLMS the performance improved with an increase in both the learning rates μ_h and μ_g. The ACLMS is therefore better suited to revealing any intricate properties of the signal generating mechanism in $\mathbb{C}$ (complex circularity).

To summarise:

- The dual univariate mode of operation (Figure 14.2a) is simple but only effective when the real and imaginary channel of a complex process are uncorrelated or weakly correlated (see Table 14.1).
- The updates for the real and imaginary parts of the weight vector within CLMS are based on both the real and imaginary errors and the real and imaginary parts of the input, thus taking into account the statistical dependence between the real and imaginary channels.
- The CLMS performs at least as well as DCRLMS for circular complex signals, whereas for noncircular processes it is suboptimal (in which case its widely linear extension – the ACLMS – is used).
- The dual channel real adaptive filter from Figure (14.1) is effectively the same filter as the widely linear complex adaptive filter (Figure 14.2b); the corresponding learning algorithms DCRLMS and ACLMS are also equivalent when the stepsize of DCRLMS is twice the stepsize of ACLMS (14.20).
- The updates for the filter coefficient vectors $\mathbf{h}_r(k)$, $\mathbf{h}_i(k)$, $\mathbf{g}_r(k)$, $\mathbf{g}_i(k)$ within ACLMS use the information from both the real and imaginary channel, that is $e_x(k)$, $e_y(k)$, $\mathbf{x}(k)$, $\mathbf{y}(k)$ (Equation 14.19). This means that the statistical dependence between the real and imaginary channel is taken into account, and the physics of the underlying processes (circularity) is catered for naturally. This gives the ACLMS potential advantages over DCRLMS, especially in noisy environments and when using individual learning rates for the coefficient vectors.
- Although it is possible to establish the duality between linear adaptive filters in the real and complex domain (DCRAF and WLCAF), the natures of nonlinearity in $\mathbb{R}^2$ and $\mathbb{C}$ are fundamentally different (for instance the only continuously differentiable function in $\mathbb{C}$ is a constant), and so complex nonlinear filters based on augmented statistics are a natural choice for the processing of the generality of complex valued signals.

15

Widely Linear Filters with Feedback

This chapter introduces 'augmented' algorithms for the training of adaptive filters with feedback, based on the widely linear autoregressive moving average (WL-ARMA) model and the augmented complex statistics discussed in Chapter 12. The algorithms are based on $\mathbb{CR}$ calculus and are derived along the lines of the Augmented Complex Least Mean Square (ACLMS) algorithm presented in Chapter 13. Two classes of such algorithms are introduced:

- *Algorithms based on a direct gradient calculation.* The stochastic gradient algorithm for training adaptive IIR filters and Complex Real Time Recurrent Learning (CRTRL) algorithm for recurrent neural networks (RNNs), derived in Chapter 7, are now reintroduced in the framework of augmented complex statistics. Since the augmented CRTRL (ACRTRL) algorithm is a direct gradient algorithm for nonlinear feedback architectures, it simplifies into an augmented recursive algorithm for the training of infinite impulse response (IIR) adaptive filters when the neurons within complex RNNs are linear; it further degenerates into ACLMS when there is no feedback present.
- *Algorithms based on nonlinear sequential state estimation.* Both the augmented complex valued Kalman filter (ACKF) and the augmented complex valued extended Kalman filter (ACEKF) algorithms are derived for the training of complex valued RNNs. The ACRTRL algorithm is used to compute the Jacobian matrix within the update of augmented Kalman filter based algorithms [95]. The 'unscented' versions of Kalman filtering algorithms (UKF) for complex RNNs are then presented in order to provide better approximations of the higher-order terms, which are neglected due to the linearisation within the extended Kalman filter (EKF); the UKF algorithms have been shown to be accurate, at least up to the second statistical moment [139, 301] (compared with first-order for the EKF).

For generality, the analysis is conducted for fully complex nonlinear activation functions of neurons. Similarly to the results in Chapter 13, it is shown that algorithms within this class offer improved performance over standard algorithms for noncircular complex signals, whereas for

Complex Valued Nonlinear Adaptive Filters: Noncircularity, Widely Linear and Neural Models
Danilo P. Mandic and Vanessa Su Lee Goh

circular signals their performances are similar to those of standard algorithms. For more detail on ARMA modelling of widely linear systems we refer to [217].

15.1 The Widely Linear ARMA (WL-ARMA) Model

The widely linear ARMA (WL-ARMA) and widely linear nonlinear ARMA (WL-NARMA) model are respectively given by[1]

$$\begin{aligned} y(k) &= \sum_{n=1}^{N} a_n y(k-n) + \sum_{m=1}^{M} b_m x(k-m) + \sum_{n=1}^{N} \alpha_n y^*(k-n) \\ &\quad + \sum_{m=1}^{M} \beta_m x^*(k-m) + b_0 x(k) + \beta_0 x^*(k) \end{aligned} \tag{15.1}$$

$$\begin{aligned} y(k) &= \Phi\left(\sum_{n=1}^{N} a_n y(k-n) + \sum_{m=1}^{M} b_m x(k-m) + \sum_{n=1}^{N} \alpha_n y^*(k-n)\right. \\ &\quad \left. + \sum_{m=1}^{M} \beta_m x^*(k-m)\right) + b_0 x(k) + \beta_0 x^*(k) \end{aligned}$$

These models are made widely linear to include the augmented complex input and feedback vectors given by

$$\begin{aligned} \mathbf{x}^a(k) &= [x(k-1), \ldots, x(k-M), x^*(k-1), \ldots, x^*(k-M)]^{\mathrm{T}} \\ \mathbf{y}^a(k) &= [y(k-1), \ldots, y(k-N), y^*(k-1), \ldots, y^*(k-N)]^{\mathrm{T}} \end{aligned} \tag{15.2}$$

together with the 'driving' input $x^a(k) = [x(k), x^*(k)]^{\mathrm{T}}$ and the corresponding fixed filter coefficients, all complex valued. This allows us to cater for second-order noncircular (improper) signals and serves as a basis for the development of augmented recursive algorithms for linear and nonlinear complex valued adaptive filters with feedback.

15.2 Widely Linear Adaptive Filters with Feedback

The coefficients of the widely linear models with memory in Equation (15.1) can be made adaptive, to give the input–output expressions for the corresponding linear and nonlinear

[1]The term *widely linear* refers to 'widely linear in the parameters'. Hence within the WL-NARMA model, the nonlinearity $\Phi(\cdot)$ is applied to the WL-ARMA term, similarly to the standard NARMA models.

adaptive filters[2] in the form

$$y(k) = \sum_{n=1}^{N} a_n(k)y(k-n) + \sum_{m=1}^{M} b_m(k)x(k-m) + \sum_{n=1}^{N} \alpha_n(k)y^*(k-n) + \sum_{m=1}^{M} \beta_m(k)x^*(k-m) \tag{15.3}$$

$$y(k) = \Phi\left(\sum_{n=1}^{N} a_n(k)y(k-n) + \sum_{m=1}^{M} b_m(k)x(k-m) + \sum_{n=1}^{N} \alpha_n(k)y^*(k-n) + \sum_{m=1}^{M} \beta_m(k)x^*(k-m)\right)$$

Adaptive filters with feedback will be analysed in the prediction setting, thus, the terms in Equation (15.1) associated with $m = 0$ are dropped. This does not influence the generality of the results as, if required, these terms can be straightforwardly included in the weight update equations.

Denote the overall input vector to an augmented feedback filter by

$$\mathbf{u}(k) = \big[x(k-1), \ldots, x(k-M), y(k-1), \ldots, y(k-N), x^*(k-1), \ldots, x^*(k-M), y^*(k-1), \ldots, y^*(k-N)\big]^{\mathrm{T}} \tag{15.4}$$

and the overall weight vector by

$$\mathbf{w}(k) = \big[b_1(k), \ldots, b_M(k), a_1(k), \ldots, a_N(k), \beta_1(k), \ldots, \beta_M(k), \alpha_1(k), \ldots, \alpha_N(k)\big]^{\mathrm{T}} \tag{15.5}$$

The output error for an IIR filter and a recurrent perceptron (NARMA filter, see Section 7.2) are given by

$$\begin{aligned} \text{Augmented IIR Filter:} \quad & e(k) = d(k) - y(k) = d(k) - \mathbf{u}^{\mathrm{T}}(k)\mathbf{w}(k) \\ \text{Recurrent Perceptron:} \quad & e(k) = d(k) - y(k) = d(k) - \Phi\left(\mathbf{u}^{\mathrm{T}}(k)\mathbf{w}(k)\right) \end{aligned} \tag{15.6}$$

where $d(k)$ is the teaching signal, and $\Phi(k)$ a nonlinear activation function, typically an elementary transcendental function given in Chapter 4.

Based on the standard cost function

$$J(k) = \frac{1}{2}|e(k)|^2 = \frac{1}{2}e(k)e^*(k) = \frac{1}{2}\left[e_r^2 + e_i^2\right] \tag{15.7}$$

the filter weights are updated based on the stochastic gradient, that is

$$\mathbf{w}(k+1) = \mathbf{w}(k) - \mu\nabla_{\mathbf{w}}J(k) \tag{15.8}$$

[2]For convenience, the bias terms within the recurrent perceptron architecture $(1 + \jmath)$ and $(1 - \jmath)$ are omitted. These are a part of the external input vector and can be included by adding one more term in $\mathbf{x}^a(k)$ in Equation (15.2).

where μ is the learning rate. We aim at minimising the error power – thus the cost function is a real function of real variables and does not admit minimisation directly in $\mathbb{C}$, as it is not $\mathbb{C}$-differentiable. Chapters 5 and 7 show that we can optimise these functions based on $\mathbb{CR}$ calculus and the use of $\mathbb{R}$-derivatives, given below

$$\begin{aligned} &\mathbb{R}\text{- derivative} \quad \frac{1}{2}\left[\frac{\partial f}{\partial x} - \jmath\frac{\partial f}{\partial y}\right] \\ &\mathbb{R}^*\text{- derivative} \quad \frac{1}{2}\left[\frac{\partial f}{\partial x} + \jmath\frac{\partial f}{\partial y}\right] \end{aligned} \tag{15.9}$$

Recall that for holomorphic functions the $\mathbb{R}^*$-derivative vanishes and the $\mathbb{R}$-derivative is equivalent to the standard complex derivative $f'(z)$.

We shall now highlight some differences between the analysis of standard feedback filters and widely linear feedback filters. These arise mainly due to the additional complex conjugate terms within the input–output models (Equation 15.3), which makes their outputs not differentiable directly in $\mathbb{C}$. In terms of complex differentiability, observe that:

- The cost function (Equation 15.7) is a real function of complex variables, and its gradient is calculated using the $\mathbb{R}^*$-derivative, as

$$\nabla_{\mathbf{w}} J(k) = \nabla_{\mathbf{w}^r} J(k) + \jmath\nabla_{\mathbf{w}^i} J(k) \tag{15.10}$$

- The output $y(k)$ is a complex function of complex variables $\{x, x^*, y, y^*\}$, and is thus not $\mathbb{C}$-differentiable – its derivatives will therefore be computed based on $\mathbb{R}^*$-derivatives;
- The nonlinear activation function $\Phi(\cdot)$ is a complex function of complex variable, it has $\mathbb{C}$-derivatives, and the Cauchy–Riemann equations hold. In addition, for the class of elementary transcendental functions

$$\Phi^*(z) = \Phi(z^*) \qquad \text{and} \qquad \left(\Phi^*\right)' = \left(\Phi'\right)^* = \Phi'^* \tag{15.11}$$

The gradient of the cost function $\nabla_{\mathbf{w}} J(k)$ can be therefore written as

$$\nabla_{\mathbf{w}} J(k) = -e^*(k)\boldsymbol{\pi}^\circ(k) - e(k)\boldsymbol{\pi}^\star(k)$$

where

$$\boldsymbol{\pi}^\circ(k) = \frac{1}{2}\left[\frac{\partial y(k)}{\partial \mathbf{w}^r(k)} + \jmath\frac{\partial y(k)}{\partial \mathbf{w}^i(k)}\right] \qquad \boldsymbol{\pi}^\star(k) = \frac{1}{2}\left[\frac{\partial y^*(k)}{\partial \mathbf{w}^r(k)} + \jmath\frac{\partial y^*(k)}{\partial \mathbf{w}^i(k)}\right] \tag{15.12}$$

that is, the terms $\partial y/\partial w$ and $\partial y^*/\partial w$ within the gradient are calculated using the $\mathbb{R}^*$-derivatives.

15.2.1 Widely Linear Adaptive IIR Filters

It has been shown in Section 7.1.1 that the sensitivities $\boldsymbol{\pi}^{\circ}(k)$ within the update of standard IIR adaptive filter vanish, whereas the sensitivities $\boldsymbol{\pi}^{\star}(k)$, are calculated based on

$$y^*(k) = \sum_{n=1}^{N} a_n^*(k) y^*(k-n) + \sum_{m=1}^{M} b_m^*(k) x^*(k-m) + \sum_{n=1}^{N} \alpha_n^*(k) y(k-n) + \sum_{m=1}^{M} \beta_m^*(k) x(k-m) \tag{15.13}$$

However, as the output $y(k)$ for widely linear IIR filters is not $\mathbb{C}$-differentiable, both the terms $\boldsymbol{\pi}^{\circ}(k)$ and $\boldsymbol{\pi}^{\star}(k)$ need to be evaluated again.

Calculation of sensitivities $\boldsymbol{\pi}^{\circ}$ (k) and $\boldsymbol{\pi}^{\star}$ (k). Consider, for instance, two particular cases

$$\pi^{\circ}_{a_n}(k) = \frac{1}{2}\left[\frac{\partial y(k)}{\partial a_n^r(k)} + \jmath \frac{\partial y(k)}{\partial a_n^i(k)}\right] \text{ and } \pi^{\circ}_{\beta_m}(k) = \frac{1}{2}\left[\frac{\partial y(k)}{\partial \beta_m^r(k)} + \jmath \frac{\partial y(k)}{\partial \beta_m^i(k)}\right].$$

These can be evaluated as

$$\begin{aligned}
\frac{\partial y(k)}{\partial a_n^r(k)} &= y(k-n) + \sum_{l=1}^{N} a_l(k) \frac{\partial y(k-l)}{\partial a_n^r(k)} + \sum_{l=1}^{N} \alpha_l(k) \frac{\partial y^*(k-l)}{\partial a_n^r(k)} \\
\frac{\partial y(k)}{\partial a_n^i(k)} &= \jmath y(k-n) + \sum_{l=1}^{N} a_l(k) \frac{\partial y(k-l)}{\partial a_n^i(k)} + \sum_{l=1}^{N} \alpha_l(k) \frac{\partial y^*(k-l)}{\partial a_n^i(k)} \\
\frac{\partial y(k)}{\partial \beta_m^r(k)} &= x^*(k-m) + \sum_{l=1}^{N} a_l(k) \frac{\partial y(k-l)}{\partial \beta_m^r(k)} + \sum_{l=1}^{N} \alpha_l(k) \frac{\partial y^*(k-l)}{\partial \beta_m^r(k)} \\
\frac{\partial y(k)}{\partial \beta_m^i(k)} &= \jmath x^*(k-m) + \sum_{l=1}^{N} a_l(k) \frac{\partial y(k-l)}{\partial \beta_m^i(k)} + \sum_{l=1}^{N} \alpha_l(k) \frac{\partial y^*(k-l)}{\partial \beta_m^i(k)}
\end{aligned} \tag{15.14}$$

Using the assumption, also used in standard adaptive IIR filters, that for small μ, for every element of $\mathbf{w}(k)$ in (15.5) we have

$$\mathbf{w}(k-1) \approx \mathbf{w}(k-2) \approx \cdots \approx \mathbf{w}(k-M) \approx \mathbf{w}(k-N) \tag{15.15}$$

This admits a recursive computation of the sensitivities in the form

$$\begin{aligned}\pi^{\circ}_{a_n}(k) &= \frac{1}{2}\left[\sum_{l=1}^{N} a_l(k)\left(\frac{\partial y(k-l)}{\partial a^r_n(k-l)} + \jmath\frac{\partial y(k-l)}{\partial a^i_n(k-l)}\right)\right.\\ &\quad \left. + \sum_{l=1}^{N} \alpha_l(k)\left(\frac{\partial y^*(k-l)}{\partial a^r_n(k-l)} + \jmath\frac{\partial y^*(k-l)}{\partial a^i_n(k-l)}\right)\right]\\ &= \sum_{l=1}^{N} a_n(k)\pi^{\circ}_{a_n}(k-l) + \sum_{l=1}^{N} \alpha_n(k)\pi^{\star}_{\alpha_n}(k-l)\\ \pi^{\circ}_{\beta_m}(k) &= \frac{1}{2}\left[\sum_{l=1}^{N} a_l(k)\left(\frac{\partial y(k-l)}{\partial \beta^r_m(k-l)} + \jmath\frac{\partial y(k-l)}{\partial \beta^i_m(k-l)}\right)\right.\\ &\quad \left. + \sum_{l=1}^{N} \alpha_l(k)\left(\frac{\partial y^*(k-l)}{\partial \beta^r_m(k-l)} + \jmath\frac{\partial y^*(k-l)}{\partial \beta^i_m(k-l)}\right)\right]\\ &= \sum_{l=1}^{N} a_l(k)\pi^{\circ}_{\beta_m}(k-l) + \sum_{l=1}^{N} \alpha_l(k)\pi^{\star}_{\beta_m}(k-l)\end{aligned} \tag{15.16}$$

Thus, for every element $w_q(k) \in \mathbf{w}(k),\ q = 1, \ldots, 2M + 2N$ in (15.5), we can write

$$\pi^{\circ}_{w_q}(k) = \sum_{l=1}^{N} a_l(k)\pi^{\circ}_{w_q}(k-l) + \sum_{l=1}^{N} \alpha_l(k)\pi^{\star}_{w_q}(k-l) \tag{15.17}$$

Similarly, for the $\pi^{\star}_{w_q}$ terms we have

$$\pi^{\star}_{w_q}(k) = u^*(k-q) + \sum_{l=1}^{N} a^*_l(k)\pi^{\star}_{w_q}(k-l) + \sum_{l=1}^{N} \alpha^*_l(k)\pi^{\circ}_{w_q}(k-l) \tag{15.18}$$

Although the updates for the terms $\pi^{\circ}_{w_q}$ represent unforced difference equations, they are also coupled with the terms $\pi^{\star}_{w_q}$ and hence do not vanish, unlike in the case of the standard complex valued adaptive IIR filters. Expressions (15.17) and (15.18), together with (15.7) and (15.12) complete the derivation of a recursive algorithm for the update of augmented complex valued adaptive IIR filters.

15.2.2 Augmented Recurrent Perceptron Learning Rule

For a recurrent perceptron (see Figure 7.2b), the sensitivities of the output to the filter weights can be calculated based on the corresponding sensitivities for the augmented IIR filter in Equations (15.17) and (15.18). The output of a recurrent perceptron is $y(k) = \Phi(net(k))$, that is

$$y(k) = \Phi\left(\sum_{n=1}^{N} a_n(k)y(k-n) + \sum_{m=1}^{M} b_m(k)x(k-m) + \sum_{n=1}^{N} \alpha_n(k)y^*(k-n) + \sum_{m=1}^{M} \beta_m(k)x^*(k-m)\right) \tag{15.19}$$

where $net(k)$ has the same form as the output of the augmented IIR filter[3]. Function $\Phi(\cdot)$ is typically a fully complex activation function, and is $\mathbb{C}$-differentiable, whereas the net input $net(k)$ is a function of both x, y and x^*, y^* and has only $\mathbb{R}^*$-derivatives. We will consider the class of nonlinearities which obeys the properties (Equation 15.11). The sensitivities within an adaptive steepest gradient algorithm for the adaptation of a recursive perceptron thus become

$$\begin{aligned}
\pi^{\circ}_{w_q}(k) &= \Phi'\big(net(k)\big)\left(\sum_{l=1}^{N} a_l(k)\pi^{\circ}_{w_q}(k-l) + \sum_{l=1}^{N} \alpha_l(k)\pi^{\star}_{w_q}(k-l)\right)\\
\pi^{\star}_{w_q}(k) &= \Phi'^*\big(net(k)\big)\left(u^*(k-q) + \sum_{l=1}^{N} a^*_l(k)\pi^{\star}_{w_q}(k-l) + \sum_{l=1}^{N} \alpha^*_l(k)\pi^{\circ}_{w_q}(k-l)\right)
\end{aligned} \tag{15.20}$$

15.3 The Augmented Complex Valued RTRL (ACRTRL) Algorithm

The overall 'augmented' input to a Recurrent Neural Network (RNN) $\mathbf{I}^a(k)$ (see Figure 7.3 in Chapter 7) can be expressed as

$$\begin{aligned}
\mathbf{I}^a(k) &= [\mathbf{I}^{\mathrm{T}}(k), \mathbf{I}^{\mathrm{H}}(k)]^{\mathrm{T}}\\
&= [x(k-1), \ldots, x(k-M), 1+\jmath, y_1(k-1), \ldots, y_N(k-1),\\
&\qquad x^*(k-1), \ldots, x^*(k-M), 1-\jmath, y_1^*(k-1), \ldots, y_N^*(k-1)]^{\mathrm{T}}
\end{aligned} \tag{15.21}$$

where for every neuron in the network

$$y_n(k) = \Phi\big(net_n(k)\big), \quad n = 1, \ldots, N \tag{15.22}$$

and

$$net_n(k) = \sum_{q=1}^{2(M+N+1)} w_{n,q}(k) I^a_q(k) \tag{15.23}$$

[3] But for a possible bias term.

is the augmented net input to nth neuron at time instant k and the vector of filter weights belonging to the nth neuron

$$\begin{aligned}\mathbf{w}_n(k) &= \big[b_{n,1}(k), \ldots, b_{n,M+1}(k), a_{n,1}(k), \ldots, a_{n,N}(k),\\ &\qquad \beta_{n,1}(k), \ldots, \beta_{n,M+1}(k), \alpha_{n,1}(k), \ldots, \alpha_{n,N}(k)\big]^{\mathrm{T}}\\ &= \big[w_{n,1}, \ldots, w_{n,2(M+N+1)}\big]^{T}\end{aligned} \tag{15.24}$$

It is convenient to combine all the weights in an RNN within a complex valued weight matrix $\mathbf{W}(k) = [\mathbf{w}_1(k), \ldots, \mathbf{w}_N(k)]$.

The error signals are available only for the L output neurons, thus the cost function, weight update, and the gradient of the cost function are given by

$$\begin{aligned}J(k) &= \frac{1}{2}\sum_{l=1}^{L} e(k)e^*(k)\\ \mathbf{w}(k+1) &= \mathbf{w}(k) - \mu\nabla_{\mathbf{w}}J(k)\\ \nabla_{w_{n,q}}J(k) &= \frac{\partial J(k)}{\partial w_{n,q}^r} + \jmath\frac{\partial J(k)}{\partial w_{n,q}^i} = -\frac{1}{2}\sum_{l=1}^{L}\left[e_l(k)\frac{\partial y_l^*(k)}{\partial w_{n,q}(k)} + \frac{\partial y_l(k)}{\partial w_{n,q}(k)}e_l^*(k)\right]\\ &= -\sum_{l=1}^{L} e_l(k)\pi_{w_{n,q}}^{\star l}(k) - \sum_{l=1}^{L} e_l^*(k)\pi_{w_{n,q}}^{\circ l}(k)\end{aligned} \tag{15.25}$$

Based on the sensitivities for recurrent perceptrons, the sensitivities within the augmented CRTRL (ACRTRL) algorithm for the training of recurrent neural networks are given by

$$\begin{aligned}\pi_{w_{n,q}}^{\circ l}(k) &= \Phi_l'\big(net_l(k)\big)\left[\sum_{p=1}^{N} a_{l,p}(k)\pi_{w_{n,q}}^{\circ p}(k-p) + \sum_{p=1}^{N}\alpha_{l,p}(k)\pi_{w_{n,q}}^{\star p}(k-p)\right]\\ \pi_{w_{n,q}}^{\star l}(k) &= \Phi_l'^{*}\big(net_l(k)\big)\left[\sum_{p=1}^{N} a_{l,p}^*(k)\pi_{w_{n,q}}^{\star p}(k-p) + \sum_{p=1}^{N}\alpha_{l,p}^*(k)\pi_{w_{n,q}}^{\circ p}(k-p) + \delta_{nl}I_q^{a^*}(k)\right]\end{aligned} \tag{15.26}$$

where δ_{nl} is the Kronecker delta function and the sensitivities are initialised with zero initial conditions.

15.4 The Augmented Kalman Filter Algorithm for RNNs

Amongst recursive filters in the domain of second-order statistics, Kalman filters are optimal sequential state estimators for nonstationary signals [112, 141]. They have also been used in several modern applications, including state estimation for car navigation systems [193, 223], parameter estimation for time series modelling [245], and the training of neural networks [112, 139, 246]. To discuss Kalman filter based algorithms for the training of complex valued RNNs, we shall first introduce an augmented state space model and the corresponding updates for the Kalman filter. Similarly to ACLMS and ACRTRL, these updates have the same generic form as the standard updates. Due to the augmentation all the vectors have two times the size and matrices four times the size of the corresponding vectors and matrices within the standard algorithm.

Consider a general state space model, given by [112]

$$\mathbf{x}_{k+1} = \mathbf{F}_{k+1}\mathbf{x}_k + \boldsymbol{\omega}_k$$
$$\mathbf{y}_k = \mathbf{H}_k\mathbf{x}_k + \boldsymbol{\nu}_k \tag{15.27}$$

where $\mathbf{x}_k$ are the states to be estimated and $\mathbf{y}_k$ is the system output (usually one or a subset of the states). Variables $\boldsymbol{\omega}_k$ and $\mathbf{v}_k$ are independent, zero mean, complex valued Gaussian noise processes with covariance matrices $\mathbf{Q}_k$ and $\mathbf{R}_k$ respectively, and $\mathbf{F}$ and $\mathbf{H}$ are the transition and measurement matrices. The augmented state space model can be written as

$$\mathbf{x}^a_{k+1} = \mathbf{F}^a_{k+1}\mathbf{x}^a_k + \boldsymbol{\omega}^a_k$$
$$\mathbf{y}^a_k = \mathbf{H}^a_k\mathbf{x}^a_k + \boldsymbol{\nu}^a_k \tag{15.28}$$

where $\mathbf{x}^a_k = \left[\mathbf{x}^{\mathrm{T}}_k, \mathbf{x}^{\mathrm{H}}_k\right]^{\mathrm{T}}, \mathbf{y}^a_k = \left[\mathbf{y}^{\mathrm{T}}_k, \mathbf{y}^{\mathrm{H}}_k\right]^{\mathrm{T}}, \mathbf{F}^a_k = \left[\mathbf{F}_k, \mathbf{F}^*_k\right], \mathbf{H}^a_k = \left[\mathbf{H}_k, \mathbf{H}^*_k\right], \boldsymbol{\omega}^a_k = \left[\boldsymbol{\omega}^{\mathrm{T}}_k, \boldsymbol{\omega}^{\mathrm{H}}_k\right]^{\mathrm{T}}$ and $\boldsymbol{\nu}^a_k = \left[\boldsymbol{\nu}^{\mathrm{T}}_k, \boldsymbol{\nu}^{\mathrm{H}}_k\right]^{\mathrm{T}}$. The augmented equivalents of $\mathbf{Q}_k$ and $\mathbf{R}_k$ are denoted respectively by $\mathbf{Q}^a_k$ and $\mathbf{R}^a_k$.

To initialise the algorithm for the time instant $k = 0$, set

$$\hat{\mathbf{x}}^a_0 = E\left[\mathbf{x}^a_0\right],$$
$$\mathbf{P}_0 = E\left[\left(\mathbf{x}^a_0 - E\left[\mathbf{x}^a_0\right]\right)\left(\mathbf{x}^a_0 - E\left[\mathbf{x}^a_0\right]\right)^{\mathrm{T}}\right] \tag{15.29}$$

The updates within the Kalman filtering algorithms are given below[4]

State estimate propagation:

$$\hat{\mathbf{x}}^{a-}_k = \mathbf{F}^a_{k,k-1}\hat{\mathbf{x}}^{a-}_{k-1} \tag{15.30}$$

Error covariance propagation:

$$\mathbf{P}^-_k = \mathbf{F}^a_{k,k-1}\mathbf{P}_{k-1}(\mathbf{F}^a_{k,k-1})^{\mathrm{H}} + \mathbf{Q}^a_{k-1} \tag{15.31}$$

Kalman gain matrix:

$$\mathbf{G}_k = \mathbf{P}^-_k(\mathbf{H}^a_k)^{\mathrm{H}}\left[\mathbf{H}^a_k\mathbf{P}_k(\mathbf{H}^a_k)^{\mathrm{H}} + \mathbf{R}^a_k\right]^{-1} \tag{15.32}$$

State estimate update:

$$\hat{\mathbf{x}}^a_k = \hat{\mathbf{x}}^{a-}_k + \mathbf{G}_k\left(\mathbf{y}^a_k - \mathbf{H}^a_k\hat{\mathbf{x}}^{a-}_k\right) \tag{15.33}$$

Error covariance update:

$$\mathbf{P}_k = \left(\mathbf{I} - \mathbf{G}_k\mathbf{H}^a_k\right)\mathbf{P}^-_k \tag{15.34}$$

This completes the description of the augmented complex valued Kalman filter (ACKF).

[4] For clarity, we use notation similar to that from [112].

15.4.1 EKF Based Training of Complex RNNs

To establish a mathematical framework for Kalman filter based training of complex RNNs, consider a nonlinear state space model[5]

$$
\begin{aligned}
\mathbf{w}_{k+1}^a &= \mathbf{w}_k^a + \boldsymbol{\omega}_k^a \\
\mathbf{y}_k^a &= \mathbf{h}(\mathbf{w}_k^a, \mathbf{u}_k^a) + \boldsymbol{\nu}_k^a
\end{aligned}
\tag{15.35}
$$

where $\mathbf{h}(\cdot)$ is a nonlinear operator associated with observations, $\mathbf{w}_k^a$ is an augmented weight vector of the network, $\mathbf{u}_k^a$ is the overall input vector to the network, and $\mathbf{y}_k^a$ is the augmented vector of observations. From the first expression in Equation (15.35), the complex weights within RNN are modelled as random walk. The idea behind the Extended Kalman Filter (EKF) is to linearise the state space model (Equation 15.35) locally (for every time instant k) based on a truncated Taylor series expansion around $\mathbf{h}$ [70, 102]. Once such a local linear model is obtained, standard ACKF updates (Equations 15.30–15.34) can be applied.

The Augmented Complex Extended Kalman Filtering Algorithm (ACEKF) for the training of complex RNNs can now be summarised as [95]

$$
\begin{aligned}
\mathbf{G}_k &= \mathbf{P}_k^-(\mathbf{H}_k^a)^{\mathrm{H}}\left[\mathbf{H}_k^a\mathbf{P}_k^-(\mathbf{H}_k^a)^{\mathrm{H}} + \mathbf{R}_k^a\right]^{-1} \\
\hat{\mathbf{w}}_k^a &= \hat{\mathbf{w}}_k^{a-} + \mathbf{G}_k\left[\mathbf{y}_k^a - \mathbf{h}(\hat{\mathbf{w}}_k^{a-}, \mathbf{u}_k^a)\right] \\
\mathbf{P}_k &= \left(\mathbf{I} - \mathbf{G}_k\mathbf{H}_k^a\right)\mathbf{P}_k^- + \mathbf{Q}_k^a
\end{aligned}
\tag{15.36}
$$

and is initialised by

$$
\begin{aligned}
\hat{\mathbf{w}}_0^a &= E\left[\mathbf{w}_0\right] \\
\mathbf{P}_0 &= E\left[\left(\mathbf{w}_0^a - E\left[\mathbf{w}_0^a\right]\right)\left(\mathbf{w}_0^a - E\left[\mathbf{w}_0^a\right]\right)^{\mathrm{T}}\right]
\end{aligned}
\tag{15.37}
$$

The augmented Jacobian[6] matrix $\mathbf{H}_k^a$ of the partial derivatives of $\mathbf{h}$ is computed using the augmented CRTRL algorithm [93] (using fully complex nonlinearities). The Kalman gain matrix $\mathbf{G}_k$ is a function of the estimated error covariance matrix $\mathbf{P}_k$, the Jacobian matrix $\mathbf{H}_k^a$ and a global scaling matrix $\mathbf{H}_k^a\mathbf{P}_k^-(\mathbf{H}_k^a)^H + \mathbf{R}_k^a$.

15.5 Augmented Complex Unscented Kalman Filter (ACUKF)

Since the higher-order terms within the Taylor series expansion in the EKF model are often not negligible, the EKF is prone to accumulating error over time (Equation 15.36). To help solve this problem, the unscented Kalman filter (UKF) [139, 301] has been proposed, whereby nonlinear transforms are used to propagate the signal statistics. This way, the information from higher-order moments of non–Gaussian processes is accounted for, and the approximations

[5]EKF based algorithms have proven successful for the training of real valued temporal neural networks [181, 222].
[6]Matrix $\mathbf{H}_k^a$ is the matrix of partial derivatives of the augmented output $\mathbf{y}_k^a$ with respect to the weights.

within the UKF scenario are accurate, at least up to second-order statistical moments[7] [139]. Within the CUKF, a series of so-called complex valued sigma vectors, that is, vectors selected to be representatives of the probability distribution, are used to calculate the cross-correlation between the error in the estimated state and error in the estimated observations, as well as the correlation matrix of the error.

Within the CUKF framework, the information about the distributions of complex random variables is propagated through the system model (Equation 15.35) using $(2L + 1)$ weighted particles, where L is the dimension of the state space of the system. The weighting for every such particle is given by

$$\begin{aligned} \mathcal{W}_0^{(m)} &= \frac{\lambda}{L+\lambda}, \\ \mathcal{W}_0^{(c)} &= \frac{\lambda}{L+\lambda} + 1 - \alpha^2 + \beta, \\ \mathcal{W}_n^{(m)} = \mathcal{W}_n^{(c)} &= \frac{\lambda}{2(L+\lambda)}, \qquad n = 1, \ldots, 2L \end{aligned} \tag{15.38}$$

where $\lambda = \alpha^2(L + \kappa) - L$ is a scaling parameter, α is set to a small value (typically of order 10^{-3}) and is related to the spread of the sample points around the mean, κ is usually set to 0, whereas parameter β incorporates knowledge from *prior* distributions (in the case of complex valued Gaussian distributions, the optimal value is $\beta = 2$).

15.5.1 State Space Equations for the Complex Unscented Kalman Filter

The CUKF effectively aims at evaluating the Jacobian matrix within CEKF through the so-called sigma-point propagation, hence not requiring any analytical calculation of the derivative. The complex valued weight vector within the network[8] and the error covariance matrix are initialised as

$$\hat{\mathbf{w}}_0 = E[\mathbf{w}], \quad \mathbf{P}_0 = E\left[(\mathbf{w} - \hat{\mathbf{w}}_0)(\mathbf{w} - \hat{\mathbf{w}}_0)^{\mathrm{T}}\right] \tag{15.39}$$

whereas the sigma-point calculation is given by [301]

$$\begin{aligned} \boldsymbol{\mathcal{S}}_k &= (L+\lambda)(\mathbf{P}_k + \mathbf{Q}_k) \\ \mathbf{W}_k &= \left[\hat{\mathbf{w}}_k, \hat{\mathbf{w}}_k + \sqrt{\boldsymbol{\mathcal{S}}_k}, \hat{\mathbf{w}}_k - \sqrt{\boldsymbol{\mathcal{S}}_k}\right] \end{aligned} \tag{15.40}$$

These sigma-point estimates are then passed through a nonlinear function $\mathbf{h}$, that is

$$\begin{aligned} \mathcal{Y}_k &= \mathbf{h}(\mathbf{W}_k, \mathbf{x}_k) \\ \mathbf{y}_k &= \mathbf{h}(\hat{\mathbf{w}}_k, \mathbf{x}_k) \end{aligned} \tag{15.41}$$

[7] The standard EKF is accurate only up to first-order statistics due to the first-order linearisation in the truncated Taylor series expansion.

[8] The ACUKF training is derived for a general case of RNNs. The algorithms can be straightforwardly simplified to IIR and FIR filters, by removing nonlinearity or feedback.

to yield the measurement update equations for the CUKF in the form

$$\mathbf{P}_{\mathbf{yy},k} = \sum_{n=0}^{2L} \mathcal{W}_n^c \left(\left(\boldsymbol{\mathcal{Y}}_{n,k} - \mathbf{y}_k \right) \left(\boldsymbol{\mathcal{Y}}_{n,k} - \mathbf{y}_k \right)^{\mathrm{T}} \right) + \mathbf{R}_k$$

$$\mathbf{P}_{\mathbf{wy},k} = \sum_{n=0}^{2L} \mathcal{W}_n^c \left(\left(\mathsf{W}_{n,k} - \hat{\mathbf{w}}_k \right) \left(\mathsf{W}_{n,k} - \hat{\mathbf{w}}_k \right)^{\mathrm{T}} \right) \tag{15.42}$$

Finally, the filter update recursions for the complex unscented Kalman filter are given by

$$\begin{aligned} \mathbf{K}_k &= \mathbf{P}_{\mathbf{wy},k} \mathbf{P}_{\mathbf{yy},k}^{-1} \\ \hat{\mathbf{w}}_{k+1} &= \hat{\mathbf{w}}_k + \mathbf{K}_k \mathbf{e}_k \\ \mathbf{P}_{k+1} &= \mathbf{P}_k - \mathbf{K}_k \mathbf{P}_{\mathbf{yy},k} \mathbf{K}_k^{\mathrm{H}} \end{aligned} \tag{15.43}$$

where the estimation error $\mathbf{e}_k = \mathbf{d}_k - \mathbf{y}_k$, and $\mathbf{d}_k$ is the desired output vector.

The conceptual differences between CUKF and complex valued EKF [95] are relatively minor, but result in significant theoretical and practical advantages. For instance, the use of sigma vectors (Equation 15.40) to improve the estimation of the statistical properties of the signal in hand facilitates the processing of non-Gaussian processes, typically found in real world applications.

15.5.2 ACUKF Based Training of Complex RNNs

Consider the augmented state space model

$$\begin{aligned} \mathbf{w}_{k+1}^a &= \mathbf{w}_k^a + \boldsymbol{\omega}_k^a \\ \mathbf{y}_k^a &= \mathbf{h}(\mathbf{w}_k^a, \mathbf{x}_k^a) + \boldsymbol{\nu}_k^a \end{aligned} \tag{15.44}$$

with the augmented complex variables as in the ACKF model. The augmented covariance matrices of zero mean complex valued Gaussian noise processes $\boldsymbol{\omega}$ and $\boldsymbol{\nu}$ are denoted respectively by $\mathbf{Q}_k^a$ and $\mathbf{R}_k^a$. After the state augmentation, based on (15.38) the $(4L+1)$ weighted particles for the augmented complex valued mean and covariance estimation become

$$\begin{aligned} \mathcal{W}_0^{(m)} &= \frac{\lambda}{2L+\lambda}, \\ \mathcal{W}_0^{(c)} &= \frac{\lambda}{2L+\lambda} + 1 - \alpha^2 + \beta, \\ \mathcal{W}_n^{(m)} = \mathcal{W}_n^{(c)} &= \frac{\lambda}{2(2L+\lambda)}, \qquad n = 1, \ldots, 4L \end{aligned}$$

where $\lambda = \alpha^2(2L+\kappa) - 2L$ is a scaling parameter.

The following expressions summarise the augmented CUKF for the training of complex valued RNNs

$$
\begin{aligned}
\hat{\mathbf{w}}_0^a &= E[\mathbf{w}^a] \\
\mathbf{P}_0 &= E\left[\left(\mathbf{w}^a - \hat{\mathbf{w}}_0^a\right)\left(\mathbf{w}^a - \hat{\mathbf{w}}_0^a\right)^{\mathrm{T}}\right] \\
\mathcal{S}_k^a &= (2L+\lambda)(\mathbf{P}_k^a + \mathbf{Q}_k^a) \\
\mathsf{W}_k &= \left[\hat{\mathbf{w}}_k, \hat{\mathbf{w}}_k + \sqrt{\mathcal{S}_k}, \hat{\mathbf{w}}_k - \sqrt{\mathcal{S}_k}\right] \\
\mathsf{W}_k^a &= [\mathsf{W}_k, \mathsf{W}_k^*]
\end{aligned}
\tag{15.45}
$$

whereby, based on Equations (15.43) and (15.44), the recursive updates within ACUKF are given by

$$
\begin{aligned}
\mathcal{Y}_k^a &= \mathbf{h}\left(\mathsf{W}_k^a, \mathbf{x}_k^a\right) \\
\mathbf{y}_k^a &= \mathbf{h}\left(\hat{\mathbf{w}}_k^a, \mathbf{x}_k^a\right) \\
\mathbf{P}_{\mathbf{yy},k}^a &= \sum_{n=0}^{4L} \mathcal{W}_n^c\left(\left(\mathcal{Y}_{n,k}^a - \mathbf{y}_k^a\right)\left(\mathcal{Y}_{n,k}^a - \mathbf{y}_k^a\right)^{\mathrm{T}}\right) + \mathbf{R}_k^a \\
\mathbf{P}_{\mathbf{wy},k}^a &= \sum_{n=0}^{4L} \mathcal{W}_n^c\left(\left(\mathsf{W}_{n,k}^a - \hat{\mathbf{w}}_k^a\right)\left(\mathsf{W}_{n,k}^a - \hat{\mathbf{w}}_k^a\right)^{\mathrm{T}}\right) \\
\mathbf{K}_k^a &= \mathbf{P}_{\mathbf{wy},k}^a\{\mathbf{P}_{\mathbf{yy},k}^a\}^{-1} \\
\hat{\mathbf{w}}_{k+1}^a &= \hat{\mathbf{w}}_k^a + \mathbf{K}_k^a\mathbf{e}_k^a \\
\mathbf{P}_{k+1}^a &= \mathbf{P}_k^a - \mathbf{K}_k^a\mathbf{P}_{\mathbf{yy},k}^a\{\mathbf{K}_k^a\}^{\mathrm{H}}
\end{aligned}
\tag{15.46}
$$

These expressions simplify straightforwardly into the corresponding learning algorithms for IIR and FIR filters.

15.6 Simulation Examples

For the simulations, the nonlinearity at the neurons was chosen to be the complex tanh function

$$
\Phi(x) = \frac{e^{\beta x} - e^{-\beta x}}{e^{\beta x} + e^{-\beta x}} \tag{15.47}
$$

with slope $\beta = 1$. In all the simulations the complex recurrent neural network (Figure 7.3) had $N = 3$ neurons and tap input length of $p = 5$. The signals used in simulations were the same as those described in Chapter 8, and the quantitative measure of the performance was the standard prediction gain $R_p = 10 \log \sigma_y^2/\sigma_e^2$.

In the first set of experiments, the performance of standard CRTRL was compared with that of ACRTRL for the one step ahead prediction of a complex radar signal. Conforming with the analysis in Chapter 12, Figure 15.1 shows that the ACRTRL algorithm was able to

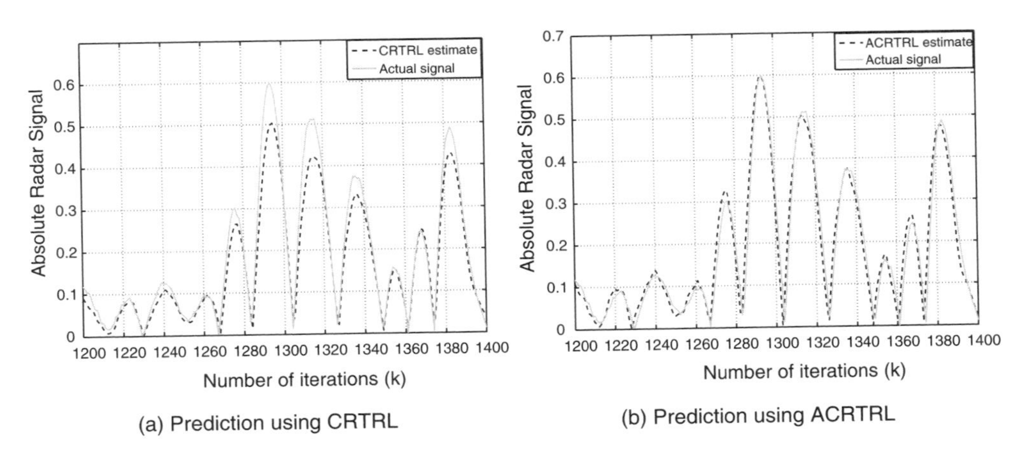

Figure 15.1 Prediction of the complex valued radar signal using CRTRL and ACRTRL

track the noncircular radar signal better than the standard CRTRL. In the next experiment, performances of the ACEKF and ACUKF algorithms were compared for the prediction of the complex AR(4) (Equation 8.22) and wind signals. Figure 15.2 illustrates that in both cases the prediction based on ACUKF nearly coincided with the original signal, whereas the ACEKF based prediction exhibited larger prediction errors. In the last set of simulations, shown in Figure 15.3, the performances of ACUKF and ACEKF were compared for the prediction of a segment of radar data; whereas both the algorithms were able to track the radar signal well, the ACUKF had better accuracy. Table 15.1 provides a comparison of the prediction gains R_p (dB) for all the algorithms introduced in this Chapter. In all the cases, the algorithms based on augmented complex statistics outperformed standard algorithms, clearly indicating the benefits of widely linear estimation. As expected, the Kalman filter based algorithms performed better

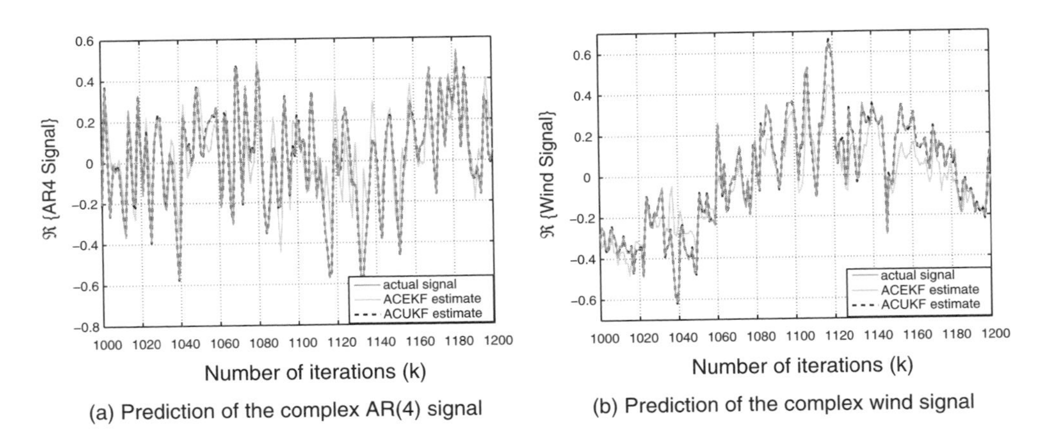

Figure 15.2 Performance comparison between ACUKF and ACEKF for the complex circular AR(4) signal (Equation 8.22) and the noncircular complex wind signal; symbol $\Re$ denotes the real part of a complex signal

Table 15.1 Prediction gains R_p [dB] for the various classes of signals

Signal	Nonlinear N1 (8.23)	AR4(8.22)	Wind	Radar
R_p (dB) (ACRTRL)	5.81	4.10	9.99	9.45
R_p (dB) (standard CRTRL)	3.76	3.54	6.12	7.22
R_p (dB) (ACEKF)	6.24	4.77	11.65	10.58
R_p (dB) (standard CEKF)	5.55	3.98	10.24	9.02
R_p (dB) (ACUKF)	7.45	6.12	12.22	11.13
R_p (dB) (standard CUKF)	6.75	4.50	9.35	9.78

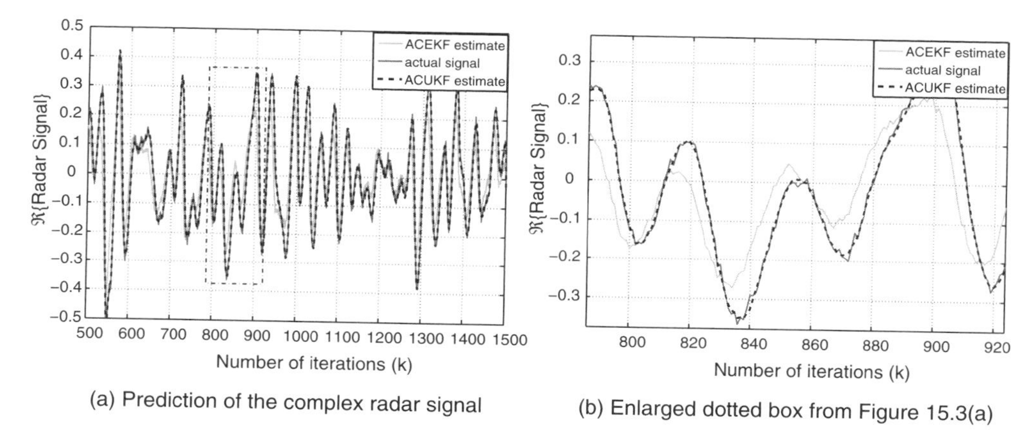

(a) Prediction of the complex radar signal

(b) Enlarged dotted box from Figure 15.3(a)

Figure 15.3 Performance comparison of ACUKF and ACEKF for the complex radar signal

than the direct gradient based CRTRL algorithms, and the algorithm based on the unscented transformation outperformed the extended Kalman filter based algorithm.

To summarise:

- The class of widely linear adaptive feedback filters has been introduced, and learning algorithms for adaptive IIR filters (adaptive WL-ARMA model) and recurrent perceptron (adaptive WL-NARMA model) have been introduced for the processing of both circular and noncircular complex valued signals.
- This has been achieved based on $\mathbb{CR}$ calculus, as the filter outputs are based on augmented inputs and do not admit differentiation directly in $\mathbb{C}$.
- The augmented complex valued real time recurrent learning (ACRTRL), augmented complex valued extended Kalman filter (ACEKF) and augmented complex valued unscented Kalman filter (ACUKF) algorithms have been introduced for the training of complex recurrent neural networks (RNNs).
- It has been shown that the training of complex RNNs in the nonlinear sequential state space estimation framework [112] improves their performance. However, this comes at the expense of significantly increased computational complexity compared with direct gradient algorithms (ACRTRL).

16

Collaborative Adaptive Filtering

Statistical, hypothesis based, techniques for the assessment of the nature of real world signals (linear, nonlinear, stochastic, deterministic – see Figure 18.2), such as the complex delay vector variance (CDVV) method described in Chapter 18 and the circularity test by Schreier *et al.* [270], are mathematically rigorous and very useful. However, they can only operate on piecewise stationary signals and in an off-line manner, whereas signal modality tracking should be performed in an online adaptive fashion. This is particularly important in adaptive prediction applications, where the information about the change in the nature of the signal in hand can be used to aid the performance of the predictor. This chapter introduces a class of online algorithms for signal modality characterisation (nonlinearity, circularity). This is achieved in a collaborative adaptive filtering framework where two filters of different natures are combined in a convex manner, hence guaranteeing the existence of a solution as long as one of the subfilters is stable. It is shown that this provides improvement in the performance (convergence, steady state properties), and that the evolution of the adaptive convex mixing parameter within this structure reflects the changes in the modality of the processed signal.

16.1 Parametric Signal Modality Characterisation

An intuitive way to judge whether the signal in question is predominantly linear or not is to perform prediction by two adaptive filters of different natures, for instance, a linear and nonlinear filter, and to compare the output errors. The degree of nonlinearity of a signal is then assessed based on a normalised ratio of the output errors of the linear and nonlinear filter [212]. Figure 16.1 illustrates the estimation of the degree of signal nonlinearity using a third-order Volterra system as the nonlinear filter and an FIR filter trained by the NLMS algorithm as a linear filter. The system input $y(k)$ was generated from a linear filtered noise signal $u(k)$ which was then passed through a nonlinear function $F(\cdot)$, as

$$u(k) = 0.5x(k) + 0.25x(k-1) + 0.125x(k-2) \tag{16.1}$$

$$y(k) = F(u(k); k) + n(k) \tag{16.2}$$

Complex Valued Nonlinear Adaptive Filters: Noncircularity, Widely Linear and Neural Models
Danilo P. Mandic and Vanessa Su Lee Goh

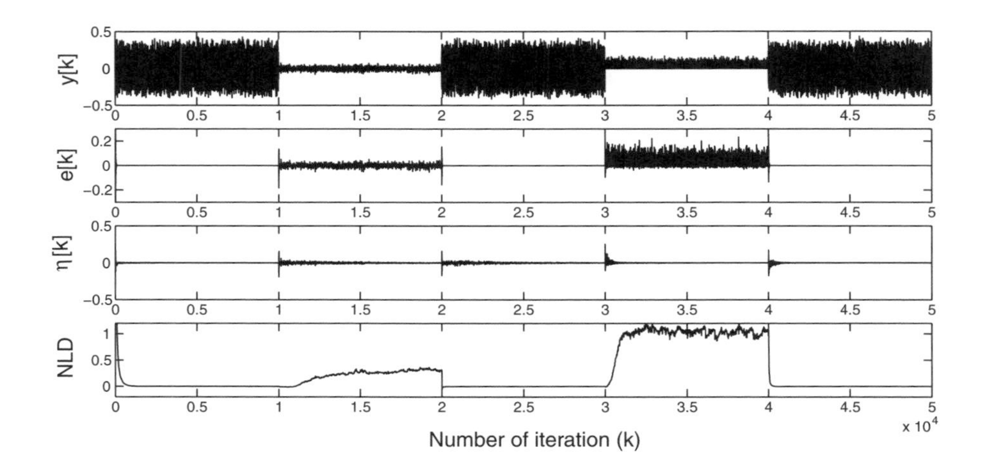

Figure 16.1 Estimated degree of signal nonlinearity for an input nature alternating between linear and nonlinear

where $x(k)$ are i.i.d. and uniformly distributed over the range $[-0.5, 0.5]$ and $n(k) \sim \mathcal{N}(0, 0.0026)$. Function $F(u(k); k)$ defines the degree of nonlinearity within different segments of signal $y(k)$, ranging from purely linear, to quadratic and cubic, that is

$$F\big(u(k); k\big) = \begin{cases} u^3(k), & \text{for } 10000 < k \leq 20000 \\ u^2(k), & \text{for } 30000 < k \leq 40000 \\ u(k), & \text{at all other times} \end{cases} \tag{16.3}$$

The generated signal $y(k)$ is shown in the first trace of Fig. 16.1, the second and third traces show respectively the residual estimation errors of the linear system and Volterra system, whereas the bottom trace shows the estimated degree of signal nonlinearity (index NLD close to zero indicates a linear signal whereas NLD close to unity indicates a nonlinear signal). Although this approach was able to correctly identify the differences in signal nonlinearity between the original linear signal $u(k)$, its square $u^2(k)$, and cube $u^3(k)$, there are also some limitations which prevent its direct extension to the complex domain. These include:

- The linear and nonlinear filter operate independently, making the results crucially dependent on the choice of filter parameters;
- In the complex domain, there are several types of nonlinearity (split-complex, fully complex) and we should check both for the linear vs nonlinear nature of the signal and for the type of nonlinearity that best models the signal;
- The only continuously differentiable function in $\mathbb{C}$ is a constant (by Liouville's theorem – see Appendix B); nonlinear functions exhibit several different types of singularities (essential, removable, isolated) and complex Volterra filters may not be best suited for all these scenarios;
- Prediction errors of the subfilters can be compared only if both of the subfilters converge, this is not always guaranteed.

To overcome these limitations, a flexible method based on collaborative adaptive filtering is next introduced [198].

16.2 Standard Hybrid Filtering in $\mathbb{R}$

The hybrid filtering configuration shown in Figure 16.2 was originally introduced with the aim of improving the performance of standard adaptive filtering algorithms [19]. It is based on the convex combination of two adaptive FIR subfilters denoted by filter1 and filter2, which operate in a collaborative fashion. The subfilters are updated independently, based on the local instantaneous output errors $e_1(k)$ and $e_2(k)$, and their respective outputs $y_1(k) = \mathbf{x}^T(k)\mathbf{w}_1(k)$ and $y_2(k) = \mathbf{x}^T(k)\mathbf{w}_2(k)$ are combined in a convex manner to give the overall output of the hybrid filter

$$y(k) = \lambda(k)y_1(k) + \big(1 - \lambda(k)\big)\, y_2(k) \tag{16.4}$$

Depending on the value of the convex mixing parameter λ, $0 \le \lambda \le 1$, the overall output of the hybrid filter $y(k)$ spans the range $\big[y_1(k), y_2(k)\big]$, as illustrated in Figure 16.3.

The idea behind the hybrid filtering architecture is as follows: assume that filter1 exhibits fast convergence, but a large steady state error, and that filter2 is slowly converging and with a small steady state error. For fast convergence, the hybrid filter should favour the fast filter1 in the beginning of the adaptation ($\lambda(k) \to 1$), whereas for good steady state properties, after the filters converge, the output of the hybrid filter should be dominated by the slow filter2 ($\lambda(k) \to 0$). For the prediction setting, this is illustrated in Figure 16.4, based on a hybrid

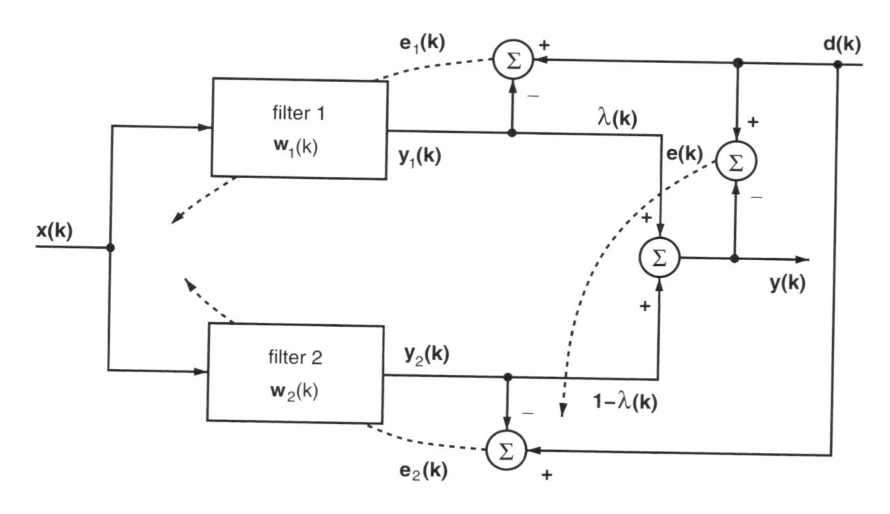

Figure 16.2 A general hybrid filtering architecture

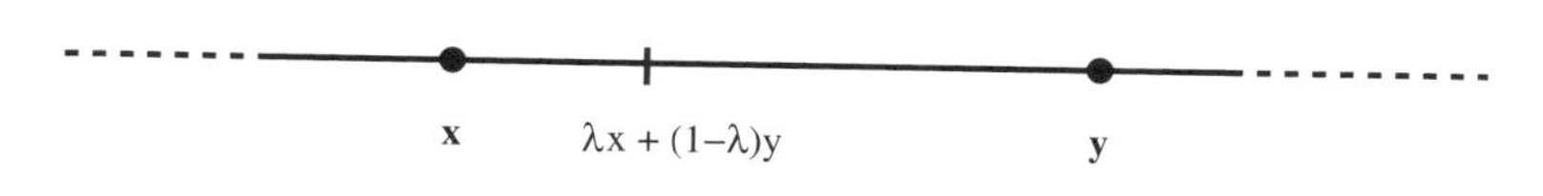

Figure 16.3 Convex combination of points x and y on a line in $\mathbb{R}$, for $0 \le \lambda(k) \le 1$

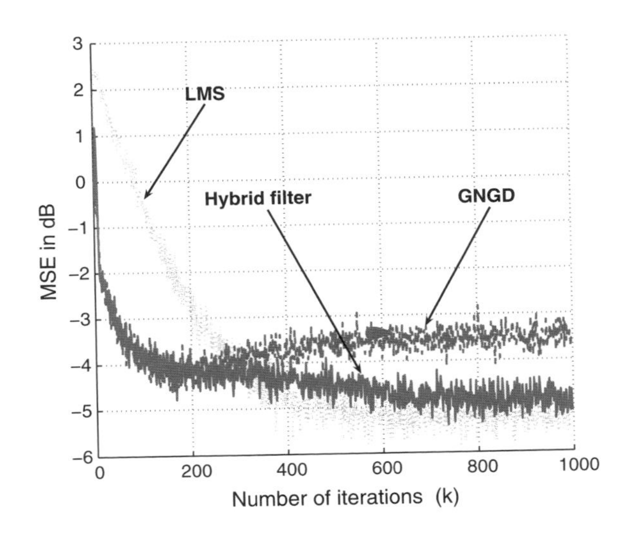

Figure 16.4 Learning curves for nonlinear signal (Equation 16.12)

filter with one linear subfilter trained with a very fast generalised normalised gradient descent (GNGD) algorithm [180] and the other linear subfilter trained by a slow LMS [197]. As desired, the learning curve follows the learning curve of the GNGD trained subfilter in the beginning of the adaptation (until about sample number 350), whereas after convergence, it approaches the learning curve of the LMS trained subfilter.

The mixing parameter λ is updated using stochastic gradient descent, based on the overall output error $e(k)$ and the cost function $J(k) = \frac{1}{2}e^2(k)$, that is

$$\begin{aligned}\lambda(k+1) &= \lambda(k) - \mu_\lambda \nabla_\lambda J(k)_{|\lambda=\lambda(k)} \\ \lambda(k+1) &= \lambda(k) + \mu_\lambda e(k)\big(y_1(k) - y_2(k)\big)\end{aligned} \tag{16.5}$$

where μ_λ is the learning rate. To ensure that the combination of adaptive filters is a convex function, it is critical that $\lambda(k)$ remains within the range $0 \leq \lambda(k) \leq 1$. This is usually achieved through the use of a sigmoid function as a post–nonlinearity to bound $\lambda(k)$ [19, 74]. Since, in order to detect the changes in the nature of a signal, our primary interest is not in the overall performance of the filter, but in the dynamics of the mixing parameter λ, the use of a nonlinear sigmoid function would interfere with true values of $\lambda(k)$. A hard limit on the set of allowed values for $\lambda(k)$ is therefore implemented; for more detail, see [133, 134].

16.3 Tracking the Linear/Nonlinear Nature of Complex Valued Signals

Consider a collaborative hybrid filtering architecture consisting of one linear and one nonlinear complex adaptive filter, shown in Figure 16.5. At every time instant k, the output of the hybrid filter is generated as a convex combination of the output of the nonlinear subfilter $y_{\mathrm{NL}}(k)$ and the output of the linear subfilter $y_{\mathrm{L}}(k)$, that is

$$y(k) = \lambda(k)y_{\mathrm{NL}}(k) + \big(1 - \lambda(k)\big)y_{\mathrm{L}}(k) \tag{16.6}$$

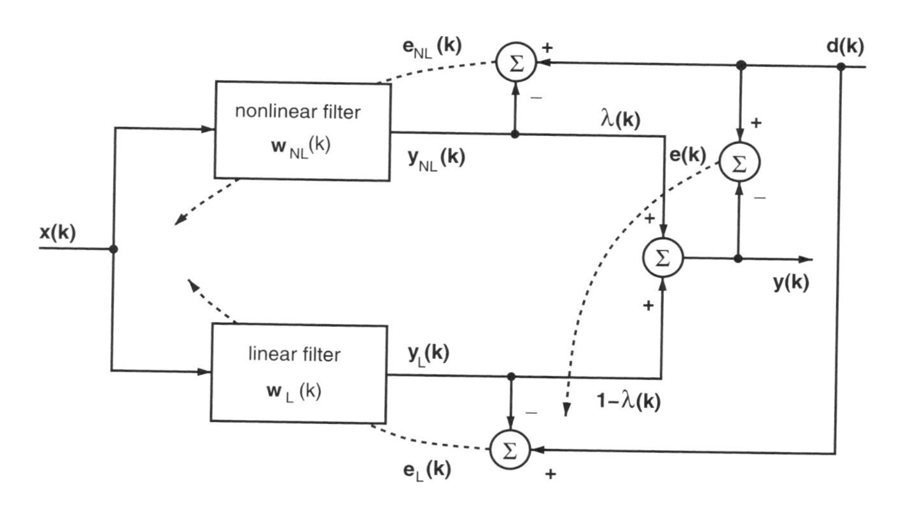

Figure 16.5 Convex combination of a linear and nonlinear complex adaptive filter

where $y_{\mathrm{NL}}(k) = \boldsymbol{x}^{\mathrm{T}}(k)\mathbf{w}_{\mathrm{NL}}(k)$, $y_{\mathrm{L}}(k) = \boldsymbol{x}^{\mathrm{T}}(k)\mathbf{w}_{\mathrm{L}}(k)$, the corresponding weight vectors are $\mathbf{w}_{\mathrm{NL}}(k)$ and $\mathbf{w}_{\mathrm{L}}(k)$, and the common input vector $\boldsymbol{x}(k) = [x_1(k), \ldots, x_N(k)]^{\mathrm{T}}$. The constituent subfilters are updated by their respective output errors $e_{\mathrm{NL}}(k)$ and $e_{\mathrm{L}}(k)$, using a common desired signal $d(k)$, whereas the mixing parameter $\lambda(k)$ is updated based on the overall output error $e(k)$. To preserve the convexity of the output of the hybrid filter from Figure 16.5, the parameter $\lambda(k)$ is kept real, and is updated as

$$\lambda(k+1) = \lambda(k) - \mu_\lambda \Re\left\{\nabla_\lambda J(k)_{|\lambda=\lambda(k)}\right\} \tag{16.7}$$

For the standard cost function $J(k) = \frac{1}{2}|e(k)|^2$ [307], the stochastic gradient update of the mixing parameter $\lambda(k)$ is calculated from

$$\nabla_\lambda J(k)_{|\lambda=\lambda(k)} = \frac{1}{2}\left\{e(k)\frac{\partial e^*(k)}{\partial\lambda(k)} + e^*(k)\frac{\partial e(k)}{\partial\lambda(k)}\right\} \tag{16.8}$$

yielding (strictly speaking the operator $\Re\{\cdot\}$ is not needed)

$$\lambda(k+1) = \lambda(k) + \mu_\lambda \Re\left\{e(k)\big(y_{\mathrm{NL}}(k) - y_{\mathrm{L}}(k)\big)^*\right\} \tag{16.9}$$

16.3.1 Signal Modality Characterisation in $\mathbb{C}$

To illustrate the ability of the collaborative hybrid filtering approach to track the nature of complex valued signals, consider two linear and two nonlinear benchmark signals:

L1. Linear complex stable $AR(1)$ process, given by

$$y(k) = 0.9y(k-1) + n(k) \tag{16.10}$$

L2. Linear complex stable $AR(4)$ process, given by

$$y(k) = 1.79y(k-1) - 1.85y(k-2) + 1.27y(k-3) - 0.41y(k-4) + n(k) \quad (16.11)$$

N1. Complex nonlinear benchmark signal [216]

$$y(k) = \frac{y(k-1)}{1 + y^2(k-1)} + n^3(k) \quad (16.12)$$

N2. Complex nonlinear benchmark signal [216]

$$y(k) = \frac{y^2(k-1)(y(k-1)+2.5)}{1 + y(k-1)^2 + y(k-2)^2} + n(k-1) \quad (16.13)$$

with complex doubly white circular Gaussian noise (CWGN) $n(k) \sim \mathcal{N}(\mathbf{0}, \mathbf{1})$ as the driving input. The CWGN can be expressed as $n(k) = n_r(k) + \jmath n_i(k)$. The real and imaginary components of CWGN are mutually independent sequences with equal variances, so that $\sigma_n^2 = \sigma_{n_r}^2 + \sigma_{n_i}^2$.

Detection of the nature of complex signals. In all the simulations, both subfilters were $N = 10$ taps long, and the simulations were performed in the prediction setting. The linear filter was trained by CLMS, whereas the nonlinear filter was trained by the Complex Nonlinear Gradient Descent (CNGD) algorithm [105], based on the fully complex logistic activation function,[1] given by

$$\Phi(z) = \frac{1}{1 + e^{-\beta z}} \qquad \beta \in \mathbb{R}^+, \ z \in \mathbb{C} \quad (16.14)$$

As illustrated in Figure 16.6, for the two nonlinear signals N1 and N2, the mixing parameter $\lambda(k)$ converged to unity, favouring the nonlinear subfilter with the output $y_{NL}(k)$, whereas for the linear signals L1 and L2, the mixing parameter converged to zero, thus illustrating the ability of the hybrid filter to detect the linear vs. nonlinear nature of the complex input. For best performance, the stepsize μ_λ was about 20 times larger than that of the constitutive subfilters.

The same experiment was repeated for the normalised versions of the learning algorithms, that is, the CNLMS for the linear subfilter, CNNGD [177] for the nonlinear subfilter, and NLMS for the adaptation of $\lambda(k)$; the evolution of the mixing parameter λ is shown in Figure 16.6(b). The normalised versions of the algorithms were less sensitive to the changes in the relative values of step sizes, and also converged faster.

Tracking the modality of complex signals. To investigate the possibility of online tracking of the nature of nonstationary signals with large dynamical changes, consider a synthetic signal consisting of alternating segments of 200 samples of linear and nonlinear benchmark signals. The same hybrid filter setting as in the previous example was used (CLMS and CNGD, Figure 16.6), and the evolution of the mixing parameter λ for the signal pairs L1–N1 and L2–N2 is shown in Figure 16.7. The behaviour of the mixing parameter $\lambda(k)$ clearly reflects the changes

[1] See Chapter 4.

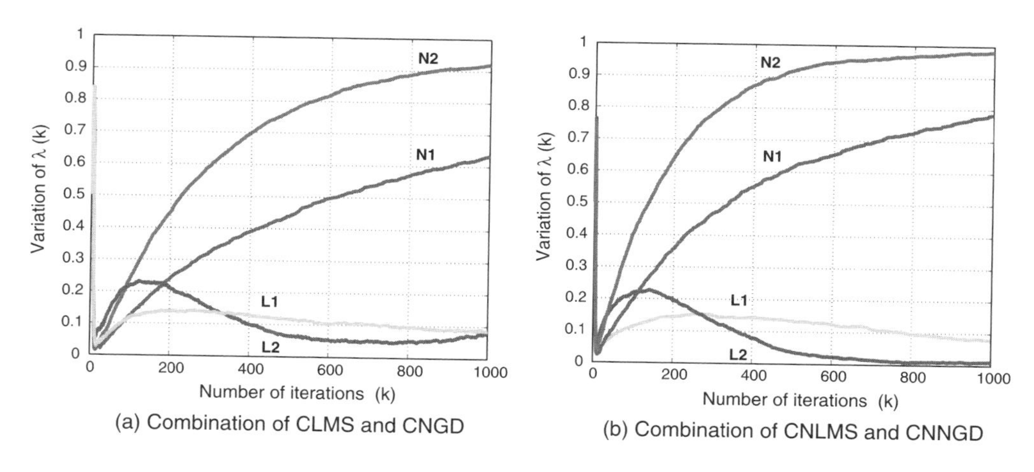

(a) Combination of CLMS and CNGD

(b) Combination of CNLMS and CNNGD

Figure 16.6 Evolution of the convex mixing parameter λ for linear inputs L1 (Equation 16.10), L2 (Equation 16.11), and nonlinear inputs N1 (Equation 16.12) and N2 (Equation 16.13), for a $N = 10$–tap hybrid filter

in the signal nature, with the high values $\lambda \approx 0.8$ and $\lambda \approx 0.9$ for the respective nonlinear segments N1 and N2, and the low values $\lambda \approx 0.2$ for the linear segments L1 and L2; the smooth transitions between the low and high values of λ are due to the sequential nature of learning.

To speed up convergence and make the hybrid filter less sensitive to the choice of filter parameters, we next consider normalised collaborative learning, whereby CNLMS and CNNGD are used for the training of the subfilters within the hybrid filter from Figure 16.5 and NLMS for the update of the mixing parameter $\lambda(k)$. Segments of linear and nonlinear data alternating every 100 and 200 samples were considered, and the dynamics of the mixing

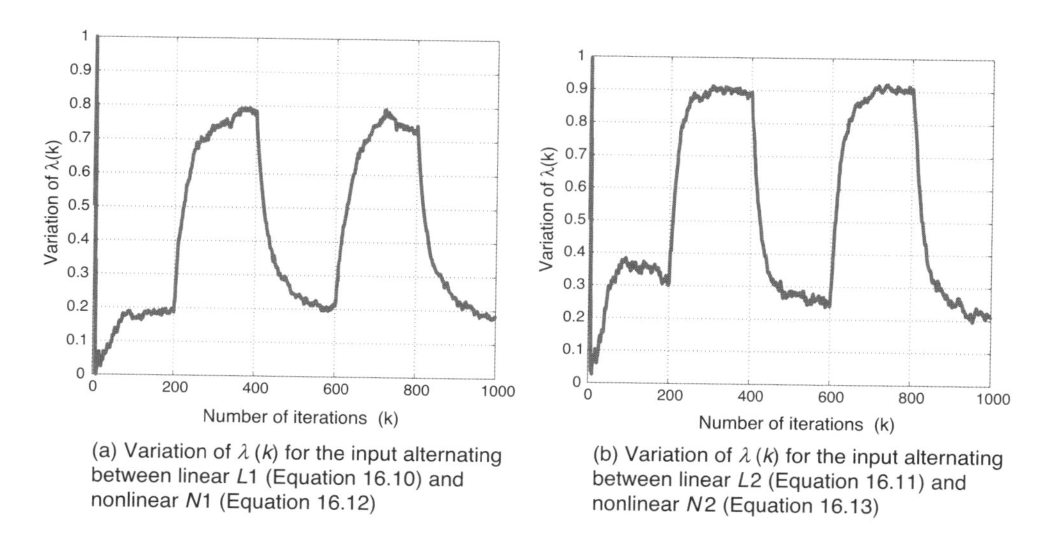

(a) Variation of $\lambda\,(k)$ for the input alternating between linear $L1$ (Equation 16.10) and nonlinear $N1$ (Equation 16.12)

(b) Variation of $\lambda\,(k)$ for the input alternating between linear $L2$ (Equation 16.11) and nonlinear $N2$ (Equation 16.13)

Figure 16.7 Evolution of the mixing parameter λ for the input nature alternating every 200 samples (based on a convex combination of CLMS and CNGD)

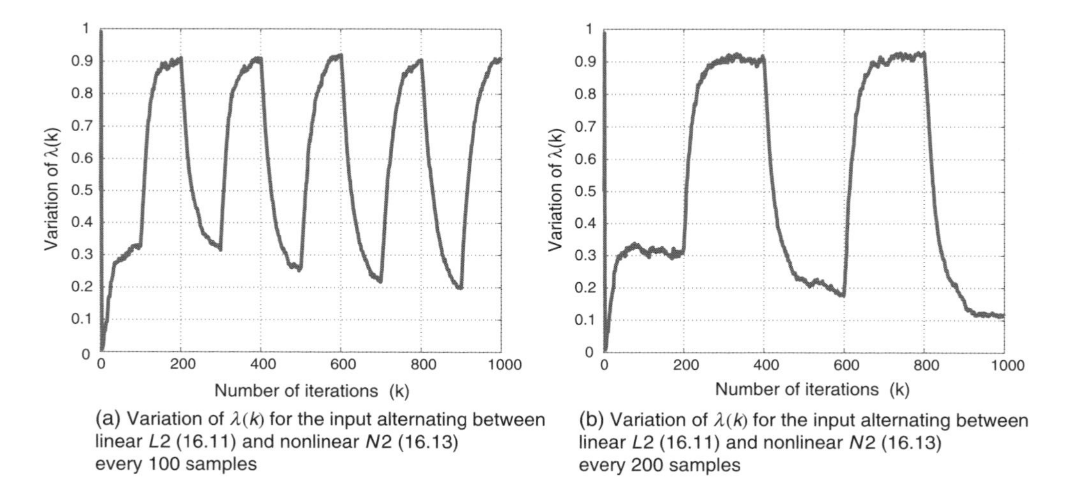

(a) Variation of $\lambda(k)$ for the input alternating between linear $L2$ (16.11) and nonlinear $N2$ (16.13) every 100 samples

(b) Variation of $\lambda(k)$ for the input alternating between linear $L2$ (16.11) and nonlinear $N2$ (16.13) every 200 samples

Figure 16.8 Evolution of the mixing parameter $\lambda(k)$ for normalised collaborative learning; the signal nature is alternating between linear L2 and nonlinear N2 every 100 samples (left) and every 200 samples (right) (based on a convex combination of CNLMS and CNNGD)

parameter $\lambda(k)$ is shown respectively in Figure 16.8(a) and (b). A comparison with the results from Figure 16.7 shows that the normalised collaborative learning exhibits faster convergence and spans a wider range of the values of the mixing parameter λ. Note the decreasing trend in the values of $\lambda(k)$ when tracking the linear signal, indicating the excellent tracking ability. The range of the values of $\lambda(k)$ for the signal with alternating 100 samples of data of different natures was smaller than that for 200 alternating samples, however, it was still possible to differentiate between the linear and nonlinear nature of the complex data.

16.4 Split vs Fully Complex Signal Natures

As has already been shown in Chapter 4, the notion of nonlinearity in $\mathbb{C}$ is quite different from that in $\mathbb{R}$. It is therefore not sufficient to check only for the linear vs nonlinear nature of the complex signal. If the signal is judged nonlinear, we also need to establish the particular type of nonlinearity best suited to the signal in question. When it comes to nonlinear adaptive filters in $\mathbb{C}$, we usually differentiate between the fully complex and split-complex nonlinearity within such filters. To test for the split- fully complex nature of a signal, we can employ a hybrid filter where the two subfilters are based on the split-complex and fully complex activation functions (SC-FC hybrid filter), as shown in Figure 16.9. The output of such a hybrid filter can be expressed as

$$y(k) = \lambda(k)y_{\text{fully}}(k) + \big(1 - \lambda(k)\big)\, y_{\text{split}}(k) \tag{16.15}$$

Values of $\lambda(k)$ approaching unity will indicate a predominantly fully complex nature of the signal in hand, whereas values of λ approaching zero will indicate a predominantly split-complex nature. To preserve the convexity (and hence the existence of the solution) of the output from Figure 16.9, the mixing parameter $\lambda(k)$ is kept real, and is hard-bounded to $0 \leq \lambda(k) \leq 1$.

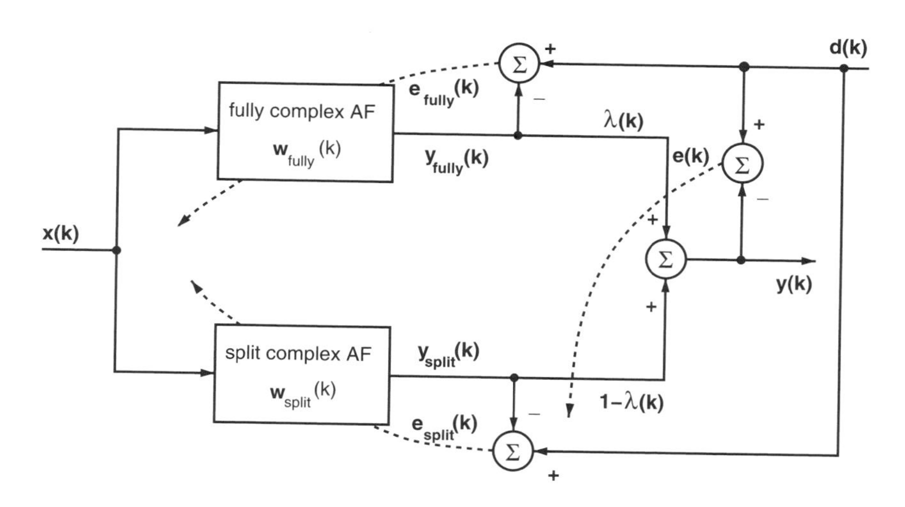

Figure 16.9 Collaborative learning based on a hybrid combination of a fully complex and split-complex (SC-FC) nonlinear filter

The update for $\lambda(k)$ is calculated based on

$$\lambda(k+1) = \lambda(k) - \mu_\lambda \nabla_\lambda J(k)_{|\lambda=\lambda(k)}$$

$$\nabla_\lambda J(k)_{|\lambda=\lambda(k)} = \frac{1}{2}\left\{e(k)\frac{\partial e^*(k)}{\partial \lambda(k)} + e^*(k)\frac{\partial e(k)}{\partial \lambda(k)}\right\} \tag{16.16}$$

yielding the stochastic gradient update

$$\lambda(k+1) = \lambda(k) + \mu_\lambda \Re\left\{e(k)\big(y_{\text{fully}}(k) - y_{\text{split}}(k)\big)^*\right\} \tag{16.17}$$

Performance on synthetically generated data. The ability of the SC-FC hybrid filter to differentiate between the split complex and fully complex nature of a signal is illustrated on the strongly nonlinear, fully complex, Ikeda map signal (Equation 13.46). Following the results from [297], the fully complex nonlinearity was the logistic sigmoid function

$$\Phi(z) = \frac{1}{1+e^{-\beta z}} \qquad z \in \mathbb{C} \tag{16.18}$$

whereas the split-complex nonlinearity was a real valued logistic function (see Appendix D)

$$\Phi(z) = \frac{1}{1+e^{-\beta z_r}} + \jmath \frac{1}{1+e^{-\beta z_i}} \qquad z \in \mathbb{C} \tag{16.19}$$

and in both cases the slope of the activation function was $\beta = 1$. The filter length was set to $N = 10$ and all the filters operated in a one step ahead prediction setting.

The nonlinear nature of the Ikeda signal (assessed using the hybrid filter from Figure 16.5), as indicated by the value of the mixing parameter $\lambda(k)$ approaching unity, is illustrated in Figure 16.10(a), whereas its fully complex nature, indicated by $\lambda(k) \to 1$ in the FC-SC setting, is illustrated in Figure 16.10(b). Since the Ikeda map is strongly nonlinear and fully complex

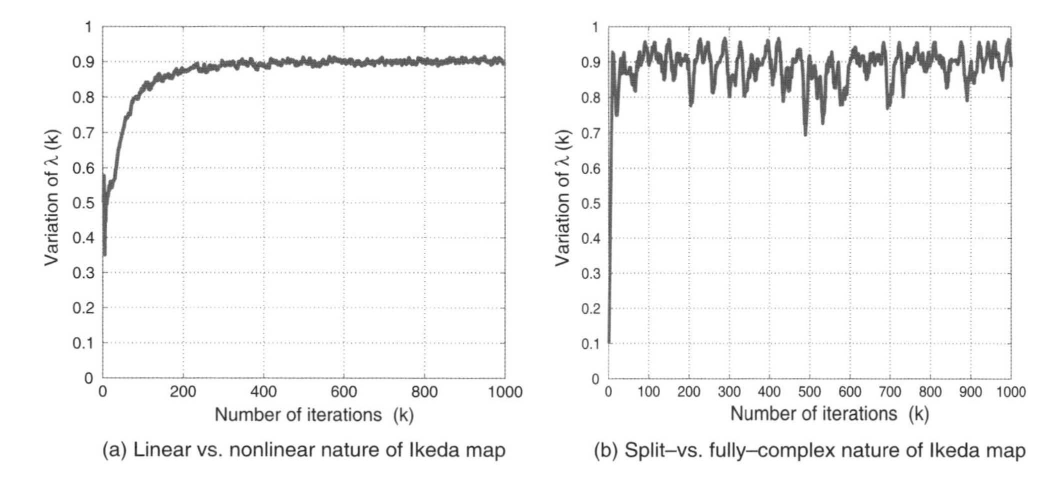

(a) Linear vs. nonlinear nature of Ikeda map

(b) Split–vs. fully–complex nature of Ikeda map

Figure 16.10 Linear vs nonlinear and fully vs split-complex nature of Ikeda map

by design, Figure 16.10 confirms the potential of the collaborative hybrid filtering approach for the modality characterisation of complex signals.

16.5 Online Assessment of the Nature of Wind Signal

We next investigate the degree of nonlinearity and the split- versus fully complex nature of the intermittent and nonstationary 'complex by convenience of representation' wind signal recorded over 24 hours in an urban environment.[2] The recording started at 2.00 PM and lasted for approximately 24 hours. Using the collaborative filtering architectures from Figures 16.5 and 16.9, the linear vs nonlinear and split- vs fully complex nature of this dataset is illustrated in Figure 16.11. Although the wind changed its nature over the 24 h period [117], it was predominantly nonlinear, as illustrated by the high values of the convex mixing parameter $\lambda(k)$ in Figure 16.11(a). The nature of nonlinearity was strongly fully complex, as indicated by the high values of the mixing parameter λ in the split- vs fully complex test shown in Figure 16.11(b). Although the online signal modality test is not as accurate as the hypothesis based tests [85, 191], and should be taken in relative rather than in absolute terms, the results conform with the rigorous hypothesis testing based statistical analysis in Chapter 18.

16.5.1 Effects of Averaging on Signal Nonlinearity

Wind recordings are very noisy and in practical applications the data are first averaged to reduce the effects of noise. When performing wind modelling for renewable energy applications, there are several reasons for employing prediction at different scales within a wind farm (WF) [117]:

- *To predict the expected production of electricity*: for large WFs this is typically achieved by using medium-range weather forecasts (several hours or longer).

[2]The wind dataset was recorded at the Institute for Industrial Science, University of Tokyo, Japan.

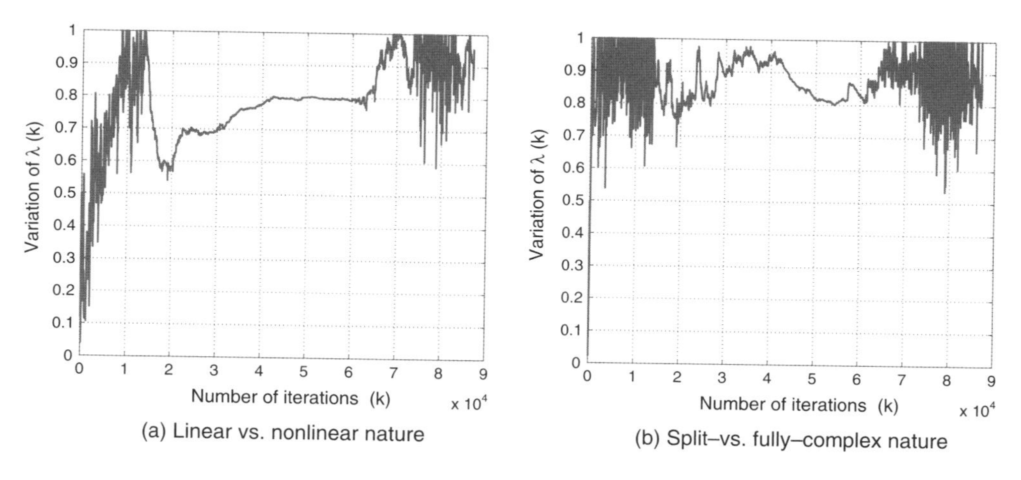

Figure 16.11 Tracking the degree of nonlinearity and the split- vs fully complex nature of a 24 h wind recording

- *For the control of wind turbines*, where short-term wind prediction (one or several steps ahead) is used to aid the control mechanism.
- *To avoid damages to wind turbines caused by gusts*: this is usually achieved based on a combination of short-term wind modelling (several steps ahead) and some sort of finite state machine.
- *To improve the efficiency of a WF*: this is achieved by adjusting the yaw of the blades and the direction of WTs so as to face the direction of the wind. This should be performed based solely on the modelling of the wind field (speed and direction), since the output power of a WT is proportional to the cube of the incident wind speed.

Since averaging is closely related to Gaussianity and linearity, it is interesting to investigate the effect that averaging has on the nature of the wind data. For that purpose, the 24 h data recording, also analysed in Figure 16.11, was used and the nonlinearity test was performed for data averaged over 5 and 10 samples. The results of simulations are shown in Figure 16.12. Compared with the nonlinearity analysis of raw data from Figure 16.11(a), the nature of the wind dataset exhibits an increase in the degree of linearity with the order of averaging, as indicated by the corresponding decrease in values of $\lambda(k)$ in Figure 16.12. This also confirms that linear modelling is adequate for heavily averaged nonlinear data, as shown in [191].

16.6 Collaborative Filters for General Complex Signals

The analysis and simulations in the preceding sections show that the benefits of using convex combinations of adaptive filters include:

- improved performance over either of the constituent subfilters, together with guaranteed stability, as long as one of the subfilters is stable;

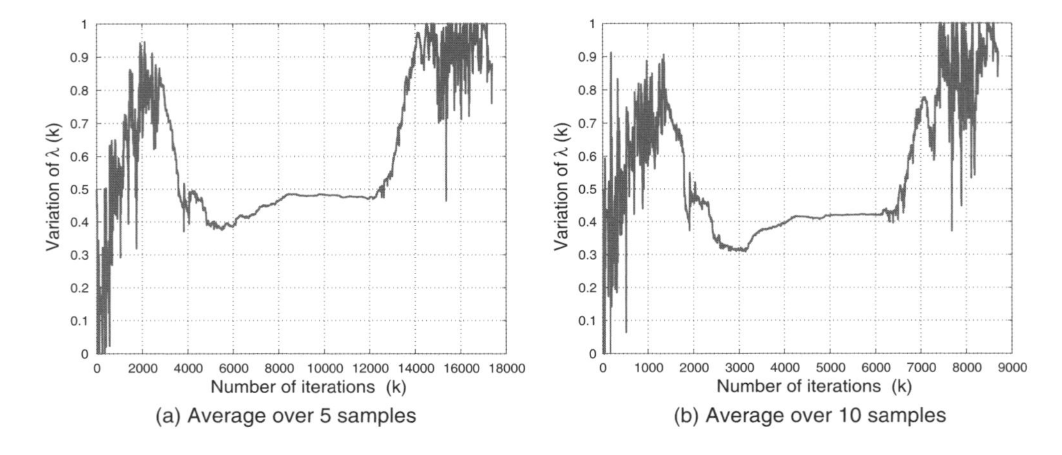

Figure 16.12 The effect of averaging on the degree of nonlinearity of wind

- insight into the nature of the signal at hand (linear, nonlinear, split-complex, fully compex) by monitoring the evolution of the convex mixing parameter $\lambda(k)$ within the hybrid filtering architecture.

We will next illustrate that collaborative adaptive filtering, based on a combination of a standard and widely linear subfilter, is suitable for the processing of the generality of complex valued signals. The evolution of the mixing parameter also provides a convenient online test for complex circularity. This is achieved at a little expense in terms of computational complexity.

16.6.1 Hybrid Filters for Noncircular Signals

It has been shown in Chapters 13 and 14 that

- The Complex Least Mean Square (CLMS) algorithm converges faster than its widely linear counterpart the Augmented Complex Least Mean Square (ACLMS) algorithm; this is due to the CLMS using half the number of coefficients compared with ACLMS;
- Both the CLMS and ACLMS converge to the same steady state solution for circular complex signals, whereas ACLMS exhibits improved performance for noncircular signals.

A hybrid filtering architecture, shown in Figure 16.13, which consists of two FIR adaptive filters trained by CLMS and ACLMS, can therefore be used to provide fast initial convergence and improved steady state properties, compared with the individual subfilters.

Similarly to Equation (16.9), the update of the convex mixing paramter $\lambda(k)$ from Figure 16.13 can be expressed as[3]

$$\lambda(k+1) = \lambda(k) + \mu_\lambda \Re\left\{e(k)\big(y_{\mathrm{ACLMS}}(k) - y_{\mathrm{CLMS}}(k)\big)^*\right\} \tag{16.20}$$

[3]Recall that the weight update for the CLMS is given by $\mathbf{w}_{\mathrm{CLMS}}(k+1) = \mathbf{w}_{\mathrm{CLMS}}(k) + \mu e(k)\boldsymbol{x}^*(k)$, whereas that for the ACLMS is $\mathbf{w}^a_{\mathrm{CLMS}}(k+1) = \mathbf{w}^a_{\mathrm{CLMS}}(k) + \mu e(k)\boldsymbol{x}^{a^*}(k)$.

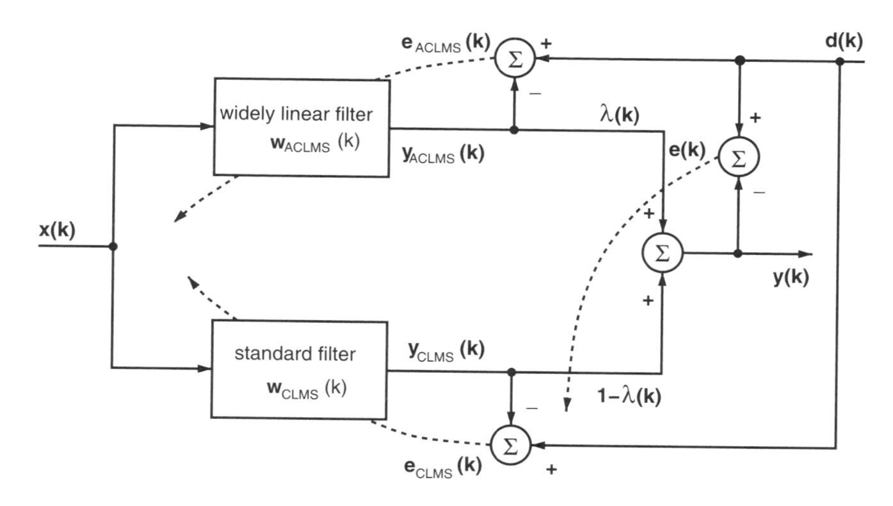

Figure 16.13 Convex combination of FIR adaptive filters in $\mathbb{C}$ trained by a widely linear (ACLMS) and standard (CLMS) adaptive filtering algorithm

To illustrate the benefits of using the hybrid filter in Figure 16.13, simulations were conducted on a synthetic circular AR(4) process (Equation 13.44) and strongly noncircular Ikeda map (Equation 13.46). Learning curves for CLMS, ACLMS, and the hybrid filter for the prediction of Ikeda map are shown in Figure 16.14(a). Conforming with the analysis, CLMS exhibited faster initial convergence than ACLMS, and ACLMS had a lower steady state error. As desired, the learning curve of the hybrid filter closely followed the CLMS learning curve in the beginning of adaptation (up to about the time instant 1000), whereas it approached the ACLMS learning curve closer to the steady state (roughly after the time instant 2000).

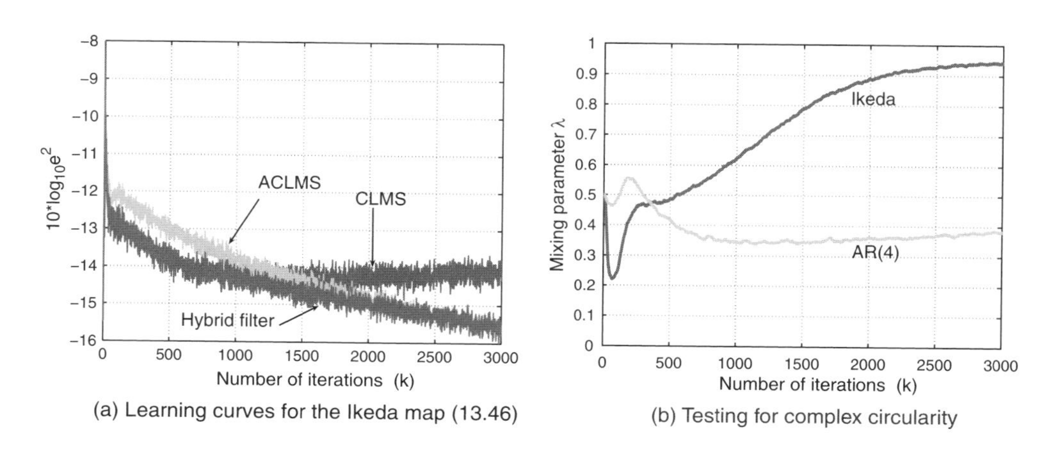

Figure 16.14 Left: performances of CLMS, ACLMS, and hybrid filter for a noncircular signal; right: evolution of $\lambda(k)$ for a circular and noncircular signal

Table 16.1 Prediction gains (dB) for the CLMS, ACLMS and hybrid filter

	AR(4)	'Calm' wind	'High' wind
CLMS	5.25	7.03	3.26
ACLMS	4.73	6.87	4.35
Hybrid	**5.66**	**7.33**	**4.48**

Quantitative performances (in terms of the prediction gain R_p) for the CLMS, ACLMS, and the hybrid filter, for the circular AR(4) signal and noncircular wind segments (see Figure 18.1) are shown in Table 16.1. In all the cases, the hybrid filter outperformed the constitutive subfilters.

16.6.2 Online Test for Complex Circularity

Similarly to the online tests for signal nonlinearity, the hybrid filter in Figure 16.13 can be used to test for complex circularity. Figure 16.14(b) shows the evolution of the convex mixing parameter $\lambda(k)$ from (Equation 16.20) for the prediction of the circular AR(4) process and the strongly noncircular Ikeda map. Observe that for the noncircular Ikeda map, the ACLMS subfilter dominated the output of the hybrid filter, as indicated by the values of $\lambda(k)$ approaching unity. For the circular AR(4) signal, after the faster initial convergence of CLMS, both ACLMS and CLMS subfilters converged to the same steady state error, as illustrated by $\lambda(k) \rightarrow 0.5$.

To summarise:

- Due to the adaptive and collaborative mode of operation, hybrid filters have great potential in the processing of nonlinear and nonstationary data.
- To test for signal modality, the subfilters within a hybrid filter are typically of different natures, e.g. linear and nonlinear. As a rule of thumb, fully complex nonlinearity is general enough to test for the lack of linear data nature.
- It has been shown that a hybrid filter, consisting of both widely linear ACLMS and standard CLMS trained subfilters, can be used for adaptive filtering of the generality of complex valued signals. It will have as fast initial convergence as CLMS, whereas its steady state properties will be similar or better than those of the subfilter most suited to the signal in hand.
- Such a hybrid filter may also be used for online testing for circularity of real world complex valued signals, as illustrated in Figure 16.14(b).
- Unlike the hypothesis based statistical tests [83, 191], which operate on a block by block basis, the tests based on collaborative learning operate in an online adaptive manner. Such tests can be incorporated into supervised or semiblind forecasting and predictive control strategies, even though they can be taken only in relative terms.

17

Adaptive Filtering Based on EMD

The performance of adaptive filtering algorithms depends strongly on the amplitude range and correlation structure of the input signal. Two frequently used approaches which help to improve the speed of convergence and stability of adaptive filtering algorithms are:

- *Normalisation*, which is performed by dividing the weight update by an instantaneous estimate of the tap input power, thus making an adaptive filter independent of the signal power [308];
- *Transform domain filtering*, whereby the signal is first transformed into a 'transform domain', typically by applying the Fast Fourier Transform (FFT) or the Discrete Cosine Transform (DCT) [69, 149], to produce a decorrelated data sequence, thus improving convergence.

Whereas normalisation in learning algorithms is well established [190, 308], transform domain filtering uses a range of techniques, each of which requires its own assumption and trade-offs. For instance, Fourier based techniques use linear superposition of trigonometric functions to represent the signal, and require rather strong assumptions of piece-wise stationarity and periodicity. Such transforms are typically applied to real valued data, which is then processed using complex adaptive filters.

Two of most important features in the interpretation of real world data are the *time scale* and *amplitude distribution* – therefore, so as best to use this information, the set of bases in transform domain representations should be:

- complete, for mathematical tractability and in order to provide good accuracy;
- orthogonal, to reduce spectral leakage and maintain a non-negative signal power;
- local, to have excellent resolution in frequency;
- adaptive, to be able to deal with nonlinear and nonstationary data.

This chapter presents a class of complex Empirical Mode Decomposition (EMD) algorithms, a fully data driven technique suitable for a time–frequency representation of complex valued nonlinear and nonstationary processes [9, 128, 196, 251]. The local and adaptive nature of

Complex Valued Nonlinear Adaptive Filters: Noncircularity, Widely Linear and Neural Models
Danilo P. Mandic and Vanessa Su Lee Goh

EMD facilitates the time–frequency representation at the level of instantaneous frequency, whereas the orthogonality of adaptive empirical modes preserves the physical meaning of the components. The benefits of using complex EMD algorithms in conjunction with complex adaptive filtering are illustrated in a nonlinear adaptive prediction setting.

17.1 The Empirical Mode Decomposition Algorithm

Empirical mode decomposition [128] is a technique to adaptively decompose a signal, by means of a process called the sifting algorithm, into a finite set of oscillatory components called intrinsic mode functions (IMFs), which represent the oscillation modes (scales) embedded in the data. The real valued EMD algorithm decomposes an arbitrary signal $x(k)$ into a sum of IMFs $\{C_i(k)\}$, $i = 1, \ldots, M$ and the residual $r(k)$, that is

$$x(k) = \sum_{i=1}^{M} C_i(k) + r(k) \tag{17.1}$$

The residual $r(k)$ is the last IMF and its physical meaning is the trend within the signal. To provide a meaningful time–frequency representation at the level of instantaneous frequency, the oscillatory modes should have the following properties:

- The upper and lower envelope are symmetric – at every point the mean value of the upper and lower envelope is zero;
- The number of zero crossings and the number of extrema are equal or they differ at most by one.

The first condition ensures that the instantaneous frequency derived from EMD will not have local fluctuations due to the asymmetry of the envelopes and IMFs can be interpreted as AM functions, whereas the second condition resembles a 'narrowband' requirement which guarantees the existence of a local oscillatory mode and hence high accuracy in the time–frequency domain.

The sifting algorithm. The IMFs are extracted from a real world signal $x(k)$ by means of an iterative algorithm called the sifting algorithm, described in Table 17.1.

Table 17.1 The sifting algorithm within the real valued EMD

1. Connect the local maxima of $x(k)$ with a spline U to form the upper envelope of the signal; connect the local minima of $x(k)$ with a spline L to form the lower envelope of the signal
2. Subtract the mean envelope $m = (U + L)/2$ from the signal to obtain a proto-IMF
3. Repeat Step 1 and Step 2 until the resulting signal is a proper IMF (meets the design criteria above)
4. Subtract the IMF from the signal $x(k)$, the residual is regarded as a new signal, that is, $r(k) \rightarrow x(k)$, and the process is repeated from Step 1
5. The sifting process is completed when the residual of Step 4 is a monotonic function.

Step 3 in Table 17.1 is checked indirectly, by evaluating a *stoppage criterion* (the standard deviation of IMFs) given by

$$\sum_{k=0}^{T} \frac{|h_{n-1}(k) - h_n(k)|^2}{h_{n-1}^2(k)} \leq SD \tag{17.2}$$

where $h_n(k)$ and $h_{n-1}(k)$ represent two successive sifting iterates. The parameter SD is set empirically and usually has the value 0.2–0.3.

17.1.1 Empirical Mode Decomposition as a Fixed Point Iteration

The sifting process within EMD can be viewed as an iterative application of a nonlinear operator T, defined as

$$h_{n+1} = T[h_n] = h_n - m_n(h_n) \tag{17.3}$$

where h_n denotes the result of the nth iteration of the sifting process and m_n is the nth local mean signal (depending on h_n). If $T[\cdot]$ is a contractive operator, since by definition IMFs have a zero local mean, the iteration (Equation 17.3) is a fixed point iteration and has a solution

$$h_n = T[h_n] \tag{17.4}$$

that is, IMFs can be considered as fixed points[1] of the iteration $h_{n+1} = T[h_n]$.

In theory, the iteration (Equation 17.3) could converge in one step, however, due to the spline approximation within the sifting algorithm and the artifacts (hidden scales) introduced by it, spurious extrema may be generated after every round of sifting. To deal with this problem, the sifting operation is carried out until the empirical stoppage criterion (Equation 17.2) is satisfied.[2]

To be able to interpret the stoppage criterion as a metric (as required by the Contraction Mapping Theorem, see Appendix P) we can modify Equation (17.2) to ensure $d(h_n, h_{n-1}) = d(h_{n-1}, h_n)$, for instance

$$d_1(h_n, h_{n-1}) = \sum_{k=0}^{T} \frac{|h_{n-1}(k) - h_n(k)|^2}{T} \tag{17.5}$$

or

$$d_2(h_n, h_{n-1}) = \sum_{k=0}^{T} \frac{|h_{n-1}(k) - h_n(k)|^2}{|h_{n-1}(k) + h_n(k)|^2} \tag{17.6}$$

---

[1] For more detail on contraction mappings and fixed point iteration, see Appendix P.

[2] A low value of SD, that is, sifting to the extreme, would remove all the amplitude modulation from the signal which would result in purely frequency modulated components (IMFs). This is not desired as IMFs would not have physical meaning [128].

This will enable to test for the Lipschitz continuity of the series $h_1, \ldots, h_\infty$, and may hence be used to determine:

- The existence of IMFs, through the examination of the series $h_1, \ldots, h_\infty$ and the use of CMT [164];
- Uniqueness, speed of convergence, and properties of the class of signals for which the sifting process converges.

17.1.2 Applications of Real Valued EMD

A time–frequency representation of a signal is produced by applying the Hilbert transform $\mathcal{H}[\cdot]$ to each IMF $C_i(k)$, $i = 1, \ldots, M$, to generate a set of analytic signals which have the IMFs as their real part and their Hilbert transforms as the imaginary part. Equation (17.1) can therefore be augmented to its analytic form given by

$$X(t) = \sum_{i=1}^{M} a_i(t) \cdot e^{\jmath \theta_i(t)} \tag{17.7}$$

where the trend $r(t)$ is omitted, due to its overwhelming power and lack of oscillatory behaviour.

This form of EMD has a time-dependent amplitude $a_i(t)$, whereas the phase function $\theta_i(t)$ provides additional computational power. The quantity

$$f_i(t) = \frac{\mathrm{d}\theta_i}{\mathrm{d}t} \tag{17.8}$$

represents the *instantaneous frequency* [52]; by plotting the amplitude $a_i(t)$ versus time t and frequency $f_i(t)$, a Time–Frequency–Amplitude (TFA) representation of the entire signal, called the Hilbert–Huang Spectrum (HHS), is obtained. It is this combination of the concept of instantaneous frequency and EMD that makes the framework so powerful as a signal decomposition tool.

Consider a signal which consists of two added sine waves with different frequencies, shown in the first row of Figure 17.1(a). By using EMD, the signal is decomposed into the intrinsic

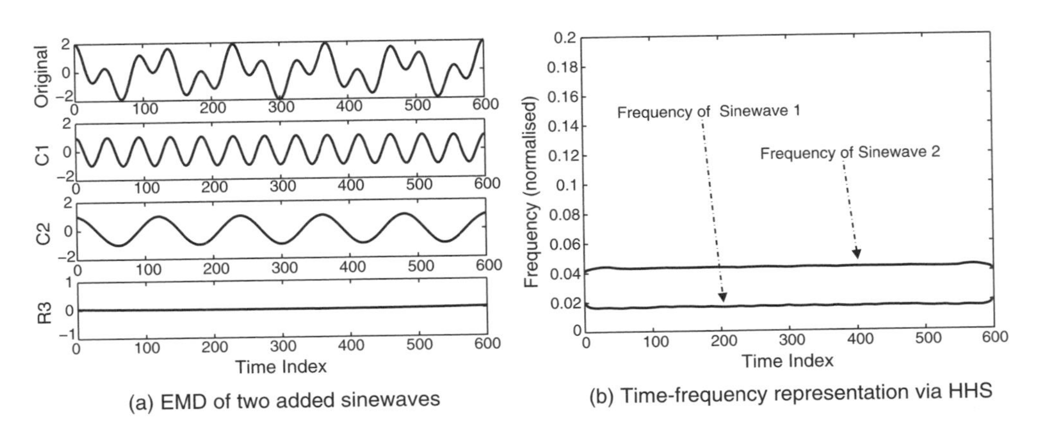

(a) EMD of two added sinewaves

(b) Time-frequency representation via HHS

Figure 17.1 Empirical mode decomposition of two added sinewaves

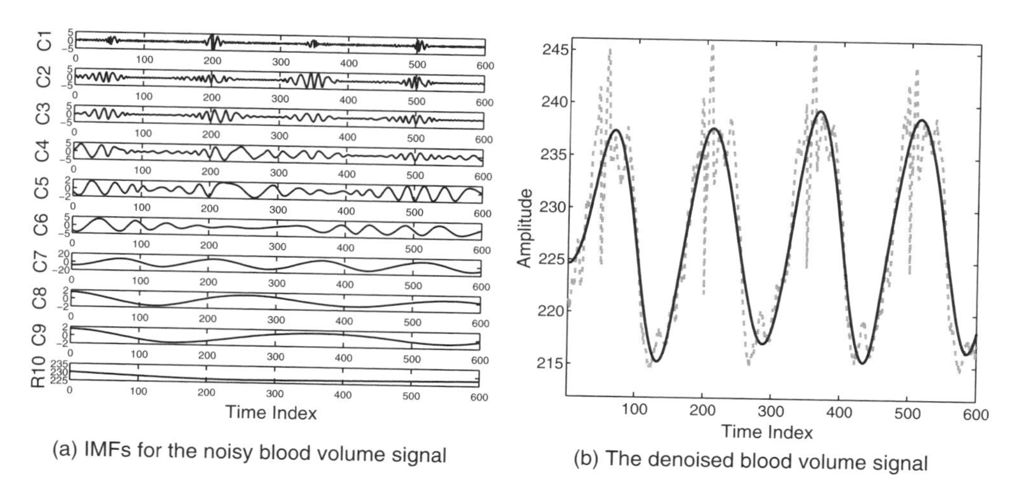

(a) IMFs for the noisy blood volume signal (b) The denoised blood volume signal

Figure 17.2 Application of EMD for the denoising of real world blood volume signal. The denoised blood volume is constructed by fusion of IMFs C_7–C_9

mode functions C_1 and C_2, and the residual (the last IMF) R_3. The Hilbert–Huang spectrum of the IMFs is shown in Figure 17.1(b). The frequencies of the sine waves comprising the original signal are clearly visible in the time–frequency spectrum in Figure 17.1(b). Figure 17.1 illustrates that EMD first identifies the highest local frequency oscillation within a signal; every subsequent IMF exhibits oscillations at a lower frequency, whereas the residual is either the mean trend or a constant.

Figure 17.2 illustrates the application of EMD to the estimation of blood volume signal during heart surgery. A sensor in the human heart records both the blood flow (a slow sine wave with drifts due to sensor movement) and the superimposed high frequency electrocardiogram (ECG). The goal is to extract a pure blood volume signal. Since the ECG and blood volume are statistically coupled, standard signal processing techniques (both supervised and blind) are bound to fail. By EMD, it has been possible to identify the IMFs in Figure 17.2(a) which represent the denoised blood volume (IMFs C_7–C_9). The original noisy blood volume (dotted line) and the denoised blood volume (solid line) are shown in Figure 17.2(b).

17.1.3 Uniqueness of the Decomposition

To this point the analysis of EMD has been based on the following assumptions:

- The signal in hand has at least one minimum and one maximum;
- The time scales are defined as the time interval between the two consecutive extremum points;
- If the signal is monotonic, that is, with no extrema, but with inflection points, such a signal can be differentiated several times to facilitate EMD.

The fully data driven, empirical nature of EMD, as described above, gives it clear advantages but also compromises the uniqueness of the decomposition. Signals with similar statistics often yield different decompositions, both in terms of the number and properties of IMFs.

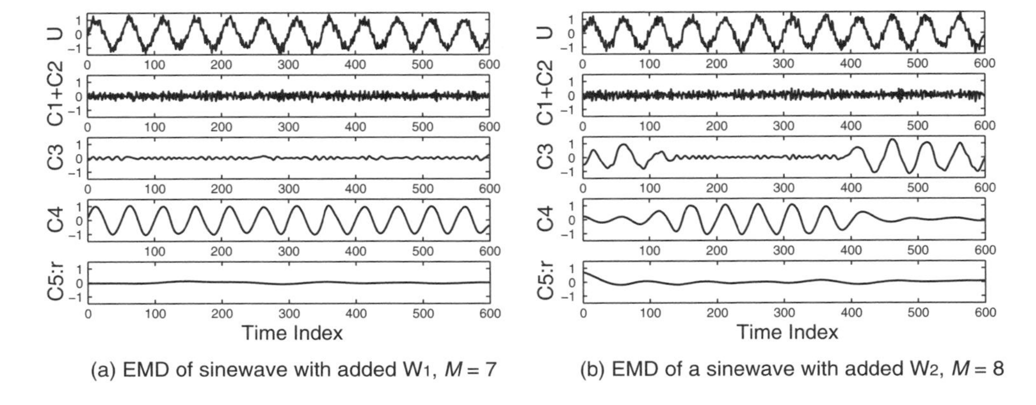

Figure 17.3 Illustration of mode mixing for a sinusoid corrupted by two different realisations of white Gaussian noise W_1 and W_2. EMD produced a different number of intrinsic mode functions, $M = 7$ in (a) and $M = 8$ in (b). The original sinusoid is located in different modes: C_4 in (a) and $C_3 + C_4$ in (b). For convenience $C_5 : r = \sum_{i=5}^{M} C_i + r$

This is manifested by so-called mode mixing, whereby two or more IMFs have the same oscillatory modes. To provide further insight into mode mixing, consider a noisy sinusoid U, corrupted by two different realisations, W_1 and W_2, of additive white Gaussian noise with the same statistical properties. The $M = 7$ extracted IMFs for the first case are shown in Figure 17.3(a). The original sinusoid corresponds to the fourth IMF C_4, whereas the noise is contained within the IMFs C_1 and C_2. Figure 17.3(b) shows the results of performing EMD on the same sinusoid corrupted by W_2, resulting in $M = 8$ IMFs. In addition to a different number of IMFs as compared with the first case, mode mixing occurred – the signal of interest is contained within two IMFs: C_3 and C_4.

The issue of mode mixing can be addressed by performing EMD over a number of independent realisations of WGN (Ensemble EMD [317]), however, this still does not guarantee a unique number of extracted IMFs. For a meaningful physical interpretation of multidimensional sources (for example the real and imaginary channel of a complex vector), the number of IMFs in all the channels should be the same.

The problem of uniqueness is addressed in Section 17.3; to this end we first need to introduce complex extensions of EMD.

17.2 Complex Extensions of Empirical Mode Decomposition

The main issue in multivariate extensions of EMD is that of envelope interpolation, as this relies on finding the extrema within the signal. Since $\mathbb{C}$ is not an ordered field (see Appendix A), it is not possible to find the local extrema directly, and several algorithms have been introduced to circumvent this problem. These include:

- The complex EMD algorithm [284] proposed by Tanaka and Mandic effectively applies real valued EMD to the signals corresponding to the positive and negative frequency component of the spectrum of analytic signals.

- The rotation invariant complex EMD (RIEMD) [9] defines the extrema as the points where the angle of the first derivative of the signal changes its sign. Since all the operations are performed directly in $\mathbb{C}$, this method provides a single set of complex IMFs.
- The bivariate EMD [251] is an extension of RIEMD, whereby envelope curves are obtained by projecting a bivariate signal in multiple directions and interpolating their extrema.

17.2.1 Complex Empirical Mode Decomposition

The first method to 'complexify' EMD was introduced in 2007, termed Complex Empirical Mode Decomposition [284]. The method is based on the inherent relationship between a complex signal and the properties of the Hilbert transform. The idea behind this approach is rather intuitive – a complex signal has a two-sided, asymmetric spectrum and can be converted into a sum of two analytic signals by first separating the positive and negative frequency components of the spectrum and then converting back into the time domain. Standard EMD is subsequently applied to the two derived signals.

More precisely, by processing a complex valued signal $x(k)$ for which the spectrum is $X(e^{\jmath w})$ with the filter given by

$$H(e^{\jmath w}) = \begin{cases} 1, & 0 < \omega \leq \pi \\ 0, & -\pi < \omega \leq 0 \end{cases} \tag{17.9}$$

two analytic signals $X_+(e^{\jmath w})$ and $X_-(e^{\jmath w})$, which correspond to the positive and the negative frequency parts of $X(e^{\jmath w})$ are generated. The subsequent application of the Inverse Fourier Transform (IFT) yields sequences $x_+(k)$ and $x_-(k)$; standard real valued EMD can then be applied to the real parts of $x_+(k)$ and $x_-(k)$, to give

$$x_+(k) = \sum_{i=1}^{M^+} x_i(k) + r_+(k), \quad x_-(k) = \sum_{i=-1}^{-M^-} x_i(k) + r_-(k) \tag{17.10}$$

where symbols M^+ and M^- denote respectively the number of IMFs for the positive and the negative frequency parts. The resulting two sets of IMFs are combined to form a complex valued signal

$$x(k) = \sum_{i=-M^-, i \neq 0}^{i=M^+} x_i(k) + r(k) \tag{17.11}$$

The ith complex IMF is therefore defined as

$$C_i(k) = \begin{cases} x_i(k) + \jmath\mathcal{H}[x_i(k)], & i = 1, \ldots, M^+, \\ \{x_i(k) + \jmath\mathcal{H}[x_i(k)]\}^*, & i = -M^-, \ldots, -1 \end{cases} \tag{17.12}$$

Although it has a straightforward mathematical derivation and preserves the dyadic filter bank property of EMD when processing complex noise [284], there is no guarantee that the positive and negative parts of the signal will yield equal numbers of IMFs. This makes the physical interpretation of the results difficult.

17.2.2 Rotation Invariant Empirical Mode Decomposition (RIEMD)

A critical aspect of the derivation of EMD in $\mathbb{C}$ is the definition of an extremum. In [9] it was proposed to use the locus where the angle of the first-order derivative (with respect to time) changes sign; this way it can be assumed that a local maximum will be followed by a local minimum (and vice versa). This criterion is equivalent to the extrema of the imaginary part of the signal, that is

$$\begin{aligned} \angle \dot{Z}(t) = 0 &\Rightarrow \angle\{\dot{x}(t) + \jmath \cdot \dot{y}(t)\} = 0 \\ &\Rightarrow \tan^{-1}\frac{\dot{y}(t)}{\dot{x}(t)} = 0 \Rightarrow \dot{y}(t) = 0 \end{aligned} \tag{17.13}$$

where $Z(t)$ is a complex signal (for convenience we here use a continuous time index t). The cubic spline interpolation is then performed directly in $\mathbb{C}$, to obtain complex valued envelopes, which are then averaged to obtain the local mean. Unlike the original CEMD [284], this method yields a single set of complex valued IMFs, and is a natural extension of real valued EMD. To illustrate RIEMD, consider a set of wind[3] speed and direction measurements, which have been made complex by convenience of representation (see Chapter 2). Figure 17.4(a) shows the 'wind rose' for the original wind signal, whereas Figure 17.4(b) illustrates the contribution of the sixth and seventh IMF ($C_6 + C_7$). It is clear that the complex IMFs have physical meaning as they reveal the dynamics of the original signal at different scales.

17.2.3 Bivariate Empirical Mode Decomposition (BEMD)

The bivariate EMD algorithm [251] calculates local mean envelopes based on the extrema of both (real and imaginary) components of a complex signal, thus yielding more accurate estimates than RIEMD. The algorithm effectively sifts rapidly rotating signal components from

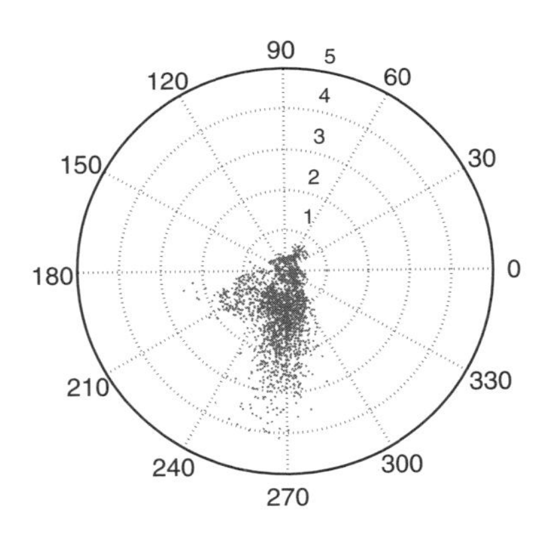

(a) A complex wind signal

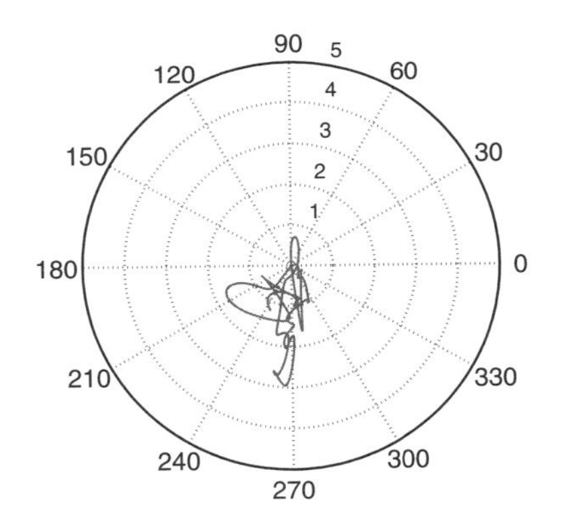

(b) Sum of the sixth and seventh complex IMF

Figure 17.4 A complex wind signal analysed by the Rotation Invariant Complex EMD

[3]Publicly available from http://mesonet.agron.iastate.edu/request/awos/1min.php.

Table 17.2 The bivariate EMD (BEMD) algorithm [251]

1. Obtain N signal projections, $\{p_{\theta_n}\}_{n=1}^N$, by projecting the complex signal $z(t)$ by means of a unit complex number $e^{j\theta_n}$, in the direction of θ_n, to obtain

$$p_{\theta_n} = \Re\left(e^{j\theta_n} z(t)\right), \quad n = 1, \ldots, N \tag{17.14}$$

   where $\theta_n = 2n\pi/N$
2. Find the locations $\{t_j^n\}_{n=1}^N$ of the maxima of $\{p_{\theta_n}\}_{n=1}^N$
3. Interpolate (using spline interpolation) between the maxima $[t_j^n,\ z(t_j^n)]$, to obtain the envelope curves $\{U_{\theta_n}\}_{n=1}^N$
4. Calculate the mean, $m(t)$, of all the envelope curves
5. Subtract $m(t)$ from the input signal $z(t)$ to yield an 'oscillatory' component, that is, $d(t) = z(t) - m(t)$. The stopping criterion is the same as for real EMD.

the slowly rotating ones and employs the same complex cubic spline interpolation scheme as RIEMD. The operation of BEMD is summarised in Table 17.2.

Bivariate EMD [251] operates in a similar fashion to rotation invariant EMD: by projecting the signal in N directions, the approach finds extrema in several directions and constructs a 3D tube by interpolating them. The local mean (centre of the tube) can be found as the barycentre of the N points or as the intersection of straight lines passing through the middle of the tangents.

Figure 17.5 illustrates the process of finding the local mean of a complex signal using RIEMD and BEMD. In both cases, envelopes are calculated in multiple directions, and are then averaged to obtain the local mean. The local mean estimate obtained by BEMD is more accurate than that using RIEMD, as it employs more directions to calculate the envelopes. This is illustrated around point 50 on the X-axis, where RIEMD could not estimate the true local mean, whereas BEMD was able to correctly differentiate between a local minimum and local maximum. For $N = 2$ BEMD and RIEMD are equivalent.

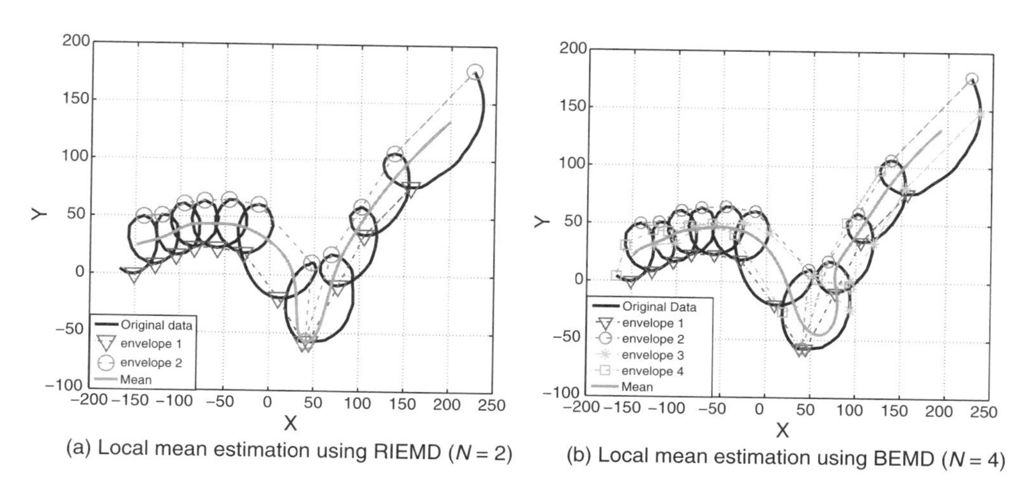

(a) Local mean estimation using RIEMD ($N = 2$)

(b) Local mean estimation using BEMD ($N = 4$)

Figure 17.5 Local mean estimation using RIEMD and BEMD

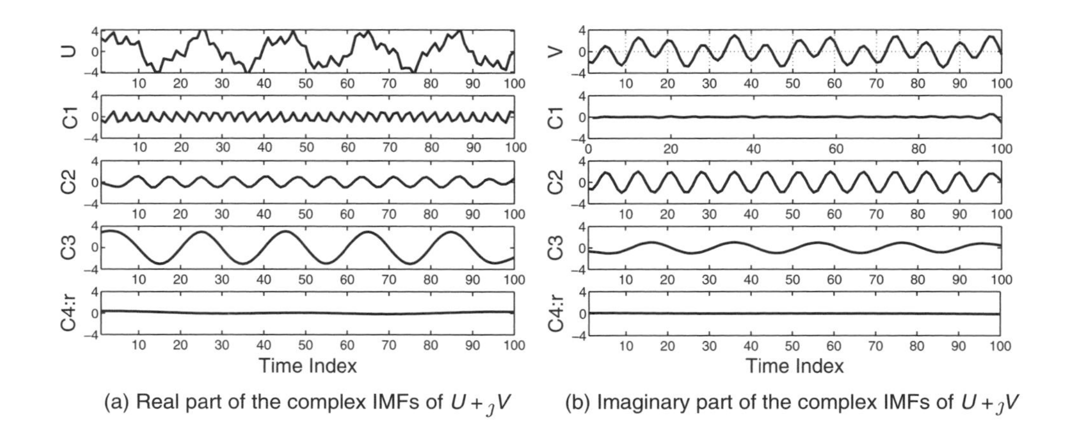

Figure 17.6 Uniqueness of the scales for RIEMD and BEMD

17.3 Addressing the Problem of Uniqueness

Both RIEMD and BEMD solve the problem of uniqueness by operating directly on the complex signal, as the algorithms produce the same number of IMFs for the real and imaginary part. To illustrate the ability of RIEMD and BEMD to produce 'common scales' within complex valued IMFs, Figure 17.6 shows IMFs for the signals $U + \jmath V$, for which the real and the imaginary part, U and V, are obtained from a set of three sine waves. Common to both U and V are the frequencies of two of the sinusoids, although their amplitude and phase were different. A third high-frequency sinusoid was added to U only. Clearly, U and V have a different number of oscillatory modes, however, by virtue of BEMD the number of IMFs for U is equal to the number of IMFs for V. The high-frequency sinusoid, contained only in U, is shown in the real part of the first IMF in Figure 17.6(a). The common frequency scales are clearly visible in IMFs C_2 and C_3. Thus, if mode mixing occurs this does not pose a problem as it occurs simultaneously in the real and imaginary part of a complex IMF.

17.4 Applications of Complex Extensions of EMD

EMD can be considered within so-called data fusion via fission framework [193], whereby the signal is first decomposed into a number of orthogonal components (fission), and then, depending on the application, the most relevant components are recombined to produce an enhanced signal (fusion), as shown in Figure 17.7(a). The standard way to use EMD in this context is to apply a binary mask to the set of IMFs, for instance, in the denoising application in Figure 17.2, IMFs C_7–C_9 are identified as information bearing components and are summed up to produce the denoised blood volume signal.

Nonlinear adaptive prediction. The use of a binary mask to identify the information bearing IMFs within EMD is not suitable for nonstationary processes. To make full use of EMD as a preprocessing step, it should be followed by an adaptive combiner, as shown in Figure 17.7(b). Any adaptive filtering architecture: linear, nonlinear, feedforward, or feedback

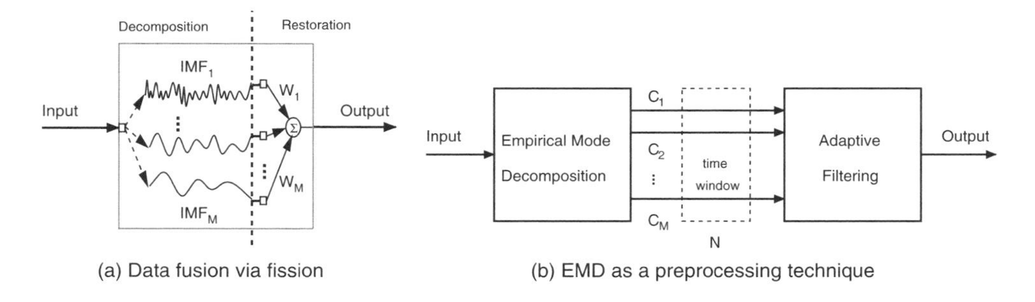

Figure 17.7 Application of EMD in the signal preconditioning stage

can be employed. By design the IMFs are locally orthogonal which helps with the speed of convergence[4] of adaptive filtering algorithms, whereas the locality of complex EMD makes this approach suitable for a generality of nonlinear and nonstationary complex valued signals [171]. Adaptive filtering can be performed across all the IMFs for a fixed time instant, or in a block fashion for a time window of length N.

To illustrate the performance, consider the task of one step ahead prediction of the complex wind signal [194], for which BEMD was used to obtain $M = 11$ complex IMFs. In the adaptive filtering stage, the normalised CLMS (NCLMS) algorithm was applied and the size of the temporal window was chosen to be $N = 3$, to give an overall block of $M \times N = 33$ data points. For a fair comparison, the filter length for the standard NCLMS algorithm was selected as $L = 33$. The prediction gain for the real part of the signal was $R_p = 23$ dB for the BEMD-NCLMS case and $R_p = 18.4$ dB when only NCLMS was applied. For the imaginary part of the wind signal these values were $R_p = 22$ dB for BEMD-NCLMS and $R_p = 17.8$ dB for NCLMS. Figure 17.8 illustrates the tracking performance of NCLMS and BEMD-NCLMS

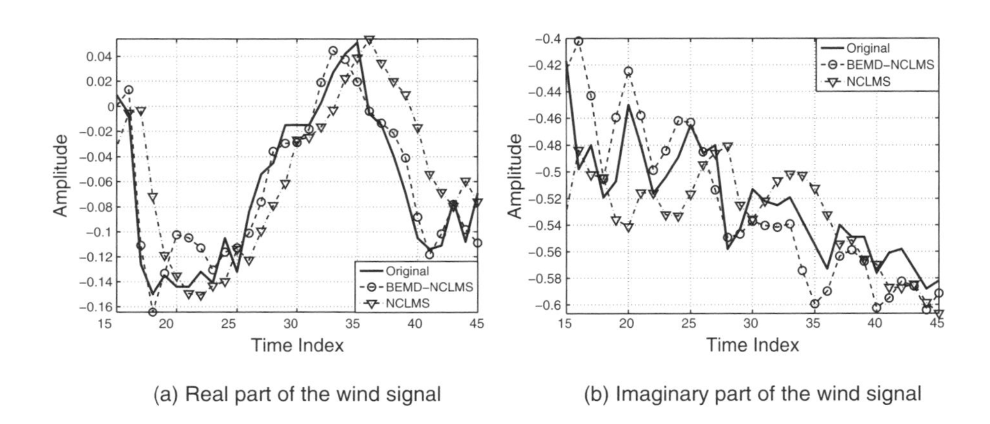

Figure 17.8 Tracking performance of BEMD-NCLMS and NCLMS

[4]In theory, since the IMFs are locally orthogonal, even a linear filter should perform well for the generality of signals.

Table 17.3 Prediction gains R_p (dB) for standard and EMD enhanced algorithms

CLMS	NCLMS	CNGD	CRTRL	ACRTRL
8.41	10.99	12.39	13.21	14.68
BEMD-CLMS	BEMD-NCLMS	BEMD-CGND	BEMD-CRTRL	BEMD-ACRTRL
11.06	13.20	13.13	15.01	16.66

algorithms for the complex wind signal; BEMD-NCLMS produced estimates which are better aligned with the original signal than those produced by the NCLMS.

In the next experiment, several complex adaptive filtering architectures, ranging from a linear feedforward filter through to a multilayer nonlinear recurrent neural network, were used to perform one step ahead prediction on a segment of raw wind data. The performances were compared with the corresponding performances for the EMD-preprocessed data. The prediction gain was calculated directly in $\mathbb{C}$, the number of complex IMFs and hence the length of the tap input delay line was $M = 10$, and recurrent neural networks had three neurons. Table 17.3 shows that the performance improved with the complexity of the algorithm, with the linear CLMS having worst performance and the augmented nonlinear CRTRL (ACRTRL) performing best. Also, due to the local orthogonality of IMFs, optimal learning rates for the decorrelated signals were on the average an order of magnitude larger.

To summarise:

- Empirical Mode Decomposition (EMD) demonstrates a considerable strength in the analysis of nonlinear and nonstationary real world data, providing a framework for information 'fusion' by performing signal 'fission' into its oscillatory components.
- The extension of EMD to the complex domain $\mathbb{C}$ enables the modelling of amplitude–phase relationships within multichannel data, and also helps with the problems of 'mode mixing' and the uniqueness of the decomposition.
- Due to the local orthogonality of intrinsic mode functions, complex extensions of EMD can be used as a preprocessing step in adaptive filtering applications.
- As signal decorrelation is achieved directly in $\mathbb{C}$, this helps increase the speed of convergence and accuracy of complex valued adaptive filtering algorithms.

18

Validation of Complex Representations – Is This Worthwhile?

So far we have addressed some inherent properties of complex processes, such as complex nonlinearity (Chapter 4), augmented complex statistics (Chapter 12), and topological properties of complex mappings (Chapter 11). It seems clear that complex valued models will be more advantageous the greater the coupling between the real and imaginary components of a process – that is, the more 'complex' the process. This is borne out by empirical evidence (Chapter 12) where it is shown that the relative benefit of complex valued modelling is related to the degree of coupling between the speed and direction components of the wind profile. In addition, signal dynamics and the degree of averaging, which affect the component coupling, will have a major influence on the choice of an appropriate signal model. Again wind data provide an example; the areas denoted by A, B and C in Figure 18.1 correspond respectively to 'high', 'medium' and 'low' dynamics.

It has been shown in Chapter 13 that the use of widely linear model is justified only if there is a statistical evidence that the signal in hand is not second-order circular. However, the pseudocovariance matrix is estimated from the data available, and such estimate will, in general, be nonzero, although the actual source is circular. Since we are mostly interested in 'complex by convenience of representation' signals (see Chapter 2), it would appear vital to establish a rigorous statistical testing framework which would reveal whether the complex representation is worthwhile – that is, does it offer theoretical and practical advantages over the bivariate[1] or dual univariate signal models?

Following on from the Delay Vector Variance (DVV) technique for statistical testing for signal nonlinearity [83, 86], one such statistical test for the 'complex nature' of real-world processes [85] is the ratio of statistical differences between realisations under the null hypothesis of 'linear bivariate' and 'linear circular.'

[1]For convenience, we use the term 'bivariate' to denote the 'real valued bivariate' signals.

Complex Valued Nonlinear Adaptive Filters: Noncircularity, Widely Linear and Neural Models
Danilo P. Mandic and Vanessa Su Lee Goh

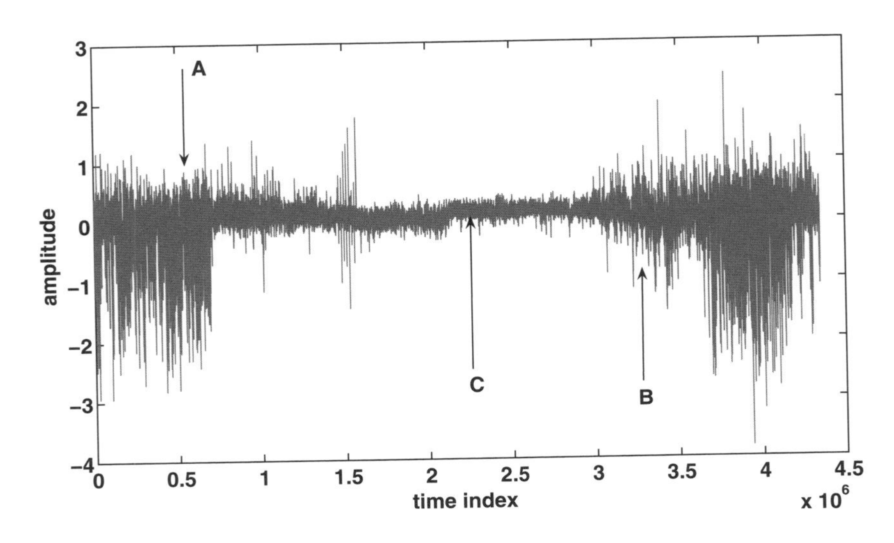

Figure 18.1 Wind speed recording over one day. Regions A, B, and C correspond to different wind regimes (high, medium, and low)

18.1 Signal Modality Characterisation in $\mathbb{R}$

Signal modality characterisation is becoming an increasingly important area of multidisciplinary research and considerable effort has been put into devising efficient algorithms for this purpose. Research in this area started in physics in the mid 1990s [303], but its applications in machine learning and signal processing are only recently becoming apparent [84]. As changes in the signal nature between, say, linear and nonlinear and deterministic and stochastic can reveal information (knowledge) which is critical in certain applications (e.g. health conditions), the accurate characterisation of the nature of signals is a key prerequisite to choosing a signal processing framework.

By the 'nature' of a signal we refer to the following fundamental properties: [82, 83, 265]:

P1. *Linear* (strict definition) – a linear signal is generated by a linear time-invariant system, driven by white Gaussian noise.

P2. *Linear* (commonly adopted) – property **P1** is relaxed somewhat by allowing the amplitude distribution of the signal to deviate from the Gaussian distribution (a linear signal from **P1** is measured by a static, possibly nonlinear, observation function).

P3. *Nonlinear* – a signal that does not meet the criteria **P1** or **P2** is considered nonlinear.

P4. *Deterministic* (predictable) – a signal is considered deterministic if it can be precisely described by a set of equations.

P5. *Stochastic* – a signal that is not deterministic.[2]

[2]The Wold decomposition theorem [314] states that any discrete stationary signal can be decomposed into its *deterministic* and *stochastic* (random) component, which are uncorrelated. This theorem forms the basis for many prediction models, since the presence of a deterministic component imposes a bound on the performance of these models.

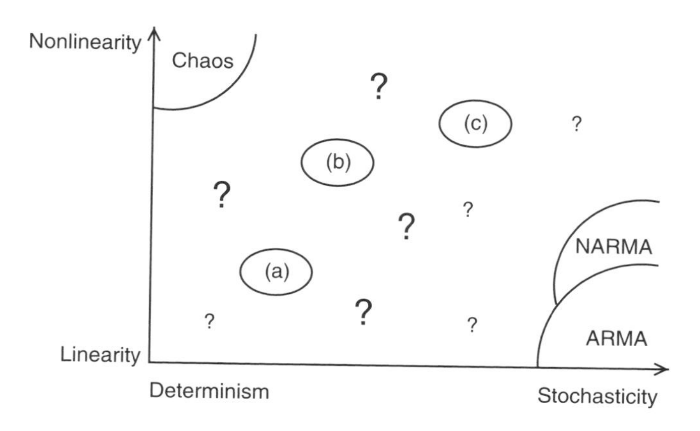

Figure 18.2 A variety of signal modalities spanned by the properties 'stochastic' and 'nonlinear'. Classes of signals for which the generating mechanisms are well understood are 'Chaos', 'ARMA', and 'NARMA'

The range of real world signals spanned by their *linear vs nonlinear* and *deterministic vs stochastic* natures is illustrated in Figure 18.2 (modified from [263]). It is interesting that the classes of signals which are well established and understood, such as the linear stochastic autoregressive moving average (ARMA) models and nonlinear deterministic chaotic signals, are at the opposite corners of Figure 18.2. Real world signals, however, are likely to belong to the areas denoted by (a), (b), (c) or '?', since they are recorded in noisy[3] environments and by nonlinear sensors; these are major signal classes about which we know little or nothing.

18.1.1 Surrogate Data Methods

The concept of 'surrogate data' was introduced by Theiler *et al.* [287], and has been extensively used in the context of statistical testing for signal nonlinearity; more detail on surrogate data methods can be found in [144, 265, 288, 290]. Hypothesis testing assesses a fundamental property of signal (say nonlinearity) by generating a large number, say 100, of independent linear realisations of the original signal (surrogates) and comparing the 'test statistic' for the surrogates against that of the original signal.

The basic principle of hypothesis based statistical testing for signal nonlinearity[4] can be summarised in the following steps:

N1. Establish a *null hypothesis* H_0, e.g. the signal is generated by a linear stochastic system driven by white Gaussian noise.

N2. Generate a number of independent surrogates, which are *linear realisations* of the original signal.

[3]The environment is also typically statistically nonstationary, and the signal modality changes with time, say from (a) to (b), or from ARMA to Chaos (heart rates, epileptic seizures), which may, e.g. indicate a health hazard.

[4]The analysis of the nonlinearity of a signal can often provide insights into the nature of the underlying signal production system. However, care should be taken in interpreting the results, since the assessment of nonlinearity within a signal does not necessarily imply that the underlying signal generation system is nonlinear: the input signal and system (transfer function) nonlinearities are confounded.

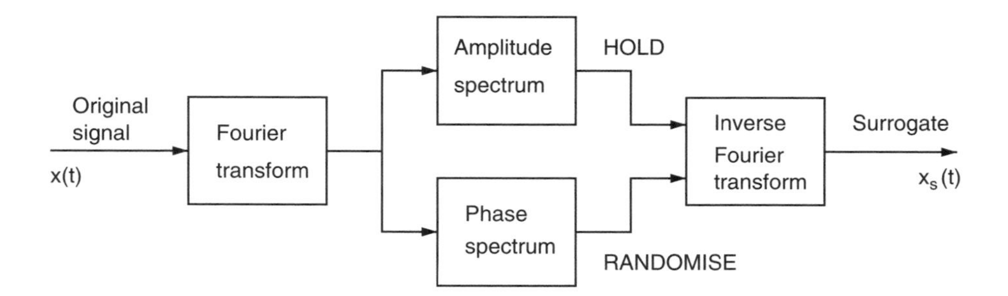

Figure 18.3 Surrogates generated based on the Fourier transform

N3. Establish a discriminating criterion between the linear and nonlinear signal, the so called *test statistic*.

N4. Based on the number of rejections of the null hypothesis from **N1**, the nature of the original signal is judged linear or nonlinear.

Since surrogate data are linear realisations of the original signal, they can be generated in many ways, for instance by ARMA modelling. It is, however, much more desirable to have nonparametric generation methods for surrogate data. By definition (**P1** and **P2**), the property of signal linearity is derived from the second-order statistics (mean, variance, autocorrelation or equivalently amplitude spectrum), and hence for a linear signal the phase spectrum and higher-order statistics (HOS) are irrelevant; one simple way to generate a number of surrogates is based on the Fourier transform (FT surrogates), as illustrated in Figure 18.3. The FT surrogates are generated by simply performing the inverse Fourier transform of a signal generated from the original amplitude spectrum and the randomised phase spectrum; this method, however, is not suitable for signals described by **P2**.

A reliable surrogate data method, capable of generating surrogates for data observed through a static nonlinearity, is the 'iterative Amplitude Adjusted Fourier Transform' (iAAFT) method [264]; it has been shown to produce superior results compared with other available surrogate data generation methods [160, 265].

The iAAFT method can be summarised as follows:

S1. Let $\{|S_k|\}$ be the Fourier amplitude spectrum of the original time series s, and $\{c_k\}$ the amplitude sorted version of the original time series

Repeat:

S2. At every iteration j generate two additional series:

(i) $r^{(j)}$, which has the same distribution as the original signal
(ii) $s^{(j)}$, which has the same amplitude spectrum as the original signal

Starting with $r^{(0)}$, a random permutation of the time samples of the original time series:

1. Compute the phase spectrum of $r^{(j-1)} \rightarrow \{\phi_k\}$

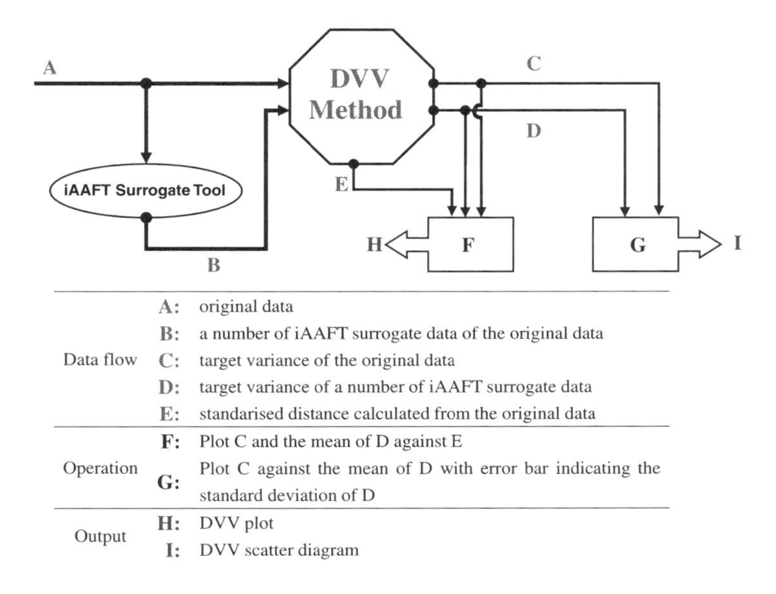

Data flow	A:	original data
	B:	a number of iAAFT surrogate data of the original data
	C:	target variance of the original data
	D:	target variance of a number of iAAFT surrogate data
	E:	standarised distance calculated from the original data
Operation	F:	Plot C and the mean of D against E
	G:	Plot C against the mean of D with error bar indicating the standard deviation of D
Output	H:	DVV plot
	I:	DVV scatter diagram

Figure 18.4 Block diagram for the Delay Vector Variance (DVV) method

2. Compute $s^{(j)}$ as the inverse transform of $\{|S_k| \exp(\jmath\, \phi_k)\}$
3. Compute $r^{(j)}$ as obtained by rank-ordering $s^{(j)}$ so as to match $\{c_k\}$

Until error convergence.

18.1.2 Test Statistics: The DVV Method

A convenient test statistic,[5] which makes use of some notions from nonlinear dynamics and chaos theory (embedding dimension and phase space) is the Delay Vector Variance (DVV) method [83]. It is based upon examining the local predictability of a signal in phase space, which when combined with the surrogate data methodology allows one to examine simultaneously the determinism and nonlinearity within a signal.

The signal flow within the DVV method is illustrated in Figure 18.4. For a given embedding dimension m, the DVV method can be summarised as follows [83, 84, 86]:

D1. The mean, μ_{d}, and standard deviation, σ_{d}, are computed over all pairwise Euclidean distances between delay vectors (DVs), $\|\mathbf{x}(i) - \mathbf{x}(j)\|$ $(i \neq j)$.

D2. The sets of 'neighbouring' delay vectors $\Omega_k(r_{\mathrm{d}})$ are generated such that $\Omega_k(r_{\mathrm{d}}) = \{\mathbf{x}(i) | \, \|\mathbf{x}(k) - \mathbf{x}(i)\| \leq r_{\mathrm{d}}\}$, that is, sets which consist of all DVs that lie closer to $\mathbf{x}(k)$ than a certain distance r_{d}, taken from the interval $[\max\{0, \mu_{\mathrm{d}} - n_{\mathrm{d}}\sigma_{\mathrm{d}}\}; \mu_{\mathrm{d}} + n_{\mathrm{d}}\sigma_{\mathrm{d}}]$,

[5] Apart from the surrogate methods, other established methods for detecting the nonlinear nature of a signal include the Deterministic versus Stochastic (DVS) plot [44] and δ–ε Method [145].

for example, N_{tv} uniformly spaced distances, where n_d is a parameter controlling the span over which to perform the DVV analysis.

D3. For every set $\Omega_k(r_d)$, the variance of the corresponding targets, $\sigma_k^2(r_d)$, is computed. The average over all sets $\Omega_k(r_d)$, normalised by the variance of the time series, σ_x^2, yields the *target variance* $\sigma^{*2}(r_d)$:

$$\sigma^{*2}(r_d) = \frac{\frac{1}{N}\sum_{k=1}^{N}\sigma_k^2(r_d)}{\sigma_x^2} \tag{18.1}$$

As a rule of thumb, we only consider a variance measurement *valid*, if the set $\Omega_k(r_d)$ contains around $N_o = 30$ DVs, since having too few points for computing a sample variance yields unreliable estimates of the true (population) variance.

As a result of the standardisation of the distance axis, the resulting 'DVV plots' (target variance, $\sigma^{*2}(r_d)$ as a function of the standardised[6] distance, $(r_d - \mu_d)/\sigma_d$) are straightforward to interpret:

- The presence of a strong deterministic component will lead to small target variances $\sigma^{*2}(r_d)$ for small spans r_d.
- The minimal target variance, $\sigma^{*2}_{min} = \min_{r_d}[\sigma^{*2}(r_d)]$, is a measure for the amount of noise which is present in the time series (the prevalence of the stochastic component).
- At the extreme right, the DVV plots smoothly converge to unity, since for maximum spans, *all* DVs belong to the same universal set, and the variance of the targets is equal to the variance of the time series.

To illustrate the operation of the DVV method, consider a linear AR(4) signal [190], given by

$$x(k) = 1.79\,x(k-1) - 1.85\,x(k-2) + 1.27\,x(k-3) - 0.41\,x(k-4) + n(k) \tag{18.2}$$

and a benchmark nonlinear signal [216], given by

$$z(k) = \frac{z(k-1)}{1 + z^2(k-1)} + x^3(k) \tag{18.3}$$

where $x(k)$ denotes the AR(4) signal defined above and $n(k) \sim \mathcal{N}(0, 1)$.

Averaged DVV plots, computed over 25 iAAFT-based surrogates for these two benchmark signals are shown respectively in Figure 18.5(a) and (b). Since the AR(4) is linear, and surrogates are also linear by design, the DVV curves for the original and averaged surrogates are very close (Figure 18.5a). This is not the case with the DVV curves for the original and averaged surrogates for the nonlinear signal in Figure 18.5(b), which are far apart, indicating that the linear surrogates were not able to model the nonlinear nature of the signal.

Due to the standardisation of the distance axis, these plots can be conveniently combined in a *scatter diagram*, where the horizontal axis corresponds to the DVV plot of the original time

[6] Note that we use the term 'standardised' in the statistical sense, namely as having zero mean and unit variance.

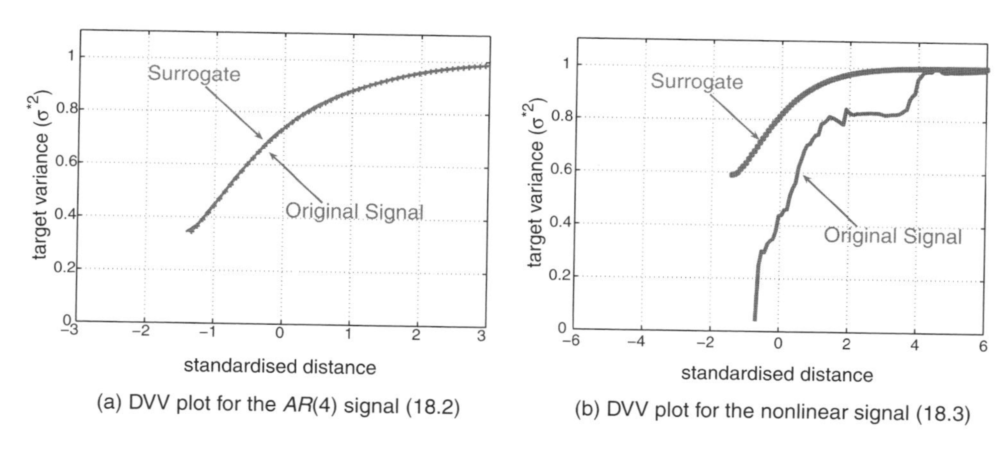

Figure 18.5 Signal modality characterisation. The DVV plots for the original and surrogates are obtained by plotting the target variance as a function of standardised distance

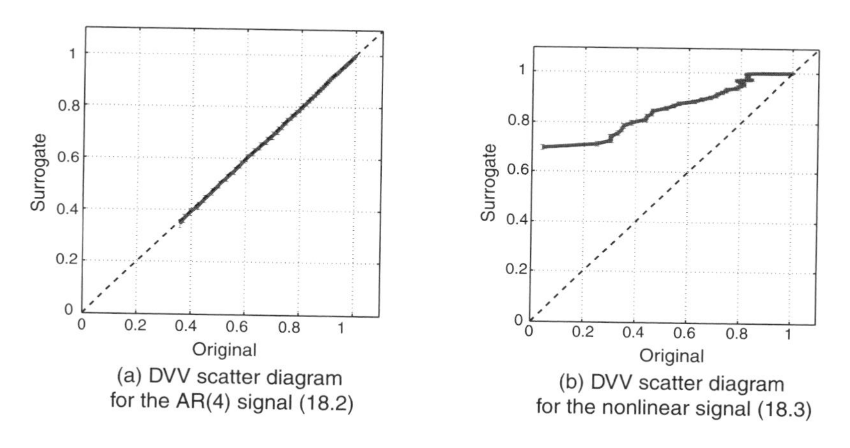

Figure 18.6 The DVV scatter diagrams obtained by plotting the target variance of the original data against the mean of the target variances of the surrogate data

series, and the vertical to that of the surrogate time series. If the averaged surrogate time series yield DVV plots similar to that of the original time series, the DVV scatter diagram coincides with the bisector line, and the original time series is judged to be linear, as illustrated in Figure 18.6(a). Conversely, as shown in Figure 18.6(b), the DVV scatter diagram for a nonlinear signal deviates from the bisector line.

18.2 Testing for the Validity of Complex Representation

When testing for the suitability of the complex valued signal representation, one convenient null hypothesis H_0 would be that the time series is generated by a linear circular complex valued process, followed by a (possibly nonlinear) static observation function, $h(\cdot)$ which operates on the *moduli* of the complex valued time samples.

To cater for complex valued signals, a straightforward extension of the (real valued) bivariate iAAFT-method [265] would be to match the amplitude spectrum of the surrogate and the amplitude spectrum of the original complex valued signal. Then, the signal distribution of the original needs to be imposed on the surrogates in the time domain (step **S3** in the iAAFT-procedure). This can be achieved by rank-ordering the real and imaginary parts of the complex valued signal separately, as this is in line with the notion of circularity. However, in practice, for complex valued signals it is more important to impose equal empirical distributions on the moduli of the samples, rather than on the real and imaginary parts separately. This way, the empirical distribution of the moduli is (approximately) identical to that of the original signal, thus retaining signal circularity (if present). This approach will be adopted in the derivation of the statistical test for the validity of complex valued representations.

Similar to the real valued iAAFT case, during the computation of complex iAAFT (CiAAFT) surrogates, $\{|S_k|\}$ denotes the Fourier amplitude spectrum of the original time series s; for every iteration j, three additional time series are generated:

- the 'modulus sorted' version of the complex valued original time series s, denoted by $\{c_k\}$;
- a time series with the same distribution (in terms of moduli) as the original complex valued time series s, denoted by $r^{(j)}$;
- a time series which has the amplitude spectrum identical to that of the original time series s, but not necessarily the same distribution of the moduli, denoted by $s^{(j)}$.

The CiAAFT procedure is summarised below; the iteration starts with $r^{(0)}$, a random permutation of the original complex valued time samples.

Repeat:

C1. Compute the phase spectrum of $r^{(j-1)} \rightarrow \{\phi_k\}$

C2. Compute $s^{(j)}$ as the inverse transform of $\{|S_k| \exp(\jmath\, \phi_k)\}$

C3. Rank-order the real and imaginary parts of $r^{(j)}$ to match the real and imaginary parts of $\{c_k\}$

C4. Rank-order the moduli of $r^{(j)}$ to match the corresponding modulus distribution of $\{c_k\}$

Until error convergence

Figure 18.7 illustrates the results of the statistical testing for the complex nature of (fully complex by design) Ikeda map. Figure 18.7(b) shows that the bivariate approach (although two-dimensional) was not able to preserve the state space properties of Ikeda map, whereas the surrogate realisation based on CiAAFT (Figure 18.7c) was much better suited for this purpose.

18.2.1 Complex Delay Vector Variance Method (CDVV)

The extension of the Delay Vector Variance method into the complex domain is straightforward. For a given embedding dimension m and a resulting time delay embedding representation (i.e., a set of complex valued delay vectors (DV), $\mathrm{s}(k) = [s_{k-m}, \ldots, s_{k-1}]^{\mathrm{T}}$), a measure of unpredictability, $\sigma^{*2}(r_{\mathrm{d}})$, is computed for a standardised range of degrees of locality, r_{d}, similar to the real case described in Section 18.1.2. Since for both the bivariate and complex time

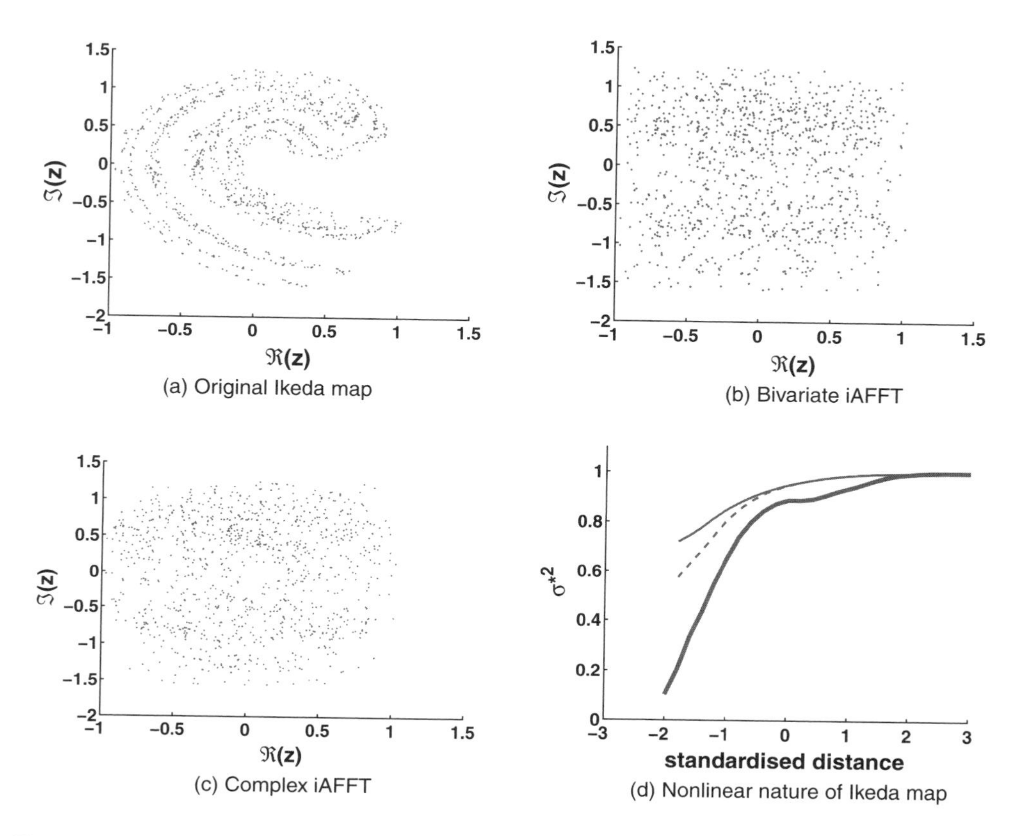

Figure 18.7 Judging the validity of complex representation of the Ikeda map time series denoted by $z = \Re(z) + \jmath\,\Im(z)$: (a) the original signal; (b) realisation using bivariate iAAFT surrogates; (c) realisation using complex iAAFT surrogates; (d) DVV plots for the Ikeda map (thick solid line), iAAFT surrogate (think dashed line), and CiAAFT surrogate (thin solid line)

series, a delay vector is generated by concatenating time delay embedded versions of the two dimensions (real and imaginary), the complex valued and real valued bivariate versions of the DVV method provide equivalent results, and the variance of such variables is computed as the sum of the variances of each variate, that is[7]

$$\sigma_s^{*2}(r_\mathrm{d}) = \sigma_{s,r}^{*2}(r_\mathrm{d}) + \sigma_{s,i}^{*2}(r_\mathrm{d})$$

where $\sigma_{s,r}^{*2}(r_\mathrm{d})$ denotes the target variance for the real part of the original signal s, and $\sigma_{s,i}^{*2}(r_\mathrm{d})$ denotes that for the imaginary part.

[7]Note that by augmenting the delay vectors within the DVV method, that is, by computing the Euclidean distances between the augmented delay vectors, would not make any difference. This is due to the deterministic relationship between a complex delay vector $\mathbf{z}(k)$ and its complex conjugate $\mathbf{z}^*(k)$ (see Chapter 12).

Test for the validity of complex valued representation. To test for the potential benefits of complex valued representations, rather than comparing the original time series to the surrogates, it is convenient to compare the surrogates generated under the fundamentally different null hypotheses of:

- a linear bivariate time series, denoted by H_0^{b}, for which the surrogates are generated using the (real valued) bivariate iAAFT [265];
- a linear and complex valued time series, denoted by H_0^{c}, for which the surrogates are generated using the CiAAFT method [85].

The DVV method can be used to characterise the natures of the two different realisations. A statistically different characterisation means that the two null hypotheses lead to different realisations of the original signal. Since the difference between the null hypotheses is the property of circularity in the linearisations of the original signal, a statistical difference can be interpreted as an indication of the presence of circularity in the original signal, which is retained in the CiAAFT realisations, and not in the iAAFT realisations. This test can be used to justify a complex valued representation over a dual univariate one.

The complex DVV based statistical test for the validity of the complex valued representation can now be summarised as follows:

T1. Generate M CiAAFT surrogates and produce the averaged DVV plot denoted by $\mathcal{D}^0$;
T2. Generate N bivariate iAAFT surrogates and produce the corresponding DVV plots, denoted by $\{\mathcal{D}^{\mathrm{b}}\}$;
T3. Generate N CiAAFT surrogates and produce the corresponding DVV plots, denoted by $\{\mathcal{D}^{\mathrm{c}}\}$;
T4. Compare $\left(\mathcal{D}^0 - \{\mathcal{D}^{\mathrm{b}}\}\right)$ and $\left(\mathcal{D}^0 - \{\mathcal{D}^{\mathrm{c}}\}\right)$.

To perform **T4** in a statistical manner, the (cumulative) empirical distributions of rootmean-square distances between $\{\mathcal{D}^{\mathrm{b}}\}$ and $\mathcal{D}^0$, and between $\{\mathcal{D}^{\mathrm{c}}\}$ and $\mathcal{D}^0$, are compared using a Kolmogorov–Smirnoff (K-S) test. This way, the different types of *linearisations* (bivariate $\{\mathcal{D}^{\mathrm{b}}\}$ from **T2**, and complex valued $\{\mathcal{D}^{\mathrm{c}}\}$ from **T3**) are compared with the 'reference' linearisation, that is, $\mathcal{D}^0$ from **T1**. If the two distributions of test statistics are significantly different at a certain confidence level α, say 95%, the original time series is judged complex valued [85].

A DVV plot of a complex signal is obtained by plotting the target variance, $\sigma^{*2}(r_{\mathrm{d}})$, as a function of the standardised distance $(r_d - \mu_d)/\sigma_d$. The DVV plots for a 1000 sample realisation of the Ikeda map and the iAAFT and CiAAFT surrogates (using $m = 3$ and $n_d = 3$) are shown in Figure 18.7(d). Figure 18.8 illustrates the convergence of CiAAFT surrogates when modelling the Ikeda map, and their behaviour in the presence of noise. The convergence curve is uniform and the CiAAFT iteration settles after about 30 iterations, as shown in Figure 18.8(a). Since the complex white Gaussian noise used in simulations was 'doubly white' (see Section 13.2), bivariate iAAFT surrogates provided a suitable model. Figure 18.8(b) shows the rejection ratios of the CDVV method for the Ikeda map contaminated with complex WGN (CWGN) over a range of power levels. The complex Ikeda map and CWGN had equal variances, and the noisy signal $\mathrm{Ikeda}_{\mathrm{noisy}}$ was generated according to

$$\mathrm{Ikeda}_{\mathrm{noisy}} = \mathrm{Ikeda}_{\mathrm{original}} + \gamma_{\mathrm{n}} \times \mathrm{CWGN} \tag{18.4}$$

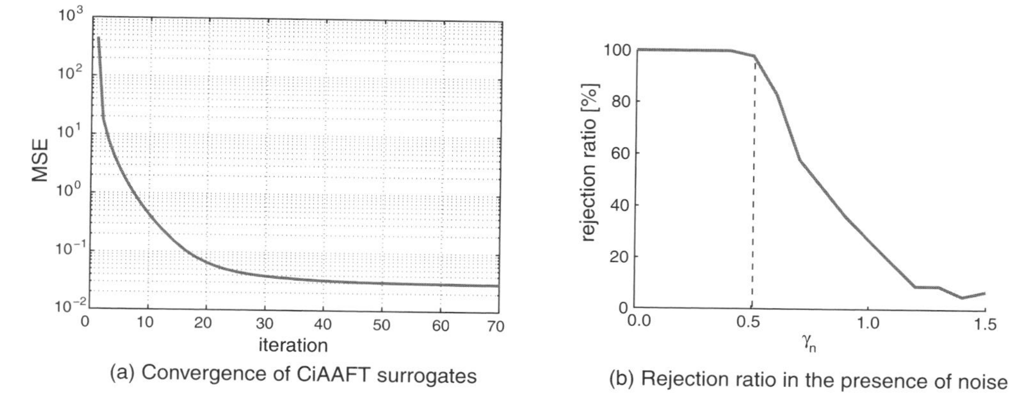

(a) Convergence of CiAAFT surrogates
(b) Rejection ratio in the presence of noise

Figure 18.8 Behaviour of CiAAFT surrogates for the example of Ikeda map. Left: convergence. Right: behaviour in the presence of noise; parameter γ_n indicates the ratio between the standard deviation of the complex Ikeda map and that of the additive white Gaussian noise – see (Equation 18.4)

where γ_n denotes the ratio of standard deviations between the Ikeda map signal and CWGN. As expected, with an increase in noise power, the validity of complex valued representation of noisy Ikeda map was less pronounced, as illustrated by a downwards trend of the curve in Figure 18.8(b).

18.3 Quantifying Benefits of Complex Valued Representation

To test for the benefits of complex valued representation of real valued processes, consider wind data recorded over 24 hours, shown in Figure 18.1. Areas denoted by A, B, and C correspond respectively to the 'high', 'medium' and 'low' dynamics of wind. It is expected that the larger the changes in wind dynamics ('high') the greater the advantage obtained by the complex valued representation of wind (see also Chapter 12). For relatively mild and slowly changing wind dynamics, it is expected that the complex valued modelling should not exhibit significant performance advantage over, say, the dual univariate one. Also, it is expected that the complex (and nonlinear) nature of wind would be less pronounced with increased averaging of the raw data.[8]

Table 18.1 illustrates the rejection ratios[9] for the null hypothesis of 'no difference between bivariate and complex linearisations', that is, for the bivariate nature of wind data from Figure 18.1, for different wind regimes and degrees of averaging.[10] Also, there are stronger indications of a complex valued nature when the wind is averaged over shorter intervals, as represented

[8]This is closely related to circularity and augmented complex statistics addressed in Chapter 12.

[9]Each result was obtained by performing the complex DVV test 100 times, and by counting the number of rejections of the null hypothesis.

[10]As expected (see Figure 2.6), there is a significant component dependence within the complex valued wind signal representation, as indicated by the rejection ratio of the null hypothesis (of a bivariate nature) being significantly greater than zero.

Table 18.1 Rejection rates (the higher the rejection rate the greater the benefit of complex valued representation) for the wind signal from Figure 18.1

Wind signal	Region A%	Region B%	Region C%
Averaged over 1 s	96	80	71
Averaged over 10 s	90	74	62
Averaged over 60 s	83	69	58

by the respective percentage values of the ratio of the rejection of the null hypothesis of a real bivariate nature.

18.3.1 Pros and Cons of the Complex DVV Method

This chapter illustrates the importance and usefulness of statistical testing for the fundamental nature of data (nonlinearity, determinism, complex valued nature). This can provide additional knowledge which can be exploited when choosing a signal processing model best suited to the data. Indeed, following from the signal modality characterisation for real valued data from Figure 18.2, the complex surrogates and complex DVV method enable us to test whether the complex valued representation of real world data is worthwhile.

Despite being well founded mathematically and extremely useful, these tests suffer from drawbacks, such as:

- Surrogate and DVV tests can only be performed in an off-line block manner;
- The DVV and surrogate data based tests are only applicable for quasistationary data segments (since we need to calculate the embedding parameters and compute the Fourier transform, these operation only apply to stationary data);
- Due to the requirement of piece-wise stationarity, it is possible to mistake the property of nonstationarity for nonlinearity;
- Hypothesis testing methods are not readily suitable for real time mode of operation, they can, however, operate in near real time on overlapping data windows.

Some of these issues have been addressed in Chapter 16, where online tests for the characterisation of the nature of complex valued data are introduced.

Appendix A

Some Distinctive Properties of Calculus in $\mathbb{C}$

Differences between real and complex algebra require us to revisit some basic notions from function analysis and topology in $\mathbb{C}$.

Ordering of numbers. A field can be ordered if and only if no sum of squares of nonzero elements is zero [16]. Since $\jmath$ and 1 are both nonzero, and $\jmath^2 + 1 = 0$, the field of complex numbers $\mathbb{C}$ *cannot be an ordered field*. We can, however, introduce 'order' in $\mathbb{C}$ by lexicographical (or dictionary) ordering [304], denoted by $\lhd$. Then

$$(a, b) \lhd (c, d) \quad \Leftrightarrow \quad a < c \quad \text{or} \quad a = c \text{ and } b < d$$

and for instance

$$(0, 1) \lhd (1, 0) \qquad (0, 0) \lhd (1, 0) \qquad (0, 0) \lhd (0, 1)$$

This ordering, however, does not posses the Archimedean property, that is, there is no positive integer n, such that $(1, 0) \lhd n(0, 1)$.

Complex probability distribution. It is usually assumed that it is not possible to assign probability distributions to complex quantities, since $\mathbb{C}$ is not an ordered field, and expressions for the cumulative and probability density function

$$F_X(x) = P(X \leq x)$$
$$f_X(x) = \frac{\partial}{\partial x} F_X(x)$$

make no sense. Whereas, it is possible to define

$$f_Z(x + \jmath y) = f_{X,Y}(x, y)$$

Complex Valued Nonlinear Adaptive Filters: Noncircularity, Widely Linear and Neural Models
Danilo P. Mandic and Vanessa Su Lee Goh

or

$$f_{X,Y}(x, y) = f_{X,Y}\left(\frac{1}{2}(z + z^*), \frac{1}{2\jmath}(z - z^*)\right) = g(z, z^*)$$

these 'distributions' are not correct, since for instance knowing z we also know z^* and thus $g(z, z^*)$ is a degenerate distribution. Some recent work by S. Olhede [224] addresses the issue of probability density functions for complex random variables, and introduces a class of such functions which are interpretable in z and z^*. This way, it is possible to state the properties of (CRV) in terms of their density, and to parameterise them in terms of the mean, covariance, and pseudocovariance matrix.

Complex mean. There are many ways to define the mean of real numbers, for instance

$$\begin{aligned} \text{Arithmetic} \quad A &= \frac{a+b}{2} \\ \text{Geometric} \quad G &= \sqrt{ab} \\ \text{Heronian} \quad H &= \frac{1}{3}\left(a + \sqrt{ab} + b\right) \\ \text{Harmonic} \quad M &= \frac{2}{\frac{1}{a} + \frac{1}{b}} \\ \text{Power} \quad P &= \left(\frac{a^n + b^n}{2}\right)^{\frac{1}{n}} \end{aligned}$$

When a and b are positive real numbers, it is easy to show that $\min\{a, b\} \le M \le G \le H \le A \le \max\{a, b\}$.

A complex mean can be defined as [272]

A complex mean is a function $m : \mathbb{C}^+ \times \mathbb{C}^+ \to \mathbb{C}^+$ *such that* $\min\{|a|, |b|\} \le |m(a, b)| \le \max\{|a|, |b|\}$, *and* $\forall a \in \mathbb{C}^+$, $m(a, a) = a$, *where* $\mathbb{C}^+ = \{x + \jmath y | x, y > 0\}$. *A complex mean* m *is said to be homogeneous if* $\forall t \in \mathbb{R}, t \ge 0$, *and* $a, b \in \mathbb{C}^+$, $m(ta, tb) = tm(a, b)$.

It can be shown [272] that when applied to complex numbers, the arithmetic mean, Heronian mean, and power mean do not satisfy the above definition. It is only the geometric mean $m(a, b) = \sqrt{ab}$ that is symmetric, homogeneous and satisfies the collinearity property.

Sign of complex numbers. When coding in Matlab, unlike for the real numbers, where the *sign* function returns 1 if the element is greater than zero, 0 if it equals zero, and -1 if it is less than zero, for a complex number $z = x + \jmath y$

$$\text{sign}(z) = \frac{z}{|z|}$$

where $|z|$ is the complex modulus of z[1].

[1] In Matlab, for nonzero complex X, $\text{sign}(X) = X/\text{abs}(X)$.

The basic arithmetic operations on complex numbers are defined as follows

$$
\begin{aligned}
(a + \jmath b) + (c + \jmath d) &= (a + c) + \jmath(b + d) \\
(a + \jmath b) - (c + \jmath d) &= (a - c) + \jmath(b - d) \\
(a + \jmath b)(c + \jmath d) &= (ac - bd) + \jmath(bc + ad) \\
\frac{a + \jmath b}{c + \jmath d} &= \frac{a + \jmath b}{c + \jmath d} \times \frac{c - \jmath d}{c - \jmath d} \\
&= \frac{ac + bd}{c^2 + d^2} + \jmath \frac{bc - ad}{c^2 + d^2}
\end{aligned} \tag{A.1}
$$

Complex conjugate. The complex conjugate is produced by changing the sign of the imaginary part of a complex number $z = x + \jmath y$. For example, the conjugate[2] of the complex number $z = x + \jmath y$ (where x and y are real numbers) is $\bar{z} = z^* = x - \jmath y$. It is common to view complex numbers as points in a plane in the Cartesian coordinate system. The complex conjugation then corresponds to mirroring of a complex number over the x-axis (see Figure A.1), whereas in the polar form, the complex conjugate of $z = re^{\jmath\theta}$ is $\bar{z} = re^{-\jmath\theta}$. Thus, in most practical settings, if a solution to a problem is the complex number z, so too is its complex conjugate $\bar{z}$; one such case are the roots of a quadratic polynomial.

For two complex numbers z and w, properties of the complex conjugation operator are:

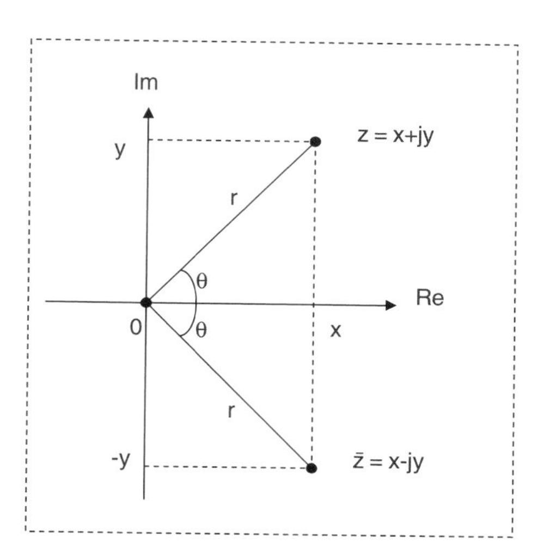

Figure A.1 Geometric representation of z and its conjugate $\bar{z}$ in the complex plane

[2]The complex conjugate of $z \in \mathbb{C}$ is commonly denoted by z^* or $\bar{z}$. For convenience, we will use the symbol $\bar{z}$ in the rest of this appendix.

1. The complex conjugate operator is distributive for complex addition and substraction, that is

$$\overline{(z+w)} = \bar{z} + \overline{w}$$

and

$$\overline{(z-w)} = \bar{z} - \overline{w}$$

2. The complex conjugate operator is distributive for complex multiplication and division, that is

$$\overline{zw} = \bar{z}\,\overline{w}$$

and

$$\frac{\bar{z}}{w} = \frac{\bar{z}}{\overline{w}}, \quad \text{if} \quad w \neq 0 \tag{A.2}$$

3. The function $\Omega(z) = \bar{z}$ from $\mathbb{C}$ to $\mathbb{C}$ is continuous. Notice that function $\Omega(z)$ is not holomorphic since it reverses the orientation, whereas holomorphic functions preserve orientation.

Generalisations of complex conjugates. The conjugate of a quarternion $a + b\imath + c\jmath + dk \in \mathbb{H}$ is $a - b\imath - c\jmath - dk \in \mathbb{H}$. It is also possible to introduce the notion of conjugation in vector spaces $\mathbf{V} \subseteq \mathbb{C}$. In this context, any (real) linear transformation $\Omega : \mathbf{V} \rightarrow \mathbf{V}$ that satisfies

1. $\Omega \neq \jmath d_V$, the identity function on $\mathbf{V}$
2. $\Omega^2 = \jmath d_V$
3. $\Omega(zv) = \bar{z}\Omega(v)$ for all $v \in \mathbf{V}$, and $z \in \mathbb{C}$

is *called a complex conjugation*. One typical example is the conjugate transpose operation for complex matrices.

Complex continuous functions. Consider a function $f : D \mapsto \mathbb{C}$ where $D \subseteq \mathbb{C}$. For $z \in \mathbb{C}$, we have,

$$f(z) = f(x, y) = u(x, y) + \jmath v(x, y) = \big(u(x, y), v(x, y)\big) \tag{A.3}$$

Limits of complex functions are defined in the same way as in elementary calculus, that is

$$\lim_{z \rightarrow z_0} f(z) = \varsigma \tag{A.4}$$

For a given an $\epsilon > 0$ there exists a $\delta > 0$ such that $|f(z) - \varsigma| < \epsilon$ when $0 < |z - z_0| < \delta$. For a complex function continuous at $z \in \mathbb{C}$, we then have

$$\lim_{z \rightarrow z_0} f(z) = f(z_0) \tag{A.5}$$

Table A.1 Classification of functions

Function type	Scalar variables $z, z^* \in \mathbb{C}$	Vector variables $\mathbf{z}, \mathbf{z}^* \in \mathbb{C}^{N\times 1}$	Matrix variables $\mathbf{Z}, \mathbf{Z}^* \in \mathbb{C}^{N\times Q}$
Scalar function $f \in \mathbb{C}$	$f(z, z^*)$ $f : \mathbb{C} \times \mathbb{C} \to \mathbb{C}$	$f(\mathbf{z}, \mathbf{z}^*)$ $f : \mathbb{C}^{N\times 1} \times \mathbb{C}^{N\times 1} \to \mathbb{C}$	$f(\mathbf{Z}, \mathbf{Z}^*)$ $f : \mathbb{C}^{N\times Q} \times \mathbb{C}^{N\times Q} \to \mathbb{C}$
Vector function $\boldsymbol{f} \in \mathbb{C}^{M\times 1}$	$\boldsymbol{f}(z, z^*)$ $\boldsymbol{f} : \mathbb{C} \times \mathbb{C} \to \mathbb{C}^{M\times 1}$	$\boldsymbol{f}(\mathbf{z}, \mathbf{z}^*)$ $\boldsymbol{f} : \mathbb{C}^{N\times 1} \times \mathbb{C}^{N\times 1} \to \mathbb{C}^{M\times 1}$	$\boldsymbol{f}(\mathbf{Z}, \mathbf{Z}^*)$ $\boldsymbol{f} : \mathbb{C}^{N\times Q} \times \mathbb{C}^{N\times Q} \to \mathbb{C}^{M\times 1}$
Matrix function $\boldsymbol{F} \in \mathbb{C}^{M\times P}$	$\boldsymbol{F}(z, z^*)$ $\boldsymbol{F} : \mathbb{C} \times \mathbb{C} \to \mathbb{C}^{M\times P}$	$\boldsymbol{F}(\mathbf{z}, \mathbf{z}^*)$ $\boldsymbol{F} : \mathbb{C}^{N\times 1} \times \mathbb{C}^{N\times 1} \to \mathbb{C}^{M\times P}$	$\boldsymbol{F}(\mathbf{Z}, \mathbf{Z}^*)$ $\boldsymbol{F} : \mathbb{C}^{N\times Q} \times \mathbb{C}^{N\times Q} \to \mathbb{C}^{M\times P}$

that is, f is continuous at z_0. If the function is continuous for all points $z \in \mathbb{C}$, then function $f(z)$ is said to be continuous. To define the derivative of a complex function, the limit

$$\lim_{\Delta z \to 0} \frac{f(z + \Delta z) - f(z)}{\Delta z} \tag{A.6}$$

must converge to a unique value regardless of how Δz approaches 0 (see also Chapter 5).

Complex matrix differentiation. In parallel with the development of augmented complex statistics, results have appeared concerning complex matrix differentiation. Thus, for instance, in [58] Price's theorem has been rederived based on the augmented complex random vector $\mathbf{z}^a$ and the augmented complex covariance matrix $\mathcal{C}_a$, to give a general result applicable for improper random vectors.

Based on the work by Hjorungnes and Gesbert [122], Table A.1 gives classes of complex functions, where scalar quantities (variables z or functions f) are denoted by lowercase symbols, vector quantities (variables $\boldsymbol{z}$ or functions $\boldsymbol{f}$) are denoted by lowercase boldface symbols, and matrix quantities (variables $\boldsymbol{Z}$ or functions $\boldsymbol{F}$) are denoted by capital boldface symbols.

Table A.2 Complex differentials

Function	Differential
$\boldsymbol{A}$	$\mathbf{0}$
$\alpha \boldsymbol{Z}$	$\alpha \mathrm{d}\boldsymbol{Z}$
$\boldsymbol{AZZ}$	$\boldsymbol{A}(\mathrm{d}\boldsymbol{Z})\boldsymbol{B}$
$\mathbf{Z}_0 + \mathbf{Z}_1$	$\mathrm{d}\mathbf{Z}_0 + \mathrm{d}\mathbf{Z}_1$
$\mathrm{Tr}\{\boldsymbol{Z}\}$	$\mathrm{Tr}\{\mathrm{d}\boldsymbol{Z}\}$
$\mathbf{Z}_0\mathbf{Z}_1$	$(\mathrm{d}\mathbf{Z}_0)\mathbf{Z}_1 + \mathbf{Z}_0(\mathrm{d}\mathbf{Z}_1)$
$\mathbf{Z}_0 \otimes \mathbf{Z}_1$	$(\mathrm{d}\mathbf{Z}_0) \otimes \mathbf{Z}_1 + \mathbf{Z}_0 \otimes (\mathrm{d}\mathbf{Z}_1)$
$\mathbf{Z}_0 \odot \mathbf{Z}_1$	$(\mathrm{d}\mathbf{Z}_0) \odot \mathbf{Z}_1 + \mathbf{Z}_0 \odot (\mathrm{d}\mathbf{Z}_1)$
$\mathbf{Z}^{-1}$	$-\mathbf{Z}^{-1}(\mathrm{d}\mathbf{Z})\mathbf{Z}^{-1}$
$\det(\boldsymbol{Z})$	$\det(\mathbf{Z})\mathrm{Tr}\left\{\mathbf{Z}^{-1} d\mathbf{Z}\right\}$
$\ln(\det \boldsymbol{Z})$	$\mathrm{Tr}\left\{\mathbf{Z}^{-1}\mathrm{d}\mathbf{Z}\right\}$
$\mathrm{reshape}(\boldsymbol{Z})$	$\mathrm{reshape}(\mathrm{d}\boldsymbol{Z})$
$\boldsymbol{Z}^*$	$\mathrm{d}\boldsymbol{Z}^*$
$\boldsymbol{Z}^H$	$\mathrm{d}\boldsymbol{Z}^H$

Table A.3 Differentials and derivatives of complex functions

Function type	Differential	Derivative with respect to z, $\boldsymbol{z}$, or $\boldsymbol{Z}$	Derivative with respect to z^*, $\boldsymbol{z}^*$, or $\boldsymbol{Z}^*$	Size of derivative
$f(z, z^*)$	$\mathrm{d}f = a_0 \mathrm{d}z + a_1 \mathrm{d}z^*$	$D_z f(z, z^*) = a_0$	$D_{z^*} f(z, z^*) = a_1$	1×1
$f(\boldsymbol{z}, \boldsymbol{z}^*)$	$\mathrm{d}f = \boldsymbol{a}_0 \mathrm{d}\boldsymbol{z} + \boldsymbol{a}_1 \mathrm{d}\boldsymbol{z}^*$	$D_{\boldsymbol{z}} f(\boldsymbol{z}, \boldsymbol{z}^*) = \boldsymbol{a}_0$	$D_{\boldsymbol{z}^*} f(\boldsymbol{z}, \boldsymbol{z}^*) = \boldsymbol{a}_1$	$1 \times N$
$f(\boldsymbol{Z}, \boldsymbol{Z}^*)$	$\mathrm{d}f = \mathrm{vec}^{\mathrm{T}}(\boldsymbol{A}_0)\mathrm{d}\, \mathrm{vec}(\boldsymbol{Z}) +$ $\mathrm{vec}^{T}(\boldsymbol{A}_1)\mathrm{d}\, \mathrm{vec}(\boldsymbol{Z}^*)$	$D_{\boldsymbol{Z}} f(\boldsymbol{Z}, \boldsymbol{Z}^*) =$ $\mathrm{vec}^{T}(\boldsymbol{A}_0)$	$D_{\boldsymbol{Z}^*} f(\boldsymbol{Z}, \boldsymbol{Z}^*) =$ $\mathrm{vec}^{T}(\boldsymbol{A}_1)$	$1 \times NQ$
$f(\boldsymbol{Z}, \boldsymbol{Z}^*)$	$\mathrm{d}f = Tr\left\{\boldsymbol{A}_0^T \mathrm{d}\boldsymbol{Z} + \boldsymbol{A}_1^T \mathrm{d}\boldsymbol{Z}^*\right\}$	$\frac{\partial}{\partial \boldsymbol{Z}^*} f(\boldsymbol{Z}, \boldsymbol{Z}^*) = \boldsymbol{A}_0$	$\frac{\partial}{\partial \boldsymbol{Z}} f(\boldsymbol{Z}, \boldsymbol{Z}^*) = \boldsymbol{A}_1$	$N \times N$
$\boldsymbol{f}(z, z^*)$	$\mathrm{d}\boldsymbol{f} = \boldsymbol{b}_0 dz + \boldsymbol{b}_1 dz^*$	$D_z \boldsymbol{f}(z, z^*) = \boldsymbol{b}_0$	$D_{z^*} \boldsymbol{f}(z, z^*) = \boldsymbol{b}_1$	$M \times 1$
$\boldsymbol{f}(\boldsymbol{z}, \boldsymbol{z}^*)$	$d\boldsymbol{f} = \boldsymbol{B}_0 d\boldsymbol{z} + \boldsymbol{B}_1 d\boldsymbol{z}^*$	$D_{\boldsymbol{z}} \boldsymbol{f}(\boldsymbol{z}, \boldsymbol{z}^*) = \boldsymbol{a}_0$	$D_{\boldsymbol{z}^*} \boldsymbol{f}(\boldsymbol{z}, \boldsymbol{z}^*) = \boldsymbol{a}_1$	$M \times N$
$\boldsymbol{f}(\boldsymbol{Z}, \boldsymbol{Z}^*)$	$d\boldsymbol{f} = \beta_0 \mathrm{d}\, \mathrm{vec}(\boldsymbol{Z}) + \beta_1 \mathrm{d}\, \mathrm{vec}(\boldsymbol{Z}^*)$	$D_{\boldsymbol{Z}} \boldsymbol{f}(\boldsymbol{Z}, \boldsymbol{Z}^*) = \beta_0$	$D_{\boldsymbol{Z}^*} \boldsymbol{f}(\boldsymbol{Z}, \boldsymbol{Z}^*) = \beta_1$	$M \times NQ$
$\boldsymbol{F}(z, z^*)$	$\mathrm{d}\, \mathrm{vec}(\boldsymbol{F}) = c_0 \mathrm{d}z + c_1 \mathrm{d}z^*$	$D_z \boldsymbol{F}(z, z^*) = c_0$	$D_{z^*} \boldsymbol{F}(z, z^*) = c_1$	$MP \times 1$
$\boldsymbol{F}(\boldsymbol{z}, \boldsymbol{z}^*)$	$\mathrm{d}\, \mathrm{vec}(\boldsymbol{F}) = C_0 \mathrm{d}\boldsymbol{z} + C_1 \mathrm{d}\boldsymbol{z}^*$	$D_{\boldsymbol{z}} \boldsymbol{F}(\boldsymbol{z}, \boldsymbol{z}^*) = C_0$	$D_{\boldsymbol{z}^*} \boldsymbol{F}(\boldsymbol{z}, \boldsymbol{z}^*) = C_1$	$MP \times N$
$\boldsymbol{F}(\boldsymbol{Z}, \boldsymbol{Z}^*)$	$\mathrm{d}\, \mathrm{vec}\boldsymbol{F} = \zeta_0 \mathrm{d}\, \mathrm{vec}(\boldsymbol{Z}) +$ $\zeta_1 \mathrm{d}\, \mathrm{vec}(\boldsymbol{Z}^*)$	$D_{\boldsymbol{Z}} \boldsymbol{F}(\boldsymbol{Z}, \boldsymbol{Z}^*) = \zeta_0$	$D_{\boldsymbol{Z}^*} \boldsymbol{F}(\boldsymbol{Z}, \boldsymbol{Z}^*) = \zeta_1$	$MP \times NQ$

Table A.2 shows the differentials associated with the classes of complex functions from Table A.1. Table A.3 lists the derivatives of the different types of function from Table A.1, this is achieved based on the differentials from Table A.2.

Appendix B

Liouville's Theorem

Theorem 1. (Liouville). *If for all z in $\mathbb{C}$, function $\Phi(z)$ is analytic and bounded by some value M, then $\Phi(z)$ is a constant.*

Proof. Using the Cauchy integral formula, we have

$$\Phi(a) = \frac{1}{2\pi \jmath} \oint_C \frac{\Phi(z)}{z - a} \mathrm{d}z \tag{B.1}$$

where $a, b \in \mathbb{C}$ and C is a circle with centre a and radius r which contains a and b. It then follows that

$$\begin{aligned} \Phi(b) - \Phi(a) &= \frac{1}{2\pi \jmath} \oint_C \frac{\Phi(z)}{z - b} \mathrm{d}z - \frac{1}{2\pi \jmath} \oint_C \frac{\Phi(z)}{z - a} \mathrm{d}z \\ &= \frac{b - a}{2\pi \jmath} \oint_C \frac{\Phi(z)}{(z - a)(z - b)} \mathrm{d}z \end{aligned} \tag{B.2}$$

Notice that $|z - a| = r$ and $|z - b| = |z - a + a - b| \geq |z - a| - |a - b| = r - |a - b|$. We can then choose r such that $|a - b| < r/2$, that is, $r - |a - b| \geq r/2$. Since for all $z \in \mathbb{C}$, $\Phi(z) \leq M$, and the length of C is $2\pi r$, we have

$$\begin{aligned} |\Phi(b) - \Phi(a)| &= \frac{|b - a|}{2\pi} \left| \oint_C \frac{\Phi(z)}{(z - a)(z - b)} \mathrm{d}z \right| \\ &\leq \frac{|b - a|\, M(2\pi r)}{2\pi(r/2)r} \\ &= \frac{2\,|b - a|\, M}{r} \end{aligned} \tag{B.3}$$

Thus, $\lim_{r \to \infty} |\Phi(b) - \Phi(a)| = 0$ and so $\Phi(a) = \Phi(b)$, which implies that $\Phi(z)$ is a constant. □

Complex Valued Nonlinear Adaptive Filters: Noncircularity, Widely Linear and Neural Models
Danilo P. Mandic and Vanessa Su Lee Goh

Appendix C

Hypercomplex and Clifford Algebras

C.1 Definitions of Algebraic Notions of Group, Ring and Field

Group. A Group $(G, *)$ is a set G, together with a binary operator $*$, defined by the following four axioms

1. *Closure:* $\forall a, b \in G, \quad a * b \in G$
2. *Identity:* $\exists i \in G$ such that $\forall a \in G \quad a * i = i * a = a$
3. *Inverse:* $\forall a \in G, \exists b \in G$, such that $a * b = b * a = i$
4. *Associativity:* $\forall a, b, c \in G \quad (a * b) * c = a * (b * c)$

Ring. A Ring $(R, +, \times)$ is a set, R, equipped with two binary operators, $+$ and $\times$ (commonly interpreted as addition and multiplication), which satisfy the following axioms

1. $(R, +)$ is a commutative group
2. $(R - 0, \times)$ satisfies the closure, identity and associativity rules
3. Operations $+$ and $\times$ satisfy the distributive law

A Ring exhibits properties of additive commutativity, additive associativity, additive identity, additive inverse, multiplicative associativity, multiplicative identity.

Field. A Field $(F, +, \times)$ is a set, F, together with two binary operators, $+$ and $\times$, defined by the following axioms

1. $(F, +)$ is a group
2. $(F - 0, \times)$ is also a group
3. Set F and operations $+$ and $\times$, satisfy the distributive law

Complex Valued Nonlinear Adaptive Filters: Noncircularity, Widely Linear and Neural Models
Danilo P. Mandic and Vanessa Su Lee Goh

In other words, a Field is any set of elements which satisfies the axioms of addition and multiplication.

C.2 Definition of a Vector Space

A vector space V over a field F is a set of elements such that $\forall X, Y, Z \in V$ and any scalars $r, s \in F$

- $(V, +)$ *is a commutative group* with properties
  - Closure: $X + Y \in V$
  - Identity: there exists a zero element 0, such that $X + 0 = X$
  - Inverse: $X + (-X) = 0$
  - Associativity: $(X + Y) + Z = X + (Y + Z)$
  - Commutativity: $X + Y = Y + X$
- *Scalar multiplication* has properties
  - Closure: $rX \in V$
  - Associativity: $r(sX) = (rs)X$
  - Distributivity of scalar sums: $(r + s)X = rX + sX$
  - Distributivity of vector sums: $r(X + Y) = rX + rY$
  - Scalar multiplication identity: $1X = X$

Basis of a Vector Space. For an n-dimensional vector space V, any n linear independent vectors $e_1, e_2, \ldots, e_n$ form a basis for the vector space. Thus, for instance, $e_1 = (1, 0)$ and $e_2 = (0, 1)$ form the basis for $\mathbb{R}^2$.

Coordinates of a vector. Any vector $X \in V$ can be uniquely expressed as a linear combination of basis vectors, that is, $X = a_1e_1 + a_2e_2 + \cdots + a_ne_n$. The symbols $a_1, a_2, \ldots, a_n$ are called coordinates of the vector X with respect to the basis $\{e_1, e_2, \ldots, e_n\}$.

C.3 Higher Dimension Algebras

Complex Numbers. Consider a set of all ordered pairs of real numbers

$$\mathbb{R}^2 = \{z = (a, b)\,|a, b \in \mathbb{R}\} \tag{C.1}$$

together with addition and multiplication operators defined for all $z_1 = (a_1, b_1), z_2 = (a_2, b_2) \in \mathbb{R}^2$, given by

$$z_1 + z_2 = (a_1 + a_2, b_1 + b_2) \tag{C.2}$$

$$z_1 \otimes z_2 = (a_1a_2 - b_1b_2, a_1b_1 + a_2b_1) \tag{C.3}$$

Then $\mathbb{C} = (\mathbb{R}^2, +, \otimes)$ is called the field of complex numbers.

The imaginary unit $\jmath$ can be expressed as $\jmath := (0, 1)$, and the relation $\jmath^2 = -1$ is then a direct consequence of (C.3). Furthermore, the usual expression for a complex number $z = a + \jmath b$ is obtained from the identity

$$z = (a, b) = (a, 0) + (0, 1) \otimes (b, 0) \tag{C.4}$$

Multiplication of complex numbers is both associative and commutative. Moreover, for all complex numbers $z = (a, b) \in \mathbb{C} \setminus (0, 0)$, we have

$$(a, b) \otimes (a/(a^2 + b^2), b/(a^2 + b^2)) = (1, 0) \tag{C.5}$$

Although the set of complex numbers $\mathbb{C}$ is primarily viewed as a field, it also comprises other algebraic structures, that is, infinitely many subfields isomorphic to $\mathbb{R}$; one obvious choice is

$$\alpha : \mathbb{R} \to \mathbb{C}, \quad a \longmapsto (a, 0) \tag{C.6}$$

For any $\lambda \in \mathbb{R}$ and any $z = (a, b) \in \mathbb{C}$ we then have

$$\alpha(\lambda) \otimes (a, b) = (\lambda, 0) \otimes (a, b) = (\lambda a, \lambda b) \tag{C.7}$$

This way, $\mathbb{C}$ becomes a real associative and commutative algebra of dimension 2 with $(1, 0)$ as the identity element.

C.4 The Algebra of Quaternions

Consider the linear space $(\mathbb{R}^4, +, \cdot)$ with standard basis $\{1 := (1_\mathbb{R}, 0, 0, 0),\ \imath := (0, 1_\mathbb{R}, 0, 0),\ \jmath := (0, 0, 1_\mathbb{R}, 0),\ k := (0, 0, 0, 1_\mathbb{R})\}$ and define a multiplication operator $\otimes$ according to Table C.1.

Then $\mathbb{H} = ((\mathbb{R}^4, +, \cdot), \otimes)$ is a real associative algebra of dimension 4, and is called the algebra of quaternions. Obviously, '1' is the identity element of $\mathbb{H}$.

A quaternion $q \in \mathbb{H}$ can be written as

$$q = q_0 + \imath q_1 + \jmath q_2 + k q_3 \tag{C.8}$$

where $q_0, q_1, q_2, q_3 \in \mathbb{R}$. Analogously to $\mathbb{C}$, the basis vectors $\{\imath, \jmath, k\}$ are often named 'imaginary units', and obey the following rules

$$\jmath k = -k\jmath = \imath \quad k\imath = -\imath k = \jmath \quad \imath\jmath = -\jmath\imath = k \tag{C.9}$$

Observe that multiplication in $\mathbb{H}$ is not commutative and there exists no other real linear structure in $\mathbb{H}$. The multiplication of quaternions is however associative (see also Chapter 11).

Table C.1 Quaternions multiplication

$\otimes$	1	$\imath$	$\jmath$	k
1	1	$\imath$	$\jmath$	k
$\imath$	$\imath$	-1	k	$-\jmath$
$\jmath$	$\jmath$	$-k$	-1	$\imath$
k	k	$\jmath$	$\imath$	-1

C.5 Clifford Algebras

There are many examples of Clifford algebras that arise naturally in mathematics, these include Quarternions, Dirac and Pauli Spin algebras. For each natural number n there exists a Clifford algebra of dimension 2^n which can be thought of as an algebra with 2^{n-1} imaginary dimensions which play a role similar to that of $\jmath = \sqrt{-1}$ in $\mathbb{C}$.

Let $\mathbf{V}$ be a real vector space equipped with a symmetric bilinear form Q. The Clifford algebra Cliff$(\mathbf{V}, Q)$ is an associative algebra; if $\mathbf{V}$ is r–dimensional and Q is negative definite, then the algebra Cliff$(\mathbf{V}, Q)$ is denoted by Cliff(r). Table C.2 summarises Clifford algebras Cliff(r) for $r < 8$. Other Clifford algebras can be computed by e.g. making use of the Bott periodicity, that is, $\text{Cliff}(r + 8) = \text{Cliff}(r) \otimes \mathbb{R}$. All real finite dimensional representations of Clifford algebras can deduced from Table C.2, because a matrix algebra $\mathbb{A}[r]$ has a unique irreducible representation, namely $\mathbb{A}^r$.

A representation of a Clifford algebra in a Euclidean space is always assumed to be compatible with the Euclidean inner product. Linear representations of the algebras $\mathbb{C} = \text{Cliff}(1)$ and $\mathbb{H} = \text{Cliff}(2)$ in $\mathbb{R}^n$ are called respectively the complex and quaternionic algebras. It is worth mentioning that all quaternions act conformally, not only as linear combinations of $1, \imath, \jmath$, and k, for more detail see [230].

Table C.2 Clifford algebras Cliff(r) for $r < 8$

r	0	1	2	3	4	5	6	7
Cliff(r)	$\mathbb{R}$	$\mathbb{C}$	$\mathbb{H}$	$\mathbb{H} \otimes \mathbb{H}$	$\mathbb{H}[2]$	$\mathbb{C}[4]$	$\mathbb{R}[8]$	$\mathbb{R}[8] \otimes \mathbb{R}[8]$

Appendix D

Real Valued Activation Functions

Two most frequently used nonlinear activation functions in $\mathbb{R}$ are the logistics function whose range is $(0, 1)$ and the tanh function whose range is $(-1, 1)$. In this Section we show simplified ways to calculate the derivatives of these two functions.

D.1 Logistic Sigmoid Activation Function

The logistic function and its derivative are given by [190]

$$F(x, \kappa, \beta) = \frac{\kappa}{1 + e^{-\beta x}} \tag{D.1}$$

and

$$F'(x, \kappa, \beta) = \frac{\partial \left[\frac{\kappa}{1+e^{-\beta x}}\right]}{\partial x} = \frac{\kappa \beta e^{-\beta x}}{\left(1 + e^{-\beta x}\right)^2} \tag{D.2}$$

where κ denotes the amplitude and β determines the steepness of the slope of the logistic function.
Letting $g(x) = F(x, \kappa, \beta)$, we have

$$\frac{\kappa}{\left(1 + e^{-\beta x}\right)^2} = \frac{g^2(x)}{\kappa} \quad \text{and} \quad e^{-\beta x} = \frac{\kappa}{g(x)} - 1 \tag{D.3}$$

which enables us to simplify the calculation of the first derivative of $F(x, \kappa, \beta)$, as follows

$$F'(x, \kappa, \beta) = \beta \frac{g^2(x)}{\kappa} \left[\frac{\kappa}{g(x)} - 1\right] = \beta g(x) \left[1 - \frac{g(x)}{\kappa}\right] \tag{D.4}$$

Complex Valued Nonlinear Adaptive Filters: Noncircularity, Widely Linear and Neural Models
Danilo P. Mandic and Vanessa Su Lee Goh

D.2 Hyperbolic Tangent Activation Function

Similarly to the case of the logistic function, for a hyperbolic tangent nonlinear activation function given by

$$F(x, \kappa, \beta) = \kappa \frac{e^{\beta x} - e^{-\beta x}}{e^{\beta x} + e^{-\beta x}} \tag{D.5}$$

and letting $g(x) = F(x, \kappa, \beta)$, we can simplify the calculation of the first derivative of (D.5), as follows

$$F'(x, \kappa, \beta) = \frac{\beta}{\kappa} \left[\kappa - g(x)\right] \left[\kappa + g(x)\right] \tag{D.6}$$

Appendix E

Elementary Transcendental Functions (ETF)

As has been shown in Appendix B, the only continuously differentiable function in $\mathbb{C}$, that is, the only function without singularities, is a constant (by Liouville's theorem). Nonlinearities within nonlinear adaptive filters in $\mathbb{C}$ can therefore be either differentiable but unbounded (fully complex) or bounded, but not differentiable (split complex and hybrid). The use of fully complex nonlinearities facilitates the development of stochastic gradient based adaptive filtering algorithms, where mathematical tractability is helped by the use of Cauchy–Riemann equations. One such fully complex nonlinearity is shown in Figure E.1; observe the periodic singularities of this function.

In order for learning algorithms not to be affected by singularities we typically standardise (scale) the data, as shown in Appendix G.4. Scaling of the range of the input signal is necessary, since, e.g. in function approximation problems we have

$$f(x) = \sum_{i=1}^{n} c_i \sigma(x - a_i) = \sum_{i=1}^{n} \frac{c_i}{1 + e^{-x} e^{a_i}} = e^x \sum_{i=1}^{n} c_i \frac{1}{e^x + e^{a_i}} \tag{E.1}$$

which becomes

$$f(x) = z \sum_{i=1}^{n} \frac{c_i}{z + \alpha_i} = r(z) \tag{E.2}$$

upon a change of variables $z = e^x$ and $\alpha_i = e^{a_i}$. The rational function $r(z)$ can be also expressed as

$$r(z) = \sum_{i=1}^{n} \frac{c_i}{z + \alpha_i} = \frac{P(z)}{Q(z)}, \quad z \in \mathbb{C} \tag{E.3}$$

Complex Valued Nonlinear Adaptive Filters: Noncircularity, Widely Linear and Neural Models
Danilo P. Mandic and Vanessa Su Lee Goh

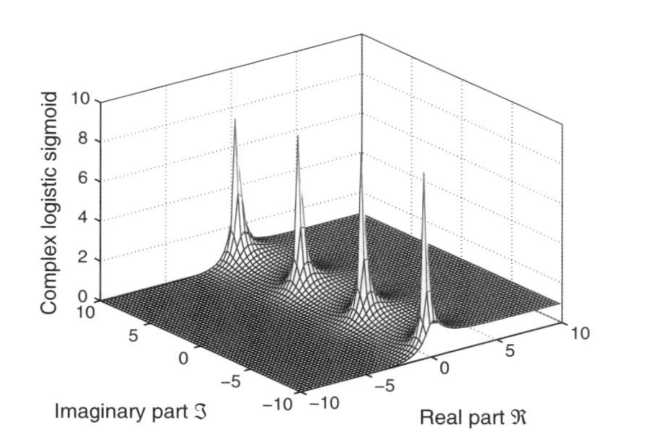

Figure E.1 Magnitude plot of the complex logistic sigmoid function $\Phi(z) = 1/1 + e^z$

Recall that any analytic function $\sigma : \mathbb{R} \rightarrow \mathbb{R}$ (such as the standard logistic sigmoid) has a convergent power series expansion $\sigma(x) = \sum_{i=0}^{\infty} \sigma_i (x - a)^i$ about a point $a \in \mathbb{R}$. When we substitute x with a complex number $z = x + \jmath y$, we obtain a series $\sum_{i=0}^{\infty} \sigma_i (z - a)^i$, which converges within a disc $|z - a| < R$, where symbol R denotes the radius of convergence of power series. Coefficients a_i correspond to the poles $\alpha_i = e^{a_i}$, whereas scaling factors c_i represent the residues of $r(z)$ at e^{a_i} [312].

Table E.1 shows some typical fully complex (differentiable but not bounded) ETFs and their corresponding type of singularity [152]. Figures E.2–E.4 show respectively the $\sin(z)$, $\arctan(z)$ and $\sinh(z)$ fully complex nonlinear activation functions.

Figure E.5 shows a 'hybrid', that is, a complex activation function which is differentiable in its real and imaginary component, but not in $\mathbb{C}$. This function belongs to the class of so called split-complex activation functions, and can be written as

$$\Phi(z) = \tanh(z_r) + \jmath \tanh(z_i), \quad z = z_r + \jmath z_i, \quad \tanh : \mathbb{R} \rightarrow \mathbb{R} \tag{E.4}$$

Table E.1 ETFs and their corresponding type of singularity

$\sigma(z)$	$\frac{d}{dz}\sigma(z)$	Type of singularity
$\tan z$	$\sec^2 z$	isolated
$\sin z$	$\cos z$	removable
$\arctan z$	$\frac{1}{1+z^2}$	isolated
$\arcsin z$	$(1 - z^2)^{-1/2}$	isolated
$\arccos z$	$-(1 - z^2)^{-1/2}$	removable
$\tanh z$	$\text{sech}^2 z$	isolated
$\sinh z$	$\cosh z$	removable
$\text{arctanh} z$	$(1 - z^2)^{-1}$	isolated
$\text{arcsinh} z$	$(1 + z^2)^{-1}$	removable

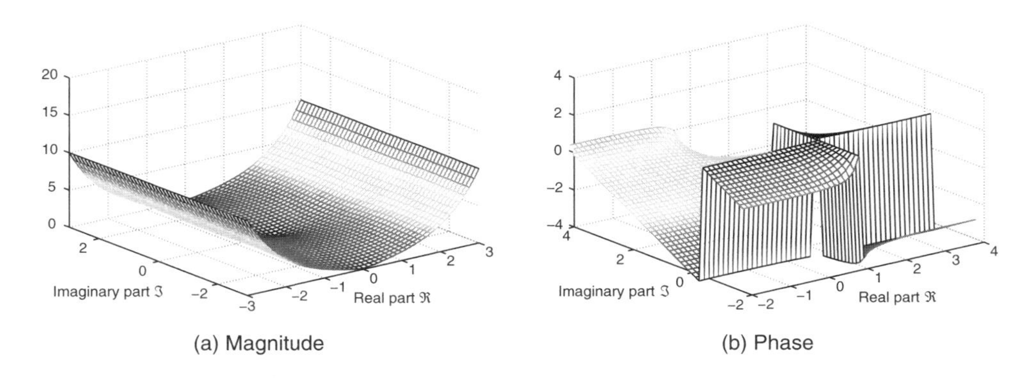

Figure E.2 The magnitude and phase plot for the sin(z) complex activation function

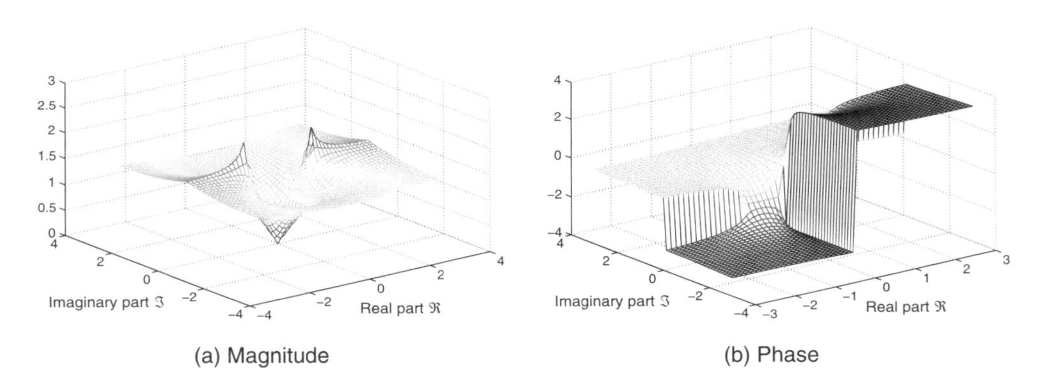

Figure E.3 The magnitude and phase plot for the arctan(z) complex activation function

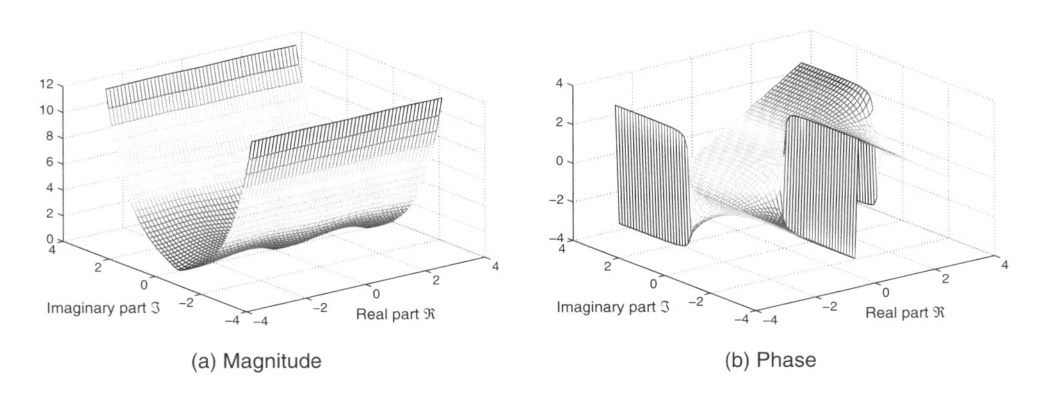

Figure E.4 The magnitude and phase plot for the sinh(z) complex activation function

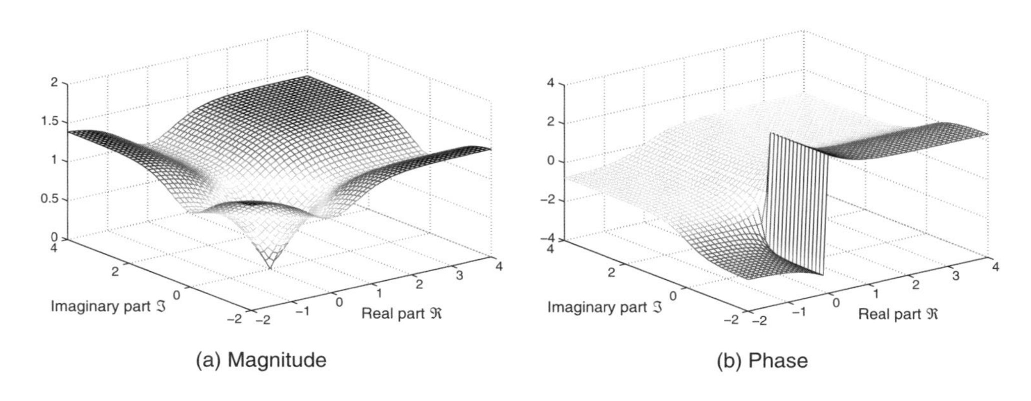

Figure E.5 Complex activation function split – tanh(z)

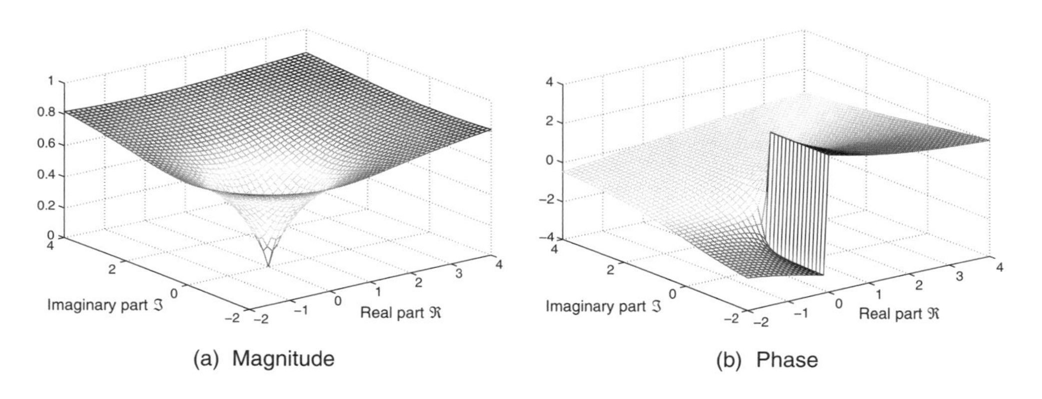

Figure E.6 Complex activation function proposed by Georgiou [88]

Despite the ease of implementation, this class of functions do not fully exploit the correlation and coupling between the real and imaginary components of a complex process.

Alternatively, we can consider hybrid 'synthetic' nonlinearities, such as the activation function proposed in [88], given by

$$\Phi(z) = \frac{z}{c + \frac{1}{r}\,|z|} \tag{E.5}$$

and shown in Figure E.6 for $C = 1$ and $r = 1$. This function is both bounded and differentiable, however, its magnitude and phase variations are rather limited.

Appendix F

The $\mathcal{O}$ Notation and Standard Vector and Matrix Differentiation

F.1 The $\mathcal{O}$ Notation

Definition F.1. *Let functions f and g be mappings $f, g : \mathbb{R}^+ \rightarrow \mathbb{R}^+$. If there exist positive numbers n_0 and c such that $f(n) \leq cg(n)$ for all $n \geq n_0$, then so called "$\mathcal{O}$ notation" is introduced as*

$$f(n) = \mathcal{O}(g(n)) \tag{F.1}$$

An algorithm is said to run in *polynomial time* if there exists $k \in \mathbb{Z}$ such that [29]

$$T(s) = \mathcal{O}(s^k) \tag{F.2}$$

F.2 Standard Vector and Matrix Differentiation

We denote vectors by lowercase bold letters and matrices by capital bold letters (see also Chapter 1). Some vector and matrix differentiation rules frequently used in this book are

- $\dfrac{\mathrm{d}\left(\mathbf{x}^{\mathrm{T}}\mathbf{A}\right)}{\mathrm{d}\mathbf{x}} = \mathbf{A}$
- $\dfrac{\mathrm{d}\left(\mathbf{x}^{\mathrm{T}}\mathbf{A}\mathbf{y}\right)}{\mathrm{d}\mathbf{A}} = \mathbf{x}\mathbf{y}^{\mathrm{T}}$

Complex Valued Nonlinear Adaptive Filters: Noncircularity, Widely Linear and Neural Models
Danilo P. Mandic and Vanessa Su Lee Goh

- $\dfrac{\mathrm{d}\,(\mathbf{Ax}+\mathbf{b})^{\mathrm{T}}\,\mathbf{C}\,(\mathbf{Dx}+\mathbf{e})}{\mathrm{d}\mathbf{x}} = \mathbf{A}^{\mathrm{T}}\mathbf{C}\,(\mathbf{Dx}+\mathbf{e}) + \mathbf{D}^{\mathrm{T}}\mathbf{C}^{\mathrm{T}}\,(\mathbf{Ax}+\mathbf{b})$
  - $\dfrac{\mathrm{d}\left(\mathbf{x}^{\mathrm{T}}\mathbf{Ax}\right)}{\mathrm{d}\mathbf{x}} = \left(\mathbf{A}+\mathbf{A}^{\mathrm{T}}\right)\mathbf{x}$
  - $\dfrac{\mathrm{d}\left(\mathbf{x}^{\mathrm{T}}\mathbf{x}\right)}{\mathrm{d}\mathbf{x}} = 2\mathbf{x}$
- $\dfrac{\mathrm{d}^2\left(\mathbf{y}^{\mathrm{T}}\mathbf{x}\right)}{\mathrm{d}\mathbf{x}^2} = 0$
- $\dfrac{\mathrm{d}^2\,(\mathbf{Ax}+\mathbf{b})\,\mathbf{C}\,(\mathbf{Dx}+\mathbf{e})}{\mathrm{d}\mathbf{x}^2} = \mathbf{A}^{\mathrm{T}}\mathbf{CD} + \mathbf{D}^{\mathrm{T}}\mathbf{C}^{\mathrm{T}}\mathbf{A}$

Appendix G

Notions From Learning Theory

The Error and the Error Function. The error at the output of an adaptive filter is defined as the difference between the target (desired output) and the filter output, that is

$$e(k) = d(k) - y(k) \tag{G.1}$$

The instantaneous error can be positive, negative or zero, and is hence not a suitable candidate for the criterion (loss, cost, objective) function which is mimimised during the training. Error functions are non-negative and are defined so that a decrease in the error function corresponds to better quality of learning; one typical error function is given by

$$E(k) = \frac{1}{2}e^2(k) \tag{G.2}$$

and aims at minimising the power of the instantaneous error of the filter.

The Objective Function is a function that we wish to minimise during training. It can be the same as the error function, but may also include other terms, for instance to penalise for high computational complexity. One such case is the regularisation type cost function, given by [289]

$$J(k) = \frac{1}{N}\sum_{i=1}^{N}\left(e^2(k-i+1) + G\left(\| \mathbf{w}(k-i+1) \|_2^2\right)\right) \tag{G.3}$$

where G is some linear or nonlinear function. We often use symbols E and J interchangeably to denote the cost function.

Complex Valued Nonlinear Adaptive Filters: Noncircularity, Widely Linear and Neural Models
Danilo P. Mandic and Vanessa Su Lee Goh

G.1 Types of Learning

Batch learning is also known as epochwise, or block learning, and is a common strategy for off-line and classification applications. Weights are updated once the whole training set has been presented, as follows

1. Initialise the weights
2. Repeat until some prescribed threshold
   - pass all the training data through the adaptive system;
   - sum the errors after each particular pattern has been presented;
   - update the weights based upon the total error.

Real time adaptive learning, also known as incremental learning, on-line, or pattern learning, operates as follows

1. Initialise the weights
2. Repeat
   - pass one pattern through the adaptive system;
   - perform the weight update.

Often, for adaptive systems which require initialisation, we perform batch learning in the initialisation stage, and online learning while the system is running [186, 190].

Constructive learning allows for the change of the architecture or interconnections within the adaptive system. Neural networks for which the topology changes over time are called *ontogenic* neural networks [73]. Two basic classes of constructive learning are *network growing* and *network pruning*. In the network growing approach, learning begins with a network with no hidden units, and if the error is larger than some threshold, new hidden units are added to the network and training resumes. One network growing algorithm is the so-called cascaded correlation algorithm [123]. The network pruning strategy starts from a large network and if the training error is smaller than some predetermined value, the network size is reduced until the desired ratio between accuracy and network size is reached [247, 283].

G.2 The Bias–Variance Dilemma

An optimal filter performance should provide a compromise between the bias and the variance of the estimation error. Consider the error measure

$$E\left[\left(d(k) - y(k)\right)^2\right] \tag{G.4}$$

The term within the square brackets in Equation (G.4) will not change if we add and subtract a dummy variable $E\left[y(k)\right]$; this yields [244]

$$E\left[\left(d(k) - y(k)\right)^2\right] = \underbrace{E\left[E\big(y(k)\big) - d(k)\right]^2}_{\text{squared bias}} + \underbrace{E\left[\big(y(k) - E[y(k)]\big)^2\right]}_{\text{variance}} \tag{G.5}$$

The first term on the right-hand side of Equation (G.5) represents the squared bias term, whereas the second term on the right-hand side represents the variance. Equation (G.5) shows that

- *Teacher forcing* (equation error) algorithms tend to produce biased estimates, since feedback is replaced by $d(k) \neq E[y(k)]$, which affects the first term of Equation (G.5) [190];
- *Supervised* (output error) algorithms tend not to produce biased estimates, but are prone to producing estimates with larger variance for insufficient filter orders or suboptimal choice of filter parameters (second term in Equation (G.5).

A thorough analysis of the bias/variance dilemma can be found in [87, 111].

G.3 Recursive and Iterative Gradient Estimation Techniques

In real time applications, at time instant k, the coefficient update $\mathbf{w}(k) \rightarrow \mathbf{w}(k+1)$ must be completed before the arrival of next input sample $x(k+1)$, making it possible to re-iterate the weight update.Based on the relationship between the iteration index l and time index k, we differentiate between

- *purely recursive algorithms*, which perform the standard update based on one gradient iteration (coefficient update) per sampling interval k, that is, $l = 1$;
- *a posteriori (data reusing) algorithms*, which perform several coefficient updates per sampling interval, that is, $l > 1$ [185].

G.4 Transformation of Input Data

Given that all the fully complex nonlinearities within complex valued nonlinear adaptive filters have singularities, it is important to ensure that the range of input data matches the useful range of the nonlinearity. There are several ways to modify the input data, these include

- *normalisation*, whereby each element of the input vector $\mathbf{x}(k)$ is divided by its squared norm, that is,

$$x_i(k) \in \mathbf{x}(k) \rightarrow \frac{x_i(k)}{\| \mathbf{x}(k) \|_2^2};$$

- *rescaling*, that is, performing an affine transformation on the data;
  - standardisation to zero mean and unit standard deviation, given by

$$\begin{aligned} \text{mean} &= \frac{\sum_i X_i}{N} \\ \text{std} &= \sqrt{\frac{\sum_i (X_i - \text{mean}^2}{N-1}} \end{aligned} \tag{G.6}$$

  the standardised input becomes $S_i = \mathrm{X_i} - \text{mean}/\text{std}$;

◦ standardisation to a different 'midrange' and 'range'. For instance, for midrange = 0 and range = 2, this is achieved as

$$\begin{aligned}\text{midrange} &= \frac{\max_i X_i + \min_i X_i}{2} \\ \text{range} &= \max_i X_i - \min_i X_i \\ S_i &= \frac{X_i - \text{midrange}}{\text{range}/2}\end{aligned} \tag{G.7}$$

- *nonlinear transformation of the data* can help when the dynamic range of the data is too high. In that case, for instance, we typically apply the log function to the input data (this is the basis for homomorphic neural networks [232]).

Appendix H

Notions from Approximation Theory

Definition H.1 [209]. *Function $\sigma : \mathbb{R} \rightarrow \mathbb{R}$ (not necessarily continuous) is called a* Kolmogorov *function if, for any integer $s \leq 1$, any compact set $K \subset \mathbb{R}^s$, any continuous function $f : K \rightarrow \mathbb{R}$ and any $\varepsilon > 0$, there exists an integer N, numbers $c_k, t_k \in \mathbb{R}$, and $\lambda_k \in \mathbb{R}^s$, $1 \leq k \leq N$ (possibly depending upon s, K, f, ε), such that*

$$\sup_{x \in K} \left| f(x) - \sum_{k=1}^{N} c_k \sigma(\lambda_k x - t_k) \right| < \varepsilon \tag{H.1}$$

Definition H.2. *The sigmoidal function $\sigma(x)$ with properties*

$$\lim_{x \rightarrow -\infty} \frac{\sigma(x)}{x^k} = 0 \tag{H.2}$$

$$\lim_{x \rightarrow -\infty} \frac{\sigma(x)}{x^k} = 1 \tag{H.3}$$

$$|\sigma(x)| \leq K\,(1 + |x|)^k\,, \quad K > 0 \tag{H.4}$$

is called the kth degree sigmoidal function (bounded on $\mathbb{R}$ by a polynomial of degree $d \leq k$).

Therefore Kolmogorov functions cannot be polynomials.

Complex Valued Nonlinear Adaptive Filters: Noncircularity, Widely Linear and Neural Models
Danilo P. Mandic and Vanessa Su Lee Goh

Theorem H.1. (Kolmogorov 1957 [155]). *There exist increasing continuous functions $\psi_{pq}(x)$, on $I = [0, 1]$ so that any continuous function f on I^n can be written in the form*

$$f(x_1, \ldots, x_n) = \sum_{q=1}^{2n+1} \Phi_q \left(\sum_{p=1}^{n} \psi_{pq}(x_p) \right) \tag{H.5}$$

where Φ_q are continuous functions of one variable.

This result asserts that every multivariate continuous function can be represented by a superposition of a small number of univariate *continuousfunctions* (13th problem of Hilbert).

Theorem H.2. (Kolmogorov–Sprecher Theorem). *For every integer $n \geq 2$, there exists a real monotonic increasing function $\psi(x)$, for which $\psi([0, 1]) = [0, 1]$, which depends on n and satisfies the following property*

> *For each preassigned number $\delta > 0$, there exists a rational number $\varepsilon, 0 < \varepsilon < \delta$, such that every real continuous function of n variables, $\phi(\mathbf{x})$, defined on I^n, can be exactly represented by*
>
> $$f(\mathbf{x}) = \sum_{j=1}^{2n+1} \chi \left[\sum_{i=1}^{n} \lambda^i \psi(x_i + \varepsilon(j-1)) + j - 1 \right] \tag{H.6}$$
>
> *where χ is a real and continuous function dependent upon f and λ is a constant independent of f.*

Since no constructive method for the determination of χ is known, a direct application of the Kolmogorov–Sprecher theorem is rather difficult.

Theorem H.3. (Weierstrass Theorem). *If f is a continuous real valued function on $[a, b] \in \mathbb{R}$, then for any $\varepsilon > 0$, there exists a polynomial P on $[a, b] \in \mathbb{R}$ such that*

$$|f(x) - P(x)| < \varepsilon, \quad \forall x \in [a, b] \tag{H.7}$$

In other words, any continuous function on a closed and bounded interval can be uniformly approximated by a polynomial.

Definition H.3. *Let Π be a probability measure on $\mathbb{R}^m$. For measurable functions $f_1, f_2 : \mathbb{R}^m \rightarrow \mathbb{R}$ we say that f_1 approximates f_2 with accuracy $\varepsilon > 0$ and confidence $\delta > 0$ in probability if*

$$\Pi\left(x \in \mathbb{R}^m \mid |f_1(x) - f_2(x)| > \varepsilon\right) < \delta \tag{H.8}$$

We say that function f_1 interpolates f_2 on p examples $x_1, \ldots, x_p$ if $f_1(x_i) = f_2(x_i), \ i = 1, \ldots, p$.

Definition H.4. *The function f that satisfies the condition*

$$|f(x) - f(y)| \leq L|x - y|, \quad x, y \in \mathbb{R}, L = \text{constant} \tag{H.9}$$

is called a Lipschitz function and L is called the Lipshitz constant.

Definition H.5. *A closure of a subset D of a topological space S, usually denoted by $\bar{D}$, is the set of points in S with the property that every neighbourhood of such a point has a nonempty intersection with D.*

The closure $\bar{S}$ of a set S is the set of all points in S together with the set of all limit points of S. A set S is closed if it is identical to its closure $\bar{S}$.

Definition H.6. *A subset D of a topological space S is called dense if $\bar{D} = S$.*

If D is dense in S then each element of S can be approximated arbitrarily well by elements of D. Examples are the set of rational numbers, which is dense $\mathbb{R}$, and the set of polynomials that is dense in the space of continuous functions.

Definition H.7. *A compact set is one in which every infinite subset contains at least one limit point. Every closed, bounded, finite dimensional set in a metric linear space is compact.*

Definition H.8. *A spline is an interpolating polynomial which uses information from neighbouring points to obtain smoothness. A cubic spline is a spline constructed of piece-wise third-order polynomials defined by a set of control points.*

Appendix I

Terminology Used in the Field of Neural Networks

The field of artificial neural networks has developed in connection with many disciplines, including neurobiology, mathematics, statistics, economics, computer science, engineering, and physics. Consequently, the terminology varies from discipline to discipline. A neural network is specified by its topology, constraints, initial state, and activation function. An initiative from the IEEE Neural Networks Council to standardise the terminology has led to several unifying definitions [66], given below.

Activation Function: algorithm for computing the activation value of a neurode as a function of its net input. The net input is typically a sum of weighted inputs to the neurode.
Feedforward Network: network ordered into layers with no feedback paths. The lowest layer is the input layer, the highest is the output layer. The outputs of a given layer are connected only to higher layers, and the inputs come only from lower layers.
Supervised Learning: learning procedure in which a network is presented with a set of input pattern – target pairs. The network can compare its output to the target and adapt itself according to the learning rules.
Unsupervised Learning: learning procedure in which the network is presented with a set of input patterns. The network then adapts itself according to the statistical associations within the input patterns.

Fiesler [72] provides further clarification of the terminology.

- *Neuron functions* (or transfer functions) specify the output of a neuron, given its inputs (this includes nonlinearity);
- *Learning rules* (or learning laws) define how weights (and offsets) are to be updated;
- *Clamping functions* determine if and when certain neurons will be insusceptible to incoming information, that is, which neurons are to retain their present activation value;
- *Ontogenic functions* specify changes in the neural network topology.

Complex Valued Nonlinear Adaptive Filters: Noncircularity, Widely Linear and Neural Models
Danilo P. Mandic and Vanessa Su Lee Goh

Table I.1 Terms related to learning strategies used in different communities

Signal processing	System ID	Neural networks	Adaptive systems
Output error	Parallel	Supervised	Unidirected
Equation error	Series – parallel	Teacher forcing	Directed

Some terms frequently used in the areas of signal processing, system identification, neural networks, and adaptive systems communities are summarised in Table I.1.

Appendix J

Complex Valued Pipelined Recurrent Neural Network (CPRNN)

To process highly nonlinear real valued nonstationary signals, Haykin and Li introduced the Pipelined Recurrent Neural Network (PRNN) [114], a computationally efficient modular nonlinear adaptive filter. Based on a concatenation of M modules, each consisting of FCRNNs with N neurons, the PRNN exhibits improved ability to track nonlinearity as compared to single RNNs, while maintaining low computational complexity ($\mathcal{O}(MN^4)$ for the PRNN with M modules compared with $\mathcal{O}((MN)^4)$ for the FCRNN). The PRNN architecture also helps to circumvent the problem of vanishing gradient, due to its spatial representation of a temporal pattern, and feedback connections within the architecture [190]. This architecture has been successfully employed for a variety of applications where complexity and nonlinearity pose major problems, including those in speech processing, and communications [200]. More insight into the PRNN performance is provided in [114, 190]. We now introduce the complex valued PRNN (CPRNN) as an extension of the real PRNN [114], and derive the complex real time recurrent learning (CRTRL) algorithm for CPRNNs.

J.1 The Complex RTRL Algorithm (CRTRL) for CPRNN

The CPRNN architecture contains M modules of FCRNNs (see Figure 7.3) connected in a nested manner, as shown in Figure J.1. The $(p \times 1)$–dimensional external complex valued signal vector $\mathbf{s}^T(k) = [s(k-1), \ldots, s(k-p)]$ is delayed by m time steps ($z^{-m}\mathbf{I}$) before feeding the module m, where z^{-m}, $m = 1, \ldots, M$ denotes the m-step time delay operator, and $\mathbf{I}$ is the $(p \times p)$ dimensional identity matrix. The complex valued weight vectors $\mathbf{w}_l$ are embodied in an $(p + N + 1) \times N$ dimensional weight matrix $\mathbf{W} = [\mathbf{w}_1, \ldots, \mathbf{w}_N]$. All the modules operate using the same weight matrix $\mathbf{W}$ (a full mathematical description of the PRNN is given

Complex Valued Nonlinear Adaptive Filters: Noncircularity, Widely Linear and Neural Models
Danilo P. Mandic and Vanessa Su Lee Goh

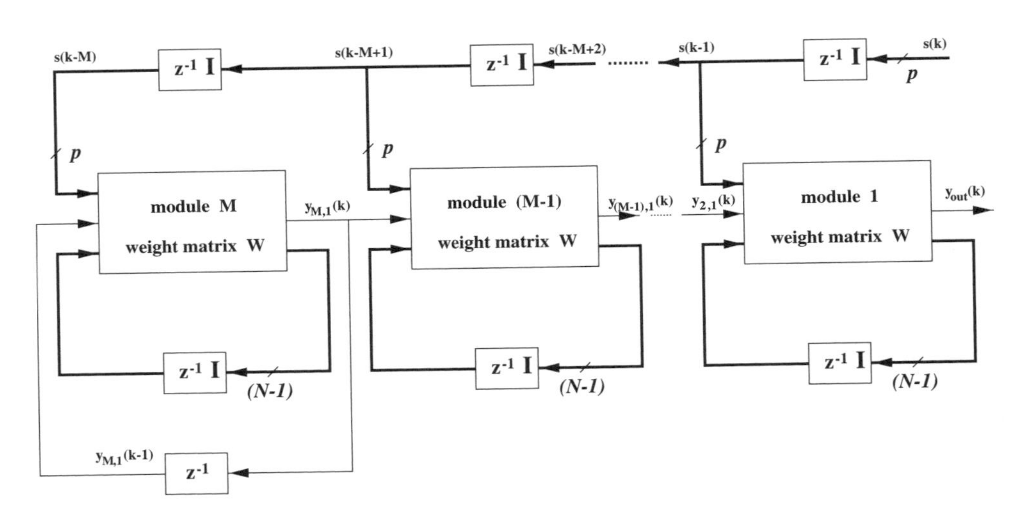

Figure J.1 Pipelined Recurrent Neural Network (PRNN)

in [114]). The following equations provide a mathematical description of the CPRNN,

$$y_{t,l}(k) = \Phi(net_{t,l}(k)), \quad t = 1, 2, \ldots, M \tag{J.1}$$

$$net_{t,l}(k) = \sum_{n=1}^{p+N+1} w_{l,n}(k) P_{t,n}(k), \quad \begin{matrix} l = 1, \ldots, N \\ n = 1, \ldots, p+N+1 \end{matrix} \tag{J.2}$$

$$\begin{aligned} \mathbf{P}_t^{\mathrm{T}}(k) = \big[&s(k-t), \ldots, s(k-t-p+1), 1+\jmath, \\ &y_{t+1,1}(k-1), y_{t,2}(k-1), \ldots, y_{t,N}(k-1)\big] \\ &\text{for} \quad 1 \le t \le M-1 \end{aligned} \tag{J.3}$$

$$\begin{aligned} \mathbf{P}_M^{\mathrm{T}}(k) = \big[&s(k-M), \ldots, s(k-M-p+1), 1+\jmath, \\ &y_{M,1}(k-1), y_{M,2}(k-1), \ldots, y_{M,N}(k-1)\big] \\ &\text{for} \quad t = M \end{aligned} \tag{J.4}$$

For simplicity we state that

$$\begin{aligned} y_{t,l}(k) &= \Phi(net_{t,l}(k)) = \Phi^r(net_{t,l}(k)) + \jmath\Phi^i(net_{t,l}(k)) \\ &= u_{t,l}(k) + \jmath v_{t,l}(k) \end{aligned} \tag{J.5}$$

The overall output of the CPRNN is $y_{1,1}(k)$, that is the output of the first neuron of the first module. At every time instant k, for every module t, $t = 1, \ldots, M$, the one step ahead instantaneous prediction error $e_t(k)$ associated with module t, is defined as

$$e_t(k) = s(k-t+1) - y_{t,1}(k) = e_t^r(k) + \jmath e_t^i(k) \tag{J.6}$$

We can split the error term into its real and imaginary parts, to yield

$$e_t^r(k) = s^r(k-t+1) - u_{t,1}(k), \quad e_t^i(k) = s^i(k-t+1) - v_{t,1}(k) \tag{J.7}$$

Since the CPRNN consists of M modules, a total of M forward prediction error signals are calculated. The cost function of the PRNN introduced in [114] can be modified to suit processing in the complex domain, for instance

$$J(k) = \sum_{t=1}^{M} \gamma^{t-1}(k)\,|e_t(k)|^2 = \sum_{t=1}^{M} \gamma^{t-1}(k)\left[e_t(k)e_t^*(k)\right] = \sum_{t=1}^{M} \gamma^{t-1}(k)\left[(e_t^r)^2 + (e_t^i)^2\right] \tag{J.8}$$

represents a weighted sum of instantaneous squared errors from outputs of the CPRNN modules, where $\gamma(k)$ is a (possible variable) forgetting factor. The forgetting factor, $\gamma < 1$ plays a very important role in nonlinear adaptive filtering of nonstationary signals and is usually set to unity for stationary signals. Since for gradient descent learning the aim is to minimise Equation (J.8) along the entire CPRNN, the weight update for the nth weight at neuron l at the time instant k is calculated as [94]

$$\Delta w_{l,n}(k) = -\mu \frac{\partial}{\partial w_{l,n}(k)} \left(\sum_{t=1}^{M} \gamma^{t-1}(k)\,|e_t(k)|^2 \right) \tag{J.9}$$

Following the derivation of the CRTRL, for $l = 1, \ldots, N$ and $n = 1, \ldots, p+N+1$, the weight update of every weight within the CPRNN can be expressed as

$$\begin{aligned}
w_{l,n}(k+1) &= w_{l,n}(k) - \mu \nabla_{w_{l,n}} J(k)|_{w_{l,n}=w_{l,n}(k)} \\
&= w_{l,n}(k) + \mu \left(\sum_{t=1}^{M} \gamma^{t-1}(k) e_t(k) \left(\pi_{l,n}^{t,1}(k)\right)^* \right) \\
&= w_{l,n}(k) + \mu \left(\sum_{t=1}^{M} \gamma^{t-1}(k) e_t(k) \left\{\Phi^*(net_{t,l}(k))\right\}' \right. \\
&\quad \times \left. \left[\sum_{\alpha=1}^{N} w_{1,\alpha+p+1}^*(k) \left(\pi_{l,n}^{t,\alpha}(k-1)\right)^* + \delta_{ln} P_{t,n}^*(k) \right] \right)
\end{aligned} \tag{J.10}$$

Notice that this weight update has the same generic form as the weight update for the real valued PRNN [114].

J.1.1 Linear Subsection Within the PRNN

The linear subsection within the CPRNN consists of an FIR filter, shown in Figure J.2. The complex valued least mean square (CLMS) algorithm is used to update the weights of this filter, for which the output is given by

$$\hat{s}(k) = \mathbf{w}_{\text{FIR}}^{\text{T}}(k)\mathbf{y}_{\text{out}}(k) \tag{J.11}$$

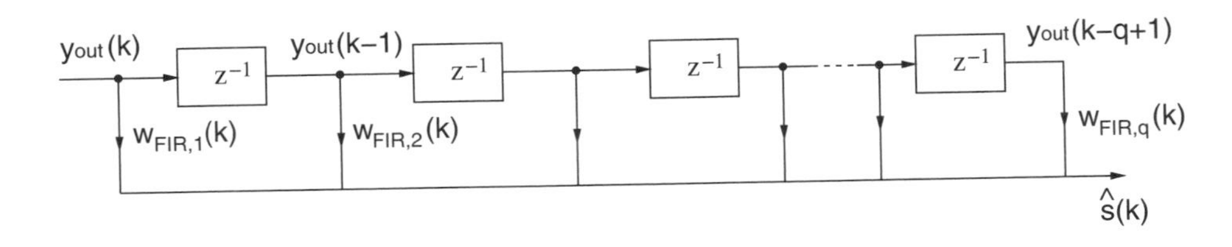

Figure J.2 Linear adaptive FIR filter

where $\mathbf{y}_{\text{out}}(k) \triangleq \left[y_{\text{out}}(k), \ldots, y_{\text{out}}(k-q+1)\right]^{\text{T}}$ is the output from the first CPRNN module ($y_{1,1}(k)$), $\mathbf{w}_{\text{FIR}}(k) \triangleq \left[w_{\text{FIR},1}(k), \ldots, w_{\text{FIR,q}}(k)\right]^{\text{T}}$ is the complex weight vector and q denotes the length of the filter. The error signal $e(k)$ required for weight adaptation is obtained as the difference between the desired response $s(k)$ and the output of the filter $\hat{s}(k)$, and is given by

$$e(k) = s(k) - \hat{s}(k) = s(k) - \mathbf{w}_{\text{FIR}}^{\text{T}}(k)\mathbf{y}_{out}(k) \tag{J.12}$$

The CLMS weight update for the FIR filter from Figure J.2 is given by [307]

$$\mathbf{w}_{FIR}(k+1) = \mathbf{w}_{FIR}(k) + \mu e(k)\mathbf{y}_{out}^{*}(k) \tag{J.13}$$

where μ is the learning rate.

Appendix K

Gradient Adaptive Step Size (GASS) Algorithms in $\mathbb{R}$

Since its introduction, some 50 years ago, the Least Mean Square (LMS) algorithm [308] has become the most frequently used algorithm for the training of adaptive finite impulse response (FIR) filters. The LMS minimises the cost function $J(k) = \frac{1}{2}e^2(k)$, and is given by [308]

$$e(k) = d(k) - \mathbf{x}^{\mathrm{T}}(k)\mathbf{w}(k) \tag{K.1}$$

$$\mathbf{w}(k+1) = \mathbf{w}(k) + \mu e(k)\mathbf{x}(k) \tag{K.2}$$

where $e(k)$ denotes the instantaneous error at the output of the filter, $d(k)$ is the desired signal, $\mathbf{x}(k) = [x(k-1), \ldots, x(k-N)]^{\mathrm{T}}$ is the input signal vector, N is the length of the filter, $(\cdot)^{\mathrm{T}}$ is the vector transpose operator, and $\mathbf{w}(k) = [w_1(k), \ldots, w_N(k)]^{\mathrm{T}}$ is the filter coefficient (weight) vector. The parameter μ is the stepsize (learning rate), which is critical to the performance, and defines how fast the algorithm is converging.

Analysis and practical experience have shown that the presence of inputs with rich dynamics and ill-conditioned input correlation matrix often leads to divergence of LMS. One modification of LMS is the normalised LMS (NLMS) algorithm [113, 308], for which the adaptive learning rate is given by

$$\eta(k) = \frac{\mu}{\| \mathbf{x}(k) \|_2^2 + \varepsilon} \tag{K.3}$$

and the term ε is included to prevent divergence for close to zero input vector $\mathbf{x}(k)$.

The performance of both LMS and NLMS is affected by the inclusion of the 'independence' assumptions in their derivation. To make this class of algorithms more robust and faster converging, a number of gradient adaptive step size (GASS) algorithms have been proposed, which include those by Benveniste [24], Mathews [207], and Farhang [13]. The aim is to ensure $dJ/d\mu = 0$, which leads to $\mu(\infty) = 0$ in the steady state. Another way of introducing more robustness into the LMS update is to make the compensation term ε in the denominator of the NLMS stepsize (Equation K.3) adaptive. One such algorithm is the Generalised Normalised Gradient Descent (GNGD) algorithm [180].

Complex Valued Nonlinear Adaptive Filters: Noncircularity, Widely Linear and Neural Models
Danilo P. Mandic and Vanessa Su Lee Goh

K.1 Gradient Adaptive Stepsize Algorithms Based on $\partial J/\partial \mu$

A gradient adaptive learning rate $\mu(k)$ can be introduced into the LMS algorithm (Equation K.2), based on

$$\mu(k+1) = \mu(k) - \rho \nabla_{\mu} J(k)_{|\mu=\mu(k-1)} \tag{K.4}$$

where parameter ρ denotes the stepsize. The gradient $\nabla_{\mu} J(k)_{|\mu=\mu(k-1)}$ can be evaluated as

$$\nabla_{\mu} J(k) = \frac{1}{2}\frac{\partial e^2(k)}{\partial e(k)}\frac{\partial e(k)}{\partial y(k)}\frac{\partial y(k)}{\partial \mathbf{w}(k)}\frac{\partial \mathbf{w}(k)}{\partial \mu(k-1)} = -e(k)\mathbf{x}^{\mathrm{T}}(k)\frac{\partial \mathbf{w}(k)}{\partial \mu(k-1)} \tag{K.5}$$

Denote $\boldsymbol{\gamma}(k) = \partial \mathbf{w}(k)/\partial \mu(k-1)$ to obtain a general update for the stepsize in the form

$$\mu(k+1) = \mu(k) + \rho e(k)\mathbf{x}^{\mathrm{T}}(k)\boldsymbol{\gamma}(k) \tag{K.6}$$

Algorithms within this class differ, depending on the way they evaluate the term $\boldsymbol{\gamma}(k)$.

Benveniste's update [24] is rigorous and evaluates the sensitivity $\boldsymbol{\gamma}(k)$ based on (K.2) as

$$\begin{aligned}\frac{\partial \mathbf{w}(k)}{\partial \mu(k-1)} &= \frac{\partial \mathbf{w}(k-1)}{\partial \mu(k-1)} + e(k-1)\mathbf{x}(k-1) + \mu(k-1)\frac{\partial e(k-1)}{\partial \mu(k-1)}\mathbf{x}(k-1) \\ &\quad + \mu(k-1)e(k-1)\frac{\partial \mathbf{x}(k-1)}{\partial \mu(k-1)}\end{aligned} \tag{K.7}$$

The last term in Equation (K.7) vanishes, since the input $\mathbf{x}(k-1)$ is independent of the learning rate $\mu(k-1)$, whereas the term $\partial e(k-1)/\partial \mu(k-1)$ becomes

$$\frac{\partial e(k-1)}{\partial \mu(k-1)} = \frac{\partial\left(d(k-1) - \mathbf{x}^{\mathrm{T}}(k-1)\mathbf{w}(k-1)\right)}{\partial \mu(k-1)} = -\mathbf{x}^{\mathrm{T}}(k-1)\frac{\partial \mathbf{w}(k-1)}{\partial \mu(k-1)} \tag{K.8}$$

The expression[1] for the gradient $\nabla_{\mu} J(k)$ in (K.5) now becomes

$$\begin{aligned}\nabla_{\mu(k-1)} J(k) &= -e(k)\mathbf{x}^{\mathrm{T}}(k)\boldsymbol{\gamma}(k) \\ \boldsymbol{\gamma}(k) &= \underbrace{\left[\mathbf{I} - \mu(k-1)\mathbf{x}(k-1)\mathbf{x}^{\mathrm{T}}(k-1)\right]}_{\text{filtering term}}\boldsymbol{\gamma}(k-1) + e(k-1)\mathbf{x}(k-1)\end{aligned} \tag{K.9}$$

The term in the square brackets represents a time-varying adaptive filter, which provides low pass filtering of the instantaneous gradient $e(k-1)\mathbf{x}(k-1)$. This way, Benveniste's update is very accurate and robust to the uncertainties due to the noisy instantaneous gradients $e(k-1)$ $\mathbf{x}(k-1)$.

[1] For a small value of μ, assume $\mu(k-1) \approx \mu(k)$ and therefore

$$\frac{\partial \mathbf{w}(k)}{\partial \mu(k-1)} \approx \frac{\partial \mathbf{w}(k)}{\partial \mu(k)} = \boldsymbol{\gamma}(k)$$

Algorithm by Farhang and Ang [13] is based on a recursive calculation of $\boldsymbol{\gamma}$ from Equation (K.9) in the form

$$\boldsymbol{\gamma}(k) = \alpha\boldsymbol{\gamma}(k-1) + e(k-1)\mathbf{x}(k-1), \quad 0 \leq \alpha \leq 1 \tag{K.10}$$

that is, a time varying instantaneous filtering term in the square brackets in Equation (K.9) is replaced by a lowpass filter with a fixed coefficient α. For each weight update $w_j(k)$, we then have

$$\gamma_j(k) = \alpha\gamma_j(k-1) + e(k-1)x_j(k-1)$$

Mathews' algorithm [207] is a simplification of the algorithm by Farhang and Ang, where $\alpha = 0$, that is, it uses noisy instantaneous estimates of the gradient, resulting in the learning rate update

$$\mu(k+1) = \mu(k) + \rho e(k)e(k-1)\mathbf{x}^{\mathrm{T}}(k)\mathbf{x}(k-1) \tag{K.11}$$

K.2 Variable Stepsize Algorithms Based on $\partial J/\partial \varepsilon$

The generalised normalised gradient descent (GNGD) algorithm [180] makes the regularisation factor ε within the NLMS algorithm (Equation K.3) gradient adaptive, and is based on

$$\varepsilon(k+1) = \varepsilon(k) - \rho\nabla_{\varepsilon(k-1)}J(k)_{|\varepsilon=\varepsilon(k-1)} \tag{K.12}$$

Similarly to the GASS algorithms based on $\partial J/\partial \mu$, the gradient $\nabla_{\varepsilon(k-1)}J(k)_{|\varepsilon=\varepsilon(k-1)}$ can be evaluated as

$$\frac{\partial J(k)}{\partial \varepsilon(k-1)} = \frac{\partial J(k)}{\partial e(k)}\frac{\partial e(k)}{\partial y(k)}\frac{\partial y(k)}{\partial \mathbf{w}(k)}\frac{\partial \mathbf{w}(k)}{\partial \varepsilon(k-1)} = -e(k)\mathbf{x}^{\mathrm{T}}(k)\frac{\partial \mathbf{w}(k)}{\partial \varepsilon(k-1)} \tag{K.13}$$

The partial derivative $\partial\mathbf{w}(k)/\partial\varepsilon(k-1)$ in Equation (K.13) now becomes

$$\frac{\partial \mathbf{w}(k)}{\partial \varepsilon(k-1)} = \frac{\partial \mathbf{w}(k-1)}{\partial \varepsilon(k-1)} - \frac{\eta e(k-1)\mathbf{x}(k-1)}{\left(\|\ \mathbf{x}(k-1)\ \|_2^2 + \varepsilon(k-1)\right)^2} + \frac{\partial e(k-1)}{\partial \varepsilon(k-1)}\eta(k-1)\mathbf{x}(k-1)$$

$$\frac{\partial e(k-1)}{\partial \varepsilon(k-1)} = \frac{\partial\left[d(k-1) - \mathbf{x}^T(k-1)\mathbf{w}(k-1)\right]}{\partial \varepsilon(k-1)} = -\mathbf{x}^T(k-1)\frac{\partial \mathbf{w}(k-1)}{\partial \varepsilon(k-1)} \tag{K.14}$$

to give

$$\frac{\partial \mathbf{w}(k)}{\partial \varepsilon(k)} = \left[\mathbf{I} - \eta(k-1)\mathbf{x}(k-1)\mathbf{x}^{\mathrm{T}}(k-1)\right]\frac{\partial \mathbf{w}(k-1)}{\partial \varepsilon(k-1)} - \frac{\eta e(k-1)\mathbf{x}(k-1)}{\left(\ \|\ \mathbf{x}(k-1)\ \|_2^2 + \varepsilon(k-1)\right)^2} \tag{K.15}$$

For simplicity, denote $\boldsymbol{\gamma}(k) = \partial\mathbf{w}(k)/\partial\varepsilon(k)$, to yield the update of the regularisation factor in the form

$$\varepsilon(k+1) = \varepsilon(k) + e(k)\mathbf{x}^{\mathrm{T}}(k)\boldsymbol{\gamma}(k) \tag{K.16}$$

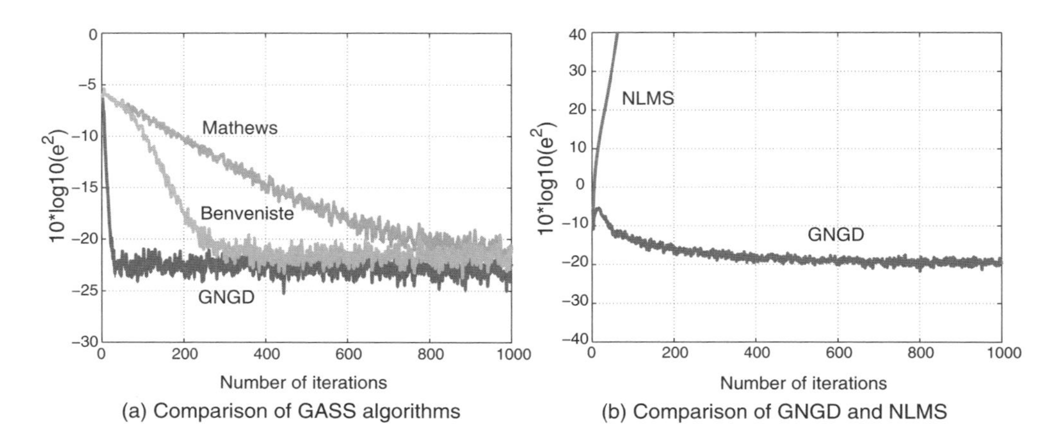

Figure K.1 Learning curves for the GASS algorithms. Left: comparison between Benveniste's, Farhang's, Mathews' algorithms and GNGD for the prediction of the coloured signal (Equation 8.22). Right: Comparison between GNGD and NLMS for prediction of the nonlinear signal (Equation 8.24), for the stepsize $\mu = 2.1$

In the standard GNGD algorithm [180], the term in the square brackets in Equation (K.15) is set to zero, giving

$$\varepsilon(k+1) = \varepsilon(k) - \rho\mu \frac{e(k)e(k-1)\mathbf{x}^{\mathrm{T}}(k)\mathbf{x}(k-1)}{\left(\| \mathbf{x}(k-1) \|_2^2 + \varepsilon(k-1)\right)^2} \tag{K.17}$$

Figure K.1(a) shows learning curves for the GASS algorithms (based on $\partial J/\partial\mu$) and the GNGD algorithm (based on $\partial J/\partial\varepsilon$) for prediction of the coloured AR(4) process (See Equation 8.22). The GNGD exhibited fastest convergence, the algorithms by Benveniste and Farhang and Ang had similar performance, whereas Mathews' algorithm was slowest to converge. Figure K.1(b) illustrates the excellent stability of GNGD compared with NLMS, for the prediction of nonlinear signal (See Equation 8.24). The learning rate was chosen to be $\mu = 2.1$, for which the NLMS diverged whereas GNGD converged.

Appendix L

Derivation of Partial Derivatives from Chapter 8

L.1 Derivation of $\partial e(k)/\partial w_n(k)$

The Cauchy–Riemann equations state that

$$\frac{\partial e_r(k)}{\partial u(k)} = \frac{\partial e_i(k)}{\partial v(k)}, \qquad \frac{\partial e_r(k)}{\partial v(k)} = -\frac{\partial e_i(k)}{\partial u(k)}$$

and

$$\frac{\partial u(k)}{\partial net_r(k)} = \frac{\partial v(k)}{\partial net_i(k)}, \qquad \frac{\partial v(k)}{\partial net_r(k)} = -\frac{\partial u(k)}{\partial net_i(k)}$$

$$\frac{\partial e(k)}{\partial w_n(k)} = \frac{\partial e_i(k)}{\partial w_n^i(k)} - \jmath \frac{\partial e_r(k)}{\partial w_n^i(k)} \tag{L.1}$$

Therefore

$$\begin{aligned}\frac{\partial e_i(k)}{\partial w_n^i(k)} &= \frac{\partial e_i(k)}{\partial v(k)}\left[\frac{\partial v(k)}{\partial net_r(k)}\frac{\partial net_r(k)}{\partial w_n^i(k)} + \frac{\partial v(k)}{\partial net_i(k)}\frac{\partial net_i(k)}{\partial w_n^i(k)}\right] \\ &= \frac{\partial e_i(k)}{\partial v(k)}\left[\frac{\partial v(k)}{\partial net_r(k)}(-x_n^i(k)) + \frac{\partial v(k)}{\partial net_i(k)}x_n^r(k)\right]\end{aligned} \tag{L.2}$$

and

$$\begin{aligned}\frac{\partial e_r(k)}{\partial w_n^i(k)} &= \frac{\partial e_r(k)}{\partial u(k)}\left[\frac{\partial u(k)}{\partial net_r(k)}\frac{\partial net_r(k)}{\partial w_n^i(k)} + \frac{\partial u(k)}{\partial net_i(k)}\frac{\partial net_i(k)}{\partial w_n^i(k)}\right] \\ &= \frac{\partial e_r(k)}{\partial u(k)}\left[\frac{\partial u(k)}{\partial net_r(k)}(-x_n^i(k)) + \frac{\partial u(k)}{\partial net_i(k)}x_n^r(k)\right]\end{aligned} \tag{L.3}$$

Complex Valued Nonlinear Adaptive Filters: Noncircularity, Widely Linear and Neural Models
Danilo P. Mandic and Vanessa Su Lee Goh

giving

$$\frac{\partial e(k)}{\partial w_n(k)} = \frac{\partial e_i(k)}{\partial v(k)} \left[\frac{\partial v(k)}{\partial net_{\mathrm{r}}(\mathrm{k})} frac\partial v(k)\partial\, net_i(k) x_n^r(k) \right] - \jmath \frac{\partial e_r(k)}{\partial u(k)} \left[\frac{\partial u(k)}{\partial\, net_r(k)}(-x_n^i(k)) + \frac{\partial u(k)}{\partial\, net_i(k)} x_n^r(k) \right] \tag{L.4}$$

From the above, we have

$$\begin{aligned} \frac{\partial e(k)}{\partial w_n(k)} &= \frac{\partial e_i(k)}{\partial v(k)} \left[\frac{\partial v(k)}{\partial\, net_r(k)}(-x_n^i(k)) \right. \\ &+ \left. \frac{\partial v(k)}{\partial net_i(k)} x_n^r(k) + \frac{\partial u(k)}{\partial\, net_r(k)}(\jmath x_n^i(k)) + \frac{\partial u(k)}{\partial net_i(k)}(-\jmath x_n^r(k)) \right] \\ &= - \left[\frac{\partial u(k)}{\partial\, net_r(k)}(x_n^r + \jmath x_n^i(k)) + \frac{\partial v(k)}{\partial net_r(k)}(-x_n^i + \jmath x_n^r(k)) \right] \\ &= -x_n(k) \left[\frac{\partial u(k)}{\partial\, net_r(k)} + \jmath \frac{\partial v(k)}{\partial net_r(k)} \right] \\ &= -\Phi'\big(net(k)\big) x_n(k) \end{aligned} \tag{L.5}$$

L.2 Derivation of $\partial e^*(k)/\partial \varepsilon(k-1)$

To calculate $\partial e^*(k)/\partial \varepsilon(k-1)$, consider the real and imaginary part separately, that as

$$\frac{\partial e^*(k)}{\partial \varepsilon(k-1)} = \frac{\partial e_r^*(k)}{\partial \varepsilon(k-1)} + \jmath \frac{\partial e_i^*(k)}{\partial \varepsilon(k-1)} \tag{L.6}$$

Knowing that

$$e^*(k) = d^*(k) - \Phi^*\big(\, net(k)\big) = d^*(k) - \Phi^*\left(\mathbf{x}^{\mathrm{T}}(k)\mathbf{w}(k)\right) \tag{L.7}$$

we have

$$e_r^*(k) = d_r^*(k) - u(k), \quad e_i^*(k) = d_i^*(k) + v(k) \tag{L.8}$$

which yields

$$\frac{\partial e_r^*(k)}{\partial u(k)} = -1, \quad \frac{\partial e_i^*(k)}{\partial v(k)} = 1 \tag{L.9}$$

Using the Cauchy–Riemann equations

$$\frac{\partial u(k)}{\partial net_r(k)} = \frac{\partial v(k)}{\partial net_i(k)}, \quad \frac{\partial v(k)}{\partial net_r(k)} = -\frac{\partial u(k)}{\partial net_i(k)} \tag{L.10}$$

we have

$$
\begin{aligned}
\frac{\partial e_r^*(k)}{\partial \varepsilon(k-1)} &= \frac{\partial e_r^*(k)}{\partial u(k)}\left[\frac{\partial u(k)}{\partial\, net_r(k)}\left(\frac{\partial net_r(k)}{\partial \mathbf{w}_r(k)}\frac{\partial \mathbf{w}_r(k)}{\partial \varepsilon(k-1)} + \frac{\partial net_r(k)}{\partial \mathbf{w}_i(k)}\frac{\partial \mathbf{w}_i(k)}{\partial \varepsilon(k-1)}\right)\right.\\
&\quad \left. + \frac{\partial u(k)}{\partial net_i(k)}\left(\frac{\partial net_i(k)}{\partial \mathbf{w}_r(k)}\frac{\partial \mathbf{w}_r(k)}{\partial \varepsilon(k-1)} + \frac{\partial net_i(k)}{\partial \mathbf{w}_i(k)}\frac{\partial \mathbf{w}_i(k)}{\partial \varepsilon(k-1)}\right)\right]\\
&= \frac{\partial u(k)}{\partial net_r(k)}\left(-\mathbf{x}_r(k)\frac{\partial \mathbf{w}_r(k)}{\partial \varepsilon(k-1)} + \mathbf{x}_i(k)\frac{\partial \mathbf{w}_i(k)}{\partial \varepsilon(k-1)}\right)\\
&\quad + \frac{\partial u(k)}{\partial net_i(k)}\left(-\mathbf{x}_i(k)\frac{\partial \mathbf{w}_r(k)}{\partial \varepsilon(k-1)} - \mathbf{x}_r(k)\frac{\partial \mathbf{w}_i(k)}{\partial \varepsilon(k-1)}\right)
\end{aligned} \tag{L.11}
$$

and

$$
\begin{aligned}
\frac{\partial e_i^*(k)}{\partial \varepsilon(k-1)} &= \frac{\partial e_i^*(k)}{\partial v(k)}\left[\frac{\partial v(k)}{\partial\, net_r(k)}\left(\frac{\partial net_r(k)}{\partial \mathbf{w}_r(k)}\frac{\partial \mathbf{w}_r(k)}{\partial \varepsilon(k-1)} + \frac{\partial net_r(k)}{\partial \mathbf{w}_i(k)}\frac{\partial \mathbf{w}_i(k)}{\partial \varepsilon(k-1)}\right)\right.\\
&\quad \left. + \frac{\partial u(k)}{\partial\, net_i(k)}\left(\frac{\partial net_i(k)}{\partial \mathbf{w}_r(k)}\frac{\partial \mathbf{w}_r(k)}{\partial \varepsilon(k-1)} + \frac{\partial net_i(k)}{\partial \mathbf{w}_i(k)}\frac{\partial \mathbf{w}_i(k)}{\partial \varepsilon(k-1)}\right)\right]\\
&= \frac{\partial v(k)}{\partial net_r(k)}\left(-\mathbf{x}_r(k)\frac{\partial \mathbf{w}_r(k)}{\partial \varepsilon(k-1)} + \mathbf{x}_i(k)\frac{\partial \mathbf{w}_i(k)}{\partial \varepsilon(k-1)}\right)\\
&\quad + \frac{\partial v(k)}{\partial net_i(k)}\left(-\mathbf{x}_i(k)\frac{\partial \mathbf{w}_r(k)}{\partial \varepsilon(k-1)} - \mathbf{x}_r(k)\frac{\partial \mathbf{w}_i(k)}{\partial \varepsilon(k-1)}\right)
\end{aligned} \tag{L.12}
$$

We can now combine the real and imaginary part of the partial derivative of the complex conjugate of the instantaneous error with respect to the regularisation parameter $\varepsilon(k)$, to obtain

$$
\begin{aligned}
\frac{\partial e^*(k)}{\partial \varepsilon(k-1)} &= \frac{\partial u(k)}{\partial net_r(k)}\left[\frac{\partial \mathbf{w}_r(k)}{\partial \varepsilon(k-1)}(-\mathbf{x}_r(k) + \jmath\mathbf{x}_i(k)) + \frac{\partial \mathbf{w}_i(k)}{\partial \varepsilon(k-1)}(\jmath\mathbf{x}_r(k) + \mathbf{x}_i(k))\right]\\
&\quad + \frac{\partial u(k)}{\partial net_i(k)}\left[\frac{\partial \mathbf{w}_r(k)}{\partial \varepsilon(k-1)}(-\jmath\mathbf{x}_r(k) - \mathbf{x}_i(k)) + \frac{\partial \mathbf{w}_i(k)}{\partial \varepsilon(k-1)}(-\mathbf{x}_r(k) + \jmath\mathbf{x}_i(k))\right]\\
&= -\left[\frac{\partial u(k)}{\partial net_r(k)}\left(\frac{\partial \mathbf{w}_r(k)}{\partial \varepsilon(k-1)} - \jmath\frac{\partial \mathbf{w}_i(k)}{\partial \varepsilon(k-1)}\right)\right.\\
&\quad \left. + \frac{\partial u(k)}{\partial net_i(k)}\left(\jmath\frac{\partial \mathbf{w}_r(k)}{\partial \varepsilon(k-1)} + \frac{\partial \mathbf{w}_i(k)}{\partial \varepsilon(k-1)}\right)\right]\mathbf{x}^*(k)\\
&= -\frac{\partial \mathbf{w}^*(k)}{\partial \varepsilon(k-1)}\left[\frac{\partial u(k)}{\partial\, net_r(k)} + \jmath\frac{\partial u(k)}{\partial\, net_i(k)}\right]\mathbf{x}^*(k) = -\frac{\partial \mathbf{w}^*(k)}{\partial \varepsilon(k-1)}\Phi'^*\big(net(k)\big)\mathbf{x}^*(k)
\end{aligned} \tag{L.13}
$$

This finally yields

$$
\frac{\partial e(k)}{\partial \varepsilon(k-1)} = -\frac{\partial \mathbf{w}(k)}{\partial \varepsilon(k-1)}\Phi'\big(net(k)\big)\mathbf{x}(k) \tag{L.14}
$$

L.3 Derivation of $\partial \mathbf{w}(k)/\partial \varepsilon(k-1)$

By a similar derivation, denote $\boldsymbol{\psi} = e(k-1)\Phi'^{*}(k-1)\mathbf{x}^{*}(k-1)$, to have

$$\frac{\partial \mathbf{w}_r(k)}{\partial \varepsilon(k-1)} = -\frac{\psi_r(k-1)}{\left[|\Phi'(net(k-1))|^2 \, \|\mathbf{x}(k-1)\|_2^2 + \varepsilon(k-1)\right]^2} \tag{L.15}$$

and

$$\frac{\partial \mathbf{w}_i(k)}{\partial \varepsilon(k-1)} = -\frac{\psi_i(k-1)}{\left[|\Phi'(net(k-1))|^2 \, \|\mathbf{x}(k-1)\|_2^2 + \varepsilon(k-1)\right]^2} \tag{L.16}$$

Appendix M

A Posteriori Learning

In the *Oxford Interactive Encyclopedia* the notions of *a priori* and *a posteriori* are defined as

> *A priori* is a term from epistemology meaning knowledge or concepts which can be gained independently of all experience. It is contrasted with *a posteriori* knowledge, in which experience plays an essential role.

Probably the oldest written study on a *posteriori* reasoning in logic was by Aristotle, sometime between 343 BC and 323 BC [20], in his famous books *Prior Analytics* and *Posterior Analytics*.[1] In the late sixteenth century, Galileo composed a manuscript, nowadays known as MS27 while he was teaching at the University of Pisa, between 1589 and 1591 [300]. The manuscript was based upon the canons introduced by Aristotle in *Posterior Analytics*, and includes *Discourse on Bodies on or in Water* and *Letters on Sunspots*. He studied the role of foreknowledge and demonstration in science in general.

> A science can give a real definition of its total subject *a posteriori* only, because the real definition is not foreknown in the science, therefore it is sought, therefore it is demonstrable [300].

> We know something either *a posteriori* or *a priori*, *a posteriori* through demonstration of the fact, *a priori* through demonstration of the reasoned fact. *A posteriori* is referred to as demonstration *froman effect*, or *conjectural* [300].

Two centuries after Galileo, in 1781, Immanuel Kant published his *Critique of Pure Reason*[2] [142], where he highlighted limitations of *a priori* reasoning.

---

[1] Aristotle used the word *Analytics* to represent today's meaning of Logic.
[2] A widely used reference is a 1929 translation by Norman Kemp Smith, then Professor of Logic and Metaphysics in the University of Edinburgh.

Complex Valued Nonlinear Adaptive Filters: Noncircularity, Widely Linear and Neural Models
Danilo P. Mandic and Vanessa Su Lee Goh

M.1 *A Posteriori* Strategies in Adaptive Learning

Consider a real valued data reusing LMS algorithm for an FIR filter, given by (see also Chapter 10)

$$\begin{aligned} \mathbf{w}_{i+1}(k) &= \mathbf{w}_i(k) + \eta e_i(k)\mathbf{x}(k), \quad i = 1, \ldots, L \\ e_i(k) &= d(k) - \mathbf{x}^{\mathrm{T}}(k)\mathbf{w}_i(k) \end{aligned} \tag{M.1}$$

where $\mathbf{w}_1(k) = \mathbf{w}(k)$ and $\mathbf{w}(k+1) = \mathbf{w}_{L+1}(k)$. For $L = 1$ equations (M.1) degenerate into the standard LMS algorithm. Time alignment for the *a priori* and *a posteriori* mode of operation is shown in Figure M.1. From Equation (M.1), it is obvious that the direction of the weight update vector $\Delta\mathbf{w}_i(k)$ is the same as that of the input vector $\mathbf{x}(k)$ (colinear), this is depicted in Figure M.2 [306].

Geometric interpretation of a posteriori learning. Figure M.3 provides a geometric interpretation of the operation of LMS, NLMS, and *a posteriori* (data reusing) LMS algorithms, where the direction of vectors $\mathbf{w}$ is given by [261]

$$\mathbf{w}(k) + \text{span}\,(\mathbf{x}(k)) \tag{M.2}$$

As the LMS algorithm is an approximative algorithm which uses instantaneous estimates instead of statistical expectations, the output error of a filter is either positive (the weight

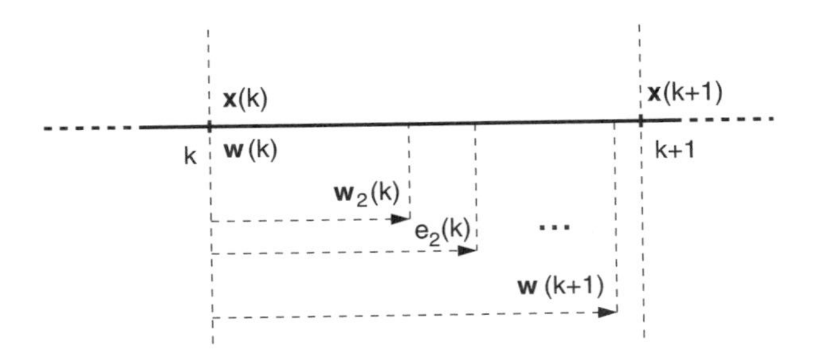

Figure M.1 Time alignment within the data reusing (a posteriori) approach

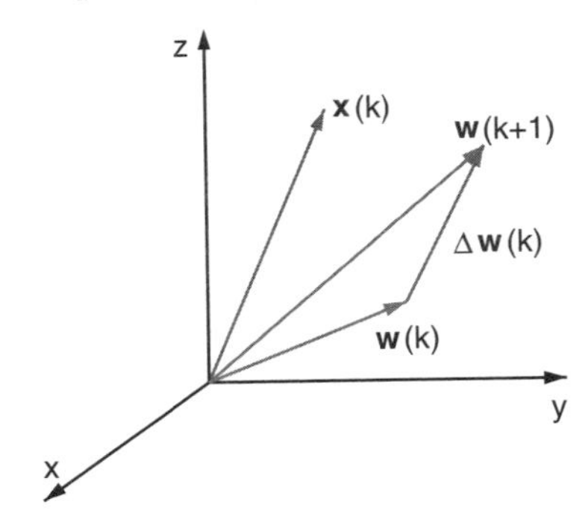

Figure M.2 Geometric view of the LMS weight update

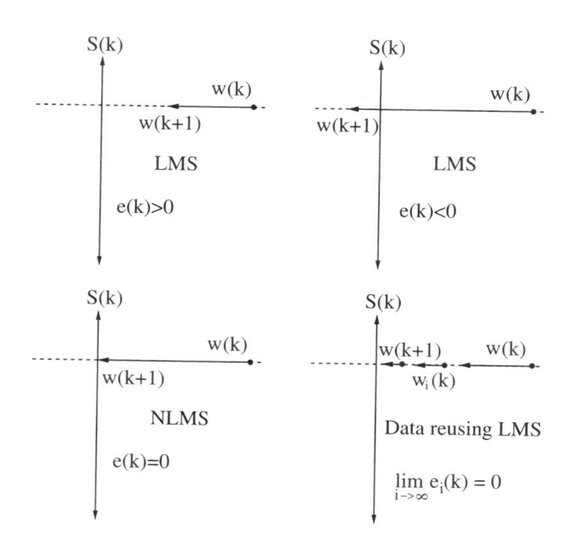

Figure M.3 Comparison of error convergence between standard and data reusing algorithms

update falling short of the optimal weight update) or negative (the weight update exceeds the optimal weight update), as illustrated in the top two diagrams of Figure M.3. *A posteriori* (data reuse) algorithms start either from the case $e(k) > 0$ or $e(k) < 0$, and the iterative data reusing weight updates approach the performance of normalised algorithms as the number of iterations increases, as shown in the bottom right diagram of Figure M.3.

Appendix N

Notions from Stability Theory

The following definitions are similar to those in [258], but are modified so as to suit this work.[1] Consider the differential equation

$$\dot{\mathbf{x}} = f(t, \mathbf{x}), \quad \mathbf{x}(t_0) = \mathbf{x}_0 \tag{N.1}$$

Definition N.1 (Autonomous system). *A system defined by Equation (N.1) is called autonomous, or time–invariant, if f does not depend on t.*

Definition N.2 (Linear system). *The system (N.1) is said to be linear if $f(t, \mathbf{x}) = A(t)\mathbf{x}$ for some $A : \mathbb{R}^+ \rightarrow \mathbb{R}^{n \times n}$.*

In terms of the scale, properties of systems defined on a closed ball $B_h \in \mathbb{R}^n$ with radius h centred at $\mathbf{0}$, can be considered

i) *locally*, if true for all $\mathbf{x}_0 \in B_h$;
ii) *globally*, if true for all $\mathbf{x}_0 \in \mathbb{R}^n$;
iii) *in any closed ball*, if true for all $\mathbf{x}_0 \in B_h$, with h arbitrary;
iv) *uniformly*, if true for all $t_0 > 0$.

Definition N.3 (Lipschitz function). *Function f is said to be Lipschitz in $\mathbf{x}$ if, for some $h > 0$, there exists $L \geq 0$ such that*

$$\| f(t, \mathbf{x}_1) - f(t, \mathbf{x}_2) \| \leq L \| \mathbf{x}_1 - \mathbf{x}_2 \| \tag{N.2}$$

for all $\mathbf{x}_1, \mathbf{x}_2 \in B_h$, $t \geq 0$. Constant L is called the Lipschitz *constant.*

[1] The roots of stability theory can be traced down to Alexander Mikhailovitch Lyapunov (1857–1918). He developed the so–called Lyapunov's Second Method in his PhD thesis *The General Problem of the Stability of Motion* in 1892. Lyapunov was interested in the problem of equilibrium figures of a rotating liquid, a problem also addressed by Maclaurin, Jacobi, and Laplace.

Complex Valued Nonlinear Adaptive Filters: Noncircularity, Widely Linear and Neural Models
Danilo P. Mandic and Vanessa Su Lee Goh

Definition N.4 (Equilibrium). *Point* $\mathbf{x}$ *is called an equilibrium point of Equation (N.1), if* $f(t, \mathbf{x}) = 0$ *for all* $t \geq 0$.

Definition N.5 (Stability). *Point* $\mathbf{x} = 0$ *is called a stable equilibrium point of Equation (N.1) if, for all* $t_0 \geq 0$ *and* $\varepsilon > 0$, *there exists* $\delta(t_0, \varepsilon)$ *such that*

$$\| \mathbf{x}_0 \| < \delta(t_0, \varepsilon) \quad \Rightarrow \quad \| \mathbf{x}(t) \| < \varepsilon, \ \ \forall \, t \geq t_0 \tag{N.3}$$

where $\mathbf{x}(t)$ *is the solution of (N.1), based on initial conditions* $\mathbf{x}_0$ *and* t_0.

Definition N.6 (Uniform stability). *Point* $\mathbf{x} = \mathbf{0}$ *is called a uniformly stable equilibrium point of Equation (N.1) if, in the above definition* δ *can be chosen independent of* t_0. *In other words, the equilibrium point is not growing progressively less stable with time.*

Definition N.7 (Asymptotic stability). *Point* $\mathbf{x} = \mathbf{0}$ *is called an asymptotically stable equilibrium point if*

a) $\mathbf{x} = \mathbf{0}$ *is a stable equilibrium point,*
b) $\mathbf{x} = \mathbf{0}$ *is attractive, that is, for all* $t_0 \geq 0$, *there exists* $\delta(t_0)$, *such that*

$$\| \mathbf{x}_0 \| < \delta \quad \Rightarrow \quad \lim_{t \to \infty} \| \mathbf{x}(t) \| = 0 \tag{N.4}$$

Definition N.8 (Uniform asymptotic stability). *Point* $\mathbf{x} = \mathbf{0}$ *is called a uniformly asymptotically stable equilibrium point of Equation (N.1) if*

a) $\mathbf{x} = \mathbf{0}$ *uniformly stable*
b) *trajectory* $\mathbf{x}(t)$ *converges to* $\mathbf{0}$ *uniformly in* t_0.

These definitions address local stability (in the neighbourhood of the equilibrium point). The following definition addresses Global Asymptotic Stability (GAS).

Definition N.9 (Global asymptotic stability). *Point* $\mathbf{x} = \mathbf{0}$ *is called a globally asymptotically stable equilibrium point of Equation (N.1), if it is asymptotically stable and* $\lim_{t \to \infty} \| \mathbf{x}(t) \| = 0$, *for all* $\mathbf{x}_0 \in \mathbb{R}^n$. *Global uniform asymptotic stability is defined similarly.*

Definition N.10 (Exponential stability). *Point* $\mathbf{x} = \mathbf{0}$ *is called an exponentially stable equilibrium point of Equation (N.1) if there exist* $m, \alpha > 0$ *such that the solution* $\mathbf{x}(t)$ *satisfies*

$$\| \mathbf{x}(t) \| \leq \ m \, e^{-\alpha(t - t_0)} \ \| \mathbf{x}_0 \| \tag{N.5}$$

for all $\mathbf{x}_0 \in B_h$, $t \geq t_0 \geq 0$. *Constant* α *is called the rate of convergence.*

In other words, GAS asserts that the system is stable for any $\mathbf{x}_0 \in \mathbb{R}^n$, whereas the definition of exponential stability is similar except that m and α may be functions of h.

Appendix O

Linear Relaxation

O.1 Vector and Matrix Norms

Vector norms. For a vector $\mathbf{x}$ of length n

- 1–norm is defined as

$$\| \mathbf{x} \|_1 = \sum_{i=1}^{n} |x_i| \tag{O.1}$$

- 2–norm is defined as

$$\| \mathbf{x} \|_2 = \sqrt{\sum_{i=1}^{n} x_i^2} \tag{O.2}$$

- ∞–norm is defined as

$$\| \mathbf{x} \|_\infty = \max_{1 \le i \le n} |x_i| \tag{O.3}$$

- p–norm is defined as

$$\| \mathbf{x} \|_p = \left\{ \sum_{i=1}^{n} |x_i|^p \right\}^{\frac{1}{p}}, \qquad \forall p \geq 1 \tag{O.4}$$

Complex Valued Nonlinear Adaptive Filters: Noncircularity, Widely Linear and Neural Models
Danilo P. Mandic and Vanessa Su Lee Goh

Matrix norms.

- 1–norm is defined as

$$\| \mathbf{A} \|_1 = \max_{1 \le j \le n} \left\{ \sum_{i=1}^{n} |a_{ij}| \right\} \tag{O.5}$$

  that is the 'maximum absolute column sum';

- 2–norm is defined as

$$\| \mathbf{A} \|_2 \quad \looparrowright \quad \text{square root of the largest eigenvalue of the matrix} \quad \mathbf{A}^{\mathrm{T}}\mathbf{A} \tag{O.6}$$

- ∞–norm is defined as

$$\| \mathbf{A} \|_\infty = \max_{1 \le i \le n} \left\{ \sum_{j=1}^{n} |a_{ij}| \right\} \tag{O.7}$$

  that is the 'maximum absolute row sum';

- Frobenius norm is defined as

$$\| \mathbf{A} \|_p = \left(\sum_{i=1}^{m} \sum_{j=1}^{n} |a_{ij}|^p \right)^{\frac{1}{p}} \quad \text{which for } p = 2 \text{ gives}$$

$$\| \mathbf{A} \|_F^2 = \sum_{i=1}^{m} \sum_{j=1}^{n} \left|a_{ij}\right|^2 = \text{trace}\left(\mathbf{A}^{\mathrm{T}}\mathbf{A}\right) = \sum_{i=1}^{\min\{m,n\}} \sigma_i \tag{O.8}$$

The Frobenius norm is usually the largest of the matrix norms and is used as a bound when analysing matrix equations.

O.2 Relaxation in Linear Systems

The problem of Global Asymptotic Stability (GAS) is important in the theory of linear systems [21, 99, 113, 140, 163]. An autonomous system which is described by an Nth-order time-variant difference equation

$$y(k) = \mathbf{a}^{\mathrm{T}}(k)\mathbf{y}(k-1) = a_1(k)y(k-1) + \cdots + a_N(k)y(k-N) \tag{O.9}$$

should ideally represent a relaxation. System (O.9) can be written in a more general form

$$\mathbf{Y}(k+1) = \mathbf{A}(k)\mathbf{Y}(k) + \mathbf{B}(k)\mathbf{u}(k) \tag{O.10}$$

which degenerates into system (O.9) when the exogenous input vector $\mathbf{u}(k) = \mathbf{0}, \ \forall k > 0$ [140, 163]. The vector–matrix form of the output of an autonomous system (O.9) now becomes

$$\begin{bmatrix} y(k+1) \\ y(k) \\ \vdots \\ y(k-N+1) \end{bmatrix} = \begin{bmatrix} a_1(k) & a_2(k) & \cdots & a_N(k) \\ 1 & 0 & \cdots & 0 \\ \vdots & \vdots & \ddots & \vdots \\ 0 & \cdots & 1 & 0 \end{bmatrix} \begin{bmatrix} y(k) \\ y(k-1) \\ \vdots \\ y(k-N) \end{bmatrix}$$

or

$$\mathbf{Y}(k+1) = \mathbf{A}(k)\mathbf{Y}(k)$$

with

$$y(k+1) = [1 \ 0 \ \cdots \ 0] \ \mathbf{Y}(k+1) \tag{O.11}$$

For simplicity, we shall only consider systems of the form (O.9) for which the coefficient vector $\mathbf{a}$ has constant coefficients. In that case, matrix $\mathbf{A}$ is a Frobenius matrix, which is a special form of the companion matrix. The fundamental theorem of matrices [124, 310] states that every matrix $\mathbf{A}$ can be reduced by a similarity transformation to a sum of Frobenius matrices [48, 310]. It is therefore important to study stability of Equation (O.11) since the results can easily generalise; for instance, the analysis of robust relaxation for nonlinear dynamical systems uses this approach and can be found in [188].

Modes of convergence of linear autonomous systems are crucially dependent on the size and norm of the coefficient vector $\mathbf{a}$ [183, 188]; this is illustrated in Figure O.1, where the solid line illustrates uniform convergence, whereas the broken line illustrates oscillatory convergence

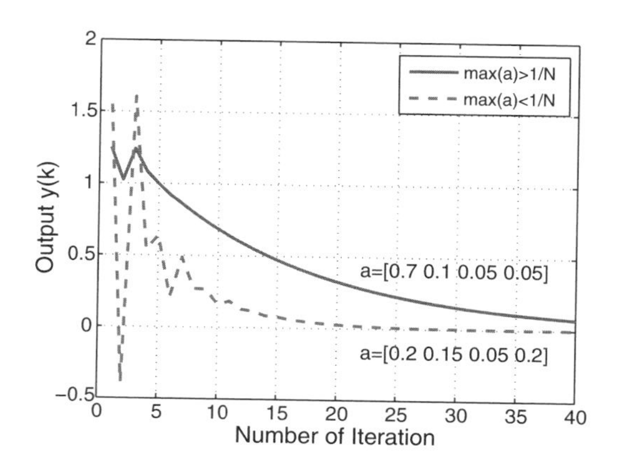

Figure O.1 Dependence of the mode of convergence on the coefficient vector. Solid line $\| \mathbf{a}(k) \| > 1/N$, broken line $\| \mathbf{a}(k) \| < 1/N$

(see also Appendix N). We can conclude that the system (O.11) with the constraint $\| \mathbf{a}(k) \|_1 < 1$ converges towards zero in the Fixed Point Iteration (FPI) sense, exhibiting linear convergence, with the convergence rate $\| \mathbf{a}(k) \|$.

Stability result for $\sum_{i=1}^{m} a_i = 1$. Consider the parameter vector $\mathbf{a}$ with constant coefficients, that is, $\mathbf{a} = [a_1, \ldots, a_N]^{\mathrm{T}}$, and $\| \mathbf{a} \|_1 = 1$. Matrix $\mathbf{A}$ from Equation (O.11) becomes a stochastic matrix [99, 169, 281], since each of its rows is a probability vector, and Equation (O.11) can be rewritten as

$$\mathbf{Y}(k+1) = \mathbf{A}\mathbf{Y}(k) = \mathbf{A}^2\mathbf{Y}(k-1) = \cdots = \mathbf{A}^k\mathbf{Y}(0) \tag{O.12}$$

Since the product of two stochastic matrices is a stochastic matrix, there exists a unique fixed vector $\mathbf{t} = [t_1, \ldots, t_N]^{\mathrm{T}}$ such that [99, 169]

$$\mathbf{tA} = \mathbf{t} \tag{O.13}$$

Since vector $\mathbf{t}$ is also a probability vector, that is, $\| \mathbf{t} \|_1 = 1$, the iteration (O.12) converges to

$$\mathbf{A}^k = \begin{bmatrix} a_1 & a_2 & \cdots & a_N \\ 1 & 0 & \cdots & 0 \\ \vdots & \vdots & \ddots & \vdots \\ 0 & \cdots & 1 & 0 \end{bmatrix}^k \xrightarrow{k \to \infty} \begin{bmatrix} t_1 & t_2 & \cdots & t_N \\ t_1 & t_2 & \cdots & t_N \\ \vdots & \vdots & \ddots & \vdots \\ t_1 & t_2 & \cdots & t_N \end{bmatrix} \tag{O.14}$$

Example. A university town has two pizza places. The statistics show that $N = 5000$ people buy one pizza every week. Tony's pizza place has the better pizza and 80% of people who buy pizza at Tony's return the following week. Mike's pizza is not that good and only 40% of the customers return the following week. Will Mike go bust?

Solution. We can describe the problem by a discrete autonomous dynamical system

$$\mathbf{x}_{n+1} = \mathbf{A}\,\mathbf{x}_n \quad \textit{where} \qquad \mathbf{A} = \begin{bmatrix} 0.8 & 0.6 \\ 0.2 & 0.4 \end{bmatrix}$$

If we assume that initially half of the customers go to Tony's and half to Mike's, that is, $\mathbf{x}_0 = [2500,\ 2500]^{\mathrm{T}}$, the above iteration gives

$$\mathbf{x}_1 = \begin{bmatrix} 3500 \\ 1500 \end{bmatrix}, \mathbf{x}_2 = \begin{bmatrix} 3700 \\ 1300 \end{bmatrix}, \mathbf{x}_3 = \begin{bmatrix} 3740 \\ 1260 \end{bmatrix}, \mathbf{x}_4 = \begin{bmatrix} 3748 \\ 1252 \end{bmatrix}, \mathbf{x}_5 = \begin{bmatrix} 3750 \\ 1250 \end{bmatrix}$$

The iteration converges very fast, and $\mathbf{x}_6 = \cdots = \mathbf{x}^* = x_\infty = [3750,\ 1250]^{\mathrm{T}}$. Clearly $\mathbf{x}^* = [3750,\ 1250]^{\mathrm{T}}$ is the fixed vector of $\mathbf{A}$, that is, $\mathbf{A}\,\mathbf{x}^* = \mathbf{x}^*$.

The eigenvectors of the system matrix are $\mathbf{v}_1 = [0.9487, 0.3162]^{\mathrm{T}}$ and $\mathbf{v}_2 = [-0.7071, 0.7071]^{\mathrm{T}}$, and the eigenvalues $\lambda_1 = 1$ and $\lambda_2 = 0.2$. Notice that the elements of $\mathbf{v}_1$ are related as $3 \div 1$, the same as the ratio of Tony's and Mike's customers. Therefore, we have reached an equilibrium, and Mike's pizza place will not go bust.

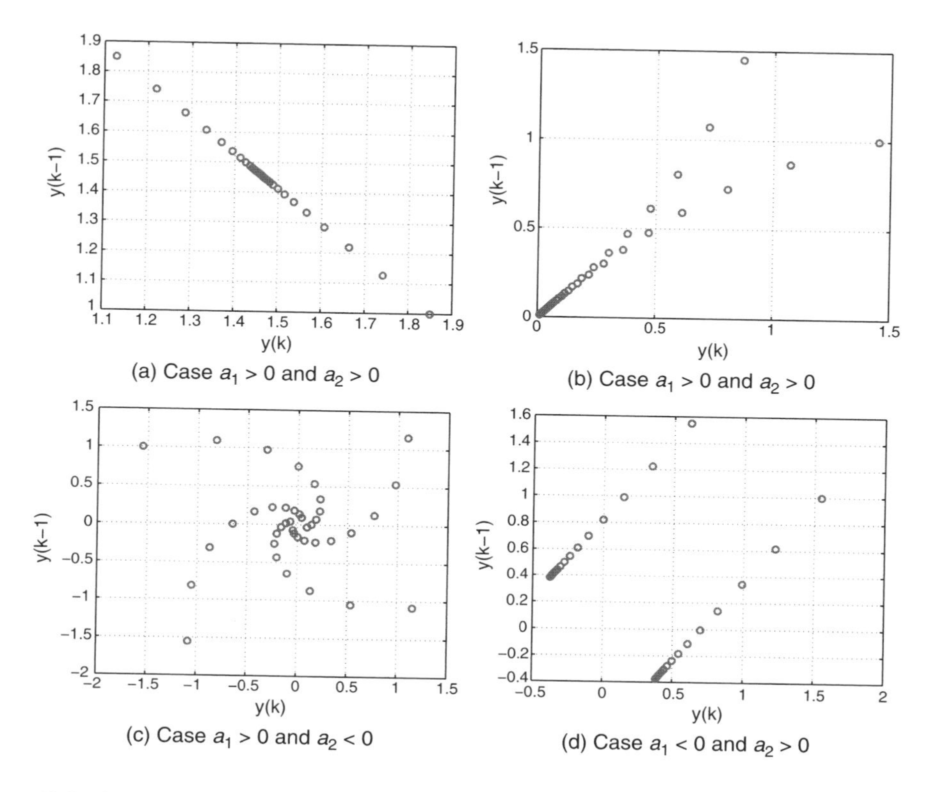

Figure O.2 State space convergence of relaxation Equation (O.9), for various combinations of the elements of the coefficient vector $\mathbf{a}$

O.2.1 Convergence in the Norm or State Space?

Let us now show that for the system (O.9) *convergence in the norm does not imply convergence in the geometric sense.*

Notice first that process (O.9), described by a vector $\mathbf{a} = [a_1, \ldots, a_N]^{\mathrm{T}}$ with non–negative constant coefficients, converges to [183]

(i) $|y_\infty| = |\sum_{i=1}^{N} t_i y(k-i)| \;\; \geq 0$ for $\| \mathbf{a} \|_1 = 1$
(ii) $y_\infty = 0$ for $\| \mathbf{a} \|_1 < 1$

from any finite initial state $\mathbf{Y}(0)$. Figure O.2 illustrates convergence in the geometric sense (state space) for various combinations of elements of the coefficient vector $\mathbf{a} = [a_1, a_2]^{\mathrm{T}}$.

Figure O.3 further depicts the need to consider the geometry of convergence; although the relaxation (O.9) converged in all the norms, it did not converge in state space.

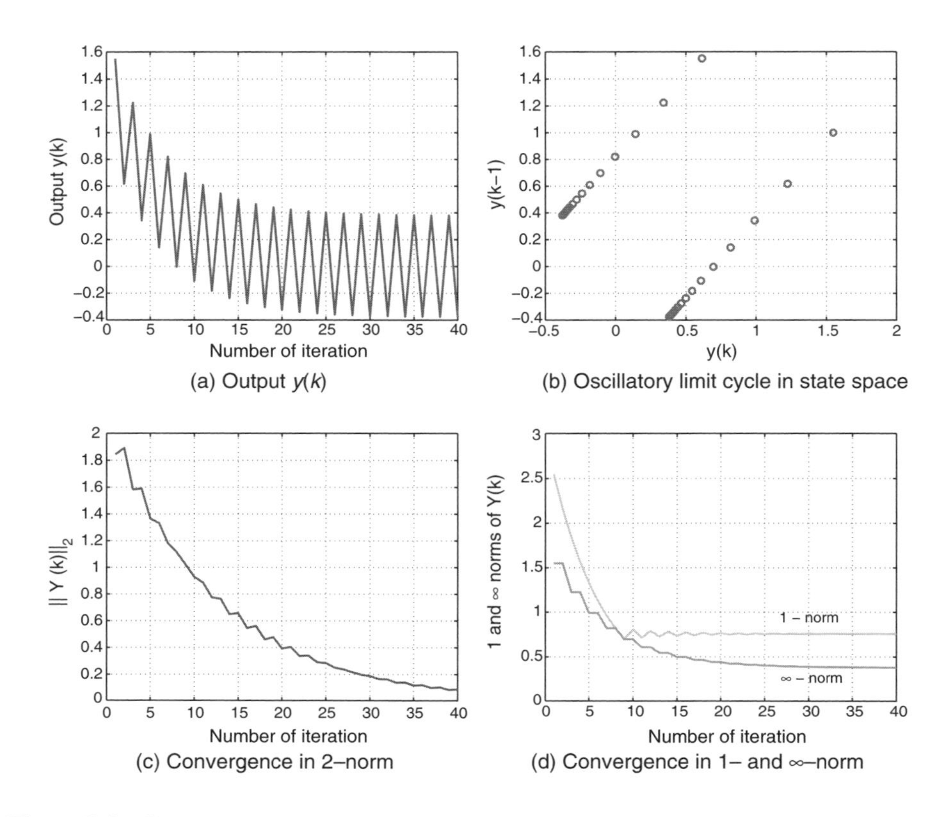

Figure O.3 Convergence in the 2-norm, 1-norm and ∞-norm for the system from Figure O.2(d)

Appendix P

Contraction Mappings, Fixed Point Iteration and Fractals

The principle of contraction mapping, and the corresponding Contraction Mapping Theorems (CMTs), have proved extremely useful in the analysis of nonlinear dynamical systems. We here provide some background material. We shall first give the definition of a *Lipschitz continuous mapping*.

Let X be a complete metric space with metric d containing a closed nonempty set Ω, and let $g : \Omega \rightarrow \Omega$. Function g is said to be *Lipschitz continuous* with *Lipschitz constant* $\gamma \in \mathbb{R}$ if

$$\forall x, y \in \Omega \qquad d\left[g(x), g(y)\right] \leq \gamma\, d(x, y)$$

We can differentiate between the following cases of Lipschitz continuity

a) for $0 \leq \gamma < 1$, g is a *contraction mapping* on Ω;
b) for $\gamma = 1$, g is a *nonexpansive* mapping;
c) for $\gamma > 1$, g is a *Lipschitz continuous mapping* on Ω.

We shall now state the *Contraction Mapping Theorem* (CMT) in $\mathbb{R}$.

CMT, Banach (1922). *Consider a continuous real valued function $K : \mathbb{R} \rightarrow \mathbb{R}$, such that*

i) For every x from the interval $[a, b]$, its image $K(x)$ also belongs to the same interval $[a, b]$ (non–expansivity – see a) and b) above);
ii) There exists a positive constant $\gamma < 1$, such that (Lipschitz continuity)

$$|K(x) - K(y)| \leq \gamma |x - y|, \quad \forall x, y \in [a, b] \tag{P.1}$$

Then the equation $x = K(x)$ has a unique solution $x^* \in [a, b]$ and the iteration (fixed point iteration (FPI))

$$x_{i+1} = K(x_i) \tag{P.2}$$

Complex Valued Nonlinear Adaptive Filters: Noncircularity, Widely Linear and Neural Models
Danilo P. Mandic and Vanessa Su Lee Goh

Figure P.1 The contraction mapping

converges to x^* from any $x_0 \in [a, b]$.

Brower's fixed point theorem gives a criterion for the *existence* of a fixed point [59, 157, 321].

Brower's fixed point theorem. Let $\Omega = [a, b]^N$ be a closed set of $\mathbb{R}^N$ and $f : \Omega \rightarrow \Omega$ be a continuous vector–valued function. Then f has at least one fixed point in Ω.

It is important to note that it is the underlying contraction mapping that provides the *existence* and *uniqueness* of the solution of the fixed point iteration (P.2).

Existence. From (i) in CMT, due to the non–expansivity, we have $a - K(a) \leq 0$ and $b - K(b) \geq 0$, as shown in Figure P.1. Then the Intermediate Value Theorem (IVT) guarantees that there exists a solution $x^* \in [a, b]$ such that $x^* = K(x^*)$

Uniqueness. If $\tilde{x} \in [a, b]$ is also a fixed point, (that is a solution of Equation P.2), then we have

$$|\tilde{x} - x^*| = |K(\tilde{x}) - K(x^*)| \leq \gamma|\tilde{x} - x^*| \tag{P.3}$$

For K a contraction, the Lipschitz constant $\gamma < 1$, and there can be only one solution to Equation (P.2), that is $\tilde{x} \equiv x^*$

Convergence. For the ith iteration of Equation (P.2) we have

$$|x_i - x^*| = |K(x_{i-1}) - K(x^*)| \leq \gamma|x_{i-1} - x^*| \tag{P.4}$$

Thus $|x_i - x^*| \leq \gamma^i|x_0 - x^*|$ and $\lim_{i\to\infty} \gamma^i = 0$. The Lipshitz constant γ therefore defines the rate of convergence towards the fixed point $\{x_i\} \xrightarrow{i} x^*$.

In practice, it is very difficult to find such γ, and the convergence is assessed by examining whether

$$|K'(x)| \leq \gamma < 1, \quad \forall x \in (a, b) \tag{P.5}$$

In order to examine the *local convergence* of the FPI (Equation P.2), we can make use of the mean value theorem (MVT). By MVT, there exists a point ξ in the open interval (a, b), for which

$$|K(x) - K(y)| = |K'(\xi)(x - y)| \Leftrightarrow |K'(\xi)||x - y| \leq \gamma|x - y|, \quad x, y \in [a, b] \tag{P.6}$$

therefore for local convergence it is sufficient to examine the gradient of K in the neighbourhood of x^*. For vector functions, the CMT can be extended to $\mathbb{R}^N$ as follows [59].

Let $\mathcal{M}$ be a closed subset of $\mathbb{R}^N$ such that

i) Function K maps set $\mathcal{M}$ onto itself, that is, $K : \mathcal{M} \rightarrow \mathcal{M}$
ii) There exists a positive constant $\gamma < 1$ such that, $\forall \mathbf{x}, \mathbf{y} \in \mathcal{M}$

$$\| \mathbf{K}(\mathbf{x}) - \mathbf{K}(\mathbf{y}) \| \leq \gamma \| \mathbf{x} - \mathbf{y} \| \tag{P.7}$$

Then the equation

$$\mathbf{x} = \mathbf{K}(\mathbf{x}) \tag{P.8}$$

has a unique solution $\mathbf{x}^* \in \mathcal{M}$ and the iteration

$$\mathbf{x}_{i+1} = \mathbf{K}(\mathbf{x}_i) \tag{P.9}$$

converges to $\mathbf{x}^*$ for any starting value $\mathbf{x}_0 \in \mathcal{M}$ (fixed point iteration).

The behaviour of state trajectories (orbits) in the vicinity of fixed points defines the character of fixed points.

- For an asymptotically stable (or *attractive*) fixed point x^* of a function K, there exists a neighborhood $\mathbb{O}(x^*)$ of x^* such that $\lim_{k\rightarrow\infty} K(x_k) = x^*$, for all $x_k \in \mathbb{O}(x^*)$. In this case, all the eigenvalues of the Jacobian of K at x^* are less than unity in magnitude, and fixed point x^* is called an *attractor*.
- Eigenvalues of the Jacobian of K which are greater than unity in magnitude indicate that K is an *expansion*, and the corresponding fixed point x^* is called a *repeller* or *repulsive* point.
- If some eigenvalues of the Jacobian of K are greater and some smaller than unity, x^* is called a *saddle point*.

One classical application of FPT is in the solution of equations, where

$$F(x) = 0 \quad \Leftrightarrow \quad x = K(x)$$

The next example shows how to set up such a fixed point iteration.

Example P.1 Find the roots of function $F(x) = x^2 - 2x - 3$.

Solution. There are several ways to set up a fixed point iteration. The roots of $F(x) = x^2 - 2x - 3$ are $x_1 = -1$ and $x_2 = 3$ and one way to find them is to rearrange

$$F(x) = 0 \quad \Leftrightarrow \quad x = K(x) = \sqrt{2x+3}$$

The FPI starting from $x_0 = 4$ gives the sequence of iterates

$$x_1 = 3.3116, x_2 = 3.1037, x_3 = 3.0344, x_4 = 3.0114, \ldots, x_8 = 3.0001, x_9 = 3.0000$$

where $x^* = 3$ is a fixed point (also in the neighborhood of $x^* = 3$, $|K'(x)| < 1$), and a solution of the equation $x^2 - 2x - 3 = 0$. The convergence of this fixed point iteration is illustrated in Figure P.2.

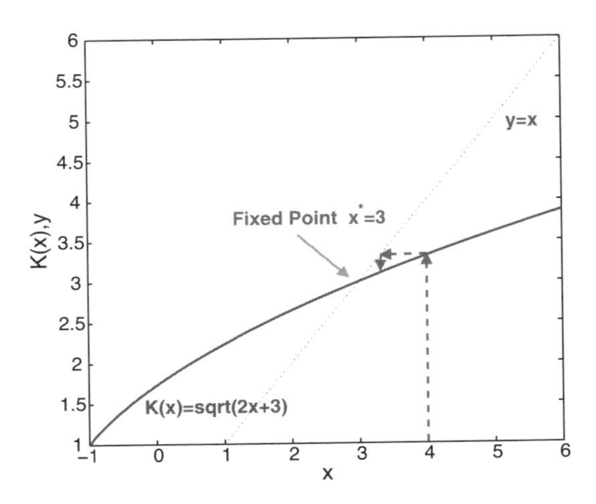

Figure P.2 Fixed point iteration for the roots of function $F(x) = x^2 - 2x - 3$

By rearranging the problem of finding zeros of F in a different way, we may encounter the other two characteristic cases, namely *oscillatory convergence* and *divergence*, as illustrated in Figures P.3(a) and (b).

For the case of divergence depicted in Figure P.3(b), rearrange $F(x) = 0 \Leftrightarrow x = K(x) = (x^2 - 3)/2$. The value of the first derivative of $K(x)$ is greater than unity for $x > x^* = 3$, hence an FPI starting from initial values $x_0 > 3$ diverges. On the other hand, an FPI starting from an initial value $x_0 < 3$ would still converge to the fixed point $x^* = 3$ (also by virtue of CMT). Hence, fixed point $x^* = 3$ is an *unstable fixed point* of function $K = (x^2 - 3)/2$.

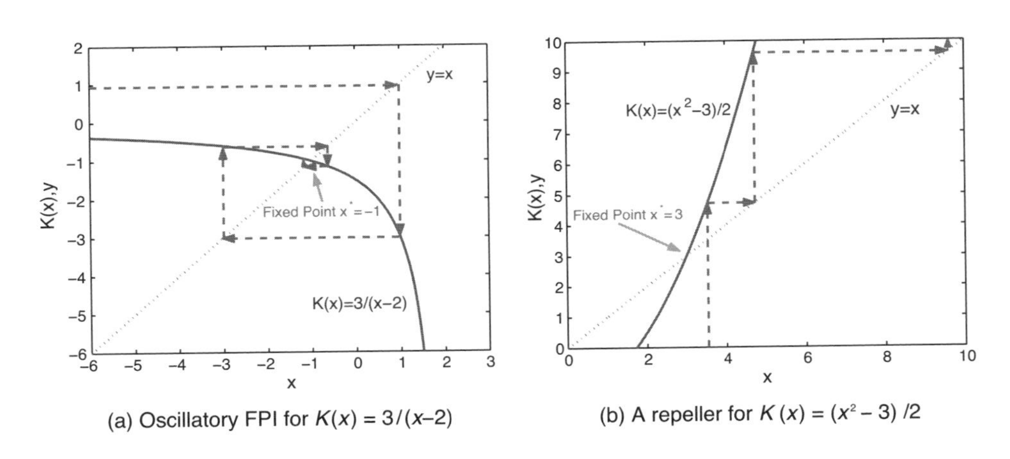

(a) Oscillatory FPI for $K(x) = 3/(x-2)$

(b) A repeller for $K(x) = (x^2 - 3)/2$

Figure P.3 Oscillatory convergence of FPI and the case of a repeller

P.1 Historical Perspective

Evidence from clay tablets in the Yale University Babylonian collection suggests that iterative techniques for finding square roots go back several thousand years [157]. One such clay tablet is YBC 7289 (ca 1800–1600 BC), shown in Figure P.4(a), which explains how to calculate the diagonal of a square with base 30. Figure P.4 illustrates the iteration procedure used by the Babylonians.[1] The Babylonians used a positional, sexagesimal (base 60) number system, and on this tablet, the 'chicken scratchings' are the sexagesimal numerals, thus for instance, as shown in Figure P.4(b), symbol V corresponds to 'one' and symbol $<$ to 'ten' [77].

Figure P.4(c) shows a simplified diagram illustrating the mathematical operation explained on YBC 7289; numbers on the diagonal give an approximation to $\sqrt{2}$ accurate to nearly six decimal places

$$1 + \frac{24}{60} + \frac{51}{60^2} + \frac{10}{60^3} \approx 1.41421296\ldots$$

whereas $30\sqrt{2}$ is calculated as (numbers along the bottom right base)

$$42 + \frac{25}{60} + \frac{35}{60^2} \approx 42.426389$$

geometrically via iteration of doubled squares.[2]

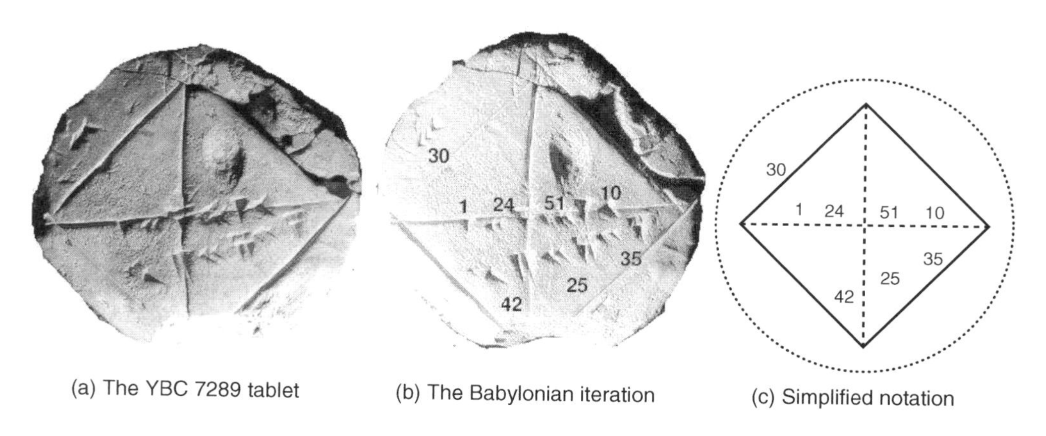

(a) The YBC 7289 tablet (b) The Babylonian iteration (c) Simplified notation

Figure P.4 The Babylonian iteration for calculating the diagonal of a square with base 30

[1] Images by Bill Casselman.

[2] Given that the Babylonians were most likely to use the knowledge of $\sqrt{2}$ in carpentry and construction, some recent authors proposed a conjecture that the Babylonians used a method of successive approximations similar to Heron's method. If the procedure starts with a guess x, then the error is calculated $e = x^2 - 2$, and the Babylonian iteration is executed until the error is very small.

In terms of modern mathematics, the Babylonians employed an iterator[3]

$$x = K(x) = \frac{x + \frac{2}{x}}{2} \tag{P.10}$$

so that, for instance, successive applications of $K(x)$, which start from $x_0 = 4.5$, give [157]

$$x_1 = K(x_0) = \frac{3}{2},\ x_2 = K(x_1) = \frac{17}{12},\ x_3 = K(x_2) = \frac{577}{408}, \ldots$$

as shown in Figure P.5. The point of intersection of curves $K(x)$ and $y = x$ can be found analytically by solving

$$\frac{1}{2}\left(x + \frac{2}{x}\right) = x$$

The solution has the value $x = \sqrt{2}$, and is a *fixed point* of K and a solution of $K(x) = x$.

The Babylonian iteration can be used to calculate the square root of any number a, by replacing 2 with 'a' in the iteration, that is

$$x_{i+1} = \frac{1}{2}\left(x_i + \frac{a}{x_i}\right)$$

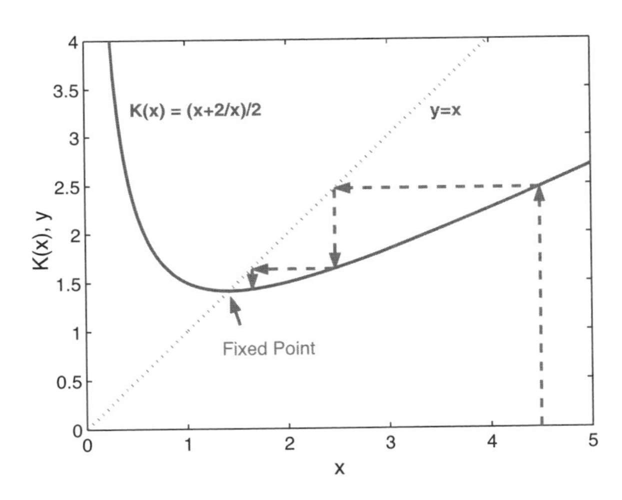

Figure P.5 The Babylonian iteration starting from $x_0 = 4.5$

[3] Observe that $x + 2/x = 2x \Leftrightarrow x^2 - 2 = 0$.

Observe that the iteration converges when the mean value between x_i and a/x_i approaches the limit $x^* = \sqrt{a}$. This is also a special case of Newton's method.[4]

P.2 More on Convergence: Modified Contraction Mapping

Despite its mathematical elegance, fixed point iteration is not guaranteed to converge from any initial condition. Let us illustrate this by an example.

Example P.2 Solve $x^2 = 3$.

Solution. Consider the fixed point iteration

$$x_{i+1} = 3/x_i$$

Starting the FPI from $x_0 = 1$, we have the sequence $x_1 = 3, x_2 = 1, x_3 = 3, x_4 = 1, \ldots$, which does not converge.

To solve this problem consider a modified FPI

$$x - Sx = Tx - Sx \qquad \leftrightarrow \qquad x_{i+1} - Sx_{i+1} = Tx_i - Sx_i$$

and rewrite

$$x = (I - S)^{-1} (Tx - Sx) \quad \leftrightarrow \quad x_{i+1} = G(x_i)$$

Choose[5] $S = -1$ and start from $x_0 = 1$, to obtain the iterates[6]

$$x_0 = 1, x_1 = 2, x_2 = 1.67, x_3 = 1.75, x_4 = 1.727, x_5 = 1.733, \ldots$$

The iteration is now convergent.

It is clear that the local convergence of $x_{i+1} = K(x_i)$ towards a fixed point x^* depends critically on the gradient (Jacobian) of K near x^*. It is therefore natural to ask ourselves whether can we modify the standard FPI to guarantee convergence for a much wider range of initial conditions. One such result has been proposed in [71]. The modified FPI from the paper by Ferrante *et al.* [71] can be presented as follows:

- assume $f(\cdot)$ is twice continuously differentiable;
- assume $x_{i+1} = g(x_i)$, the usual choice being $g = f$;
- if the spectrum (eigenvalues) $\sigma\left[J(x^*)\right]$ of the Jacobian

$$J(x) = \frac{\partial f}{\partial x}(x)$$

[4]For a differentiable function f, the Newton iteration can be expressed as

$$x_{i+1} = x_i - \frac{f(x_i)}{f'(x_i)}.$$

[5]Notice $S = 1$ is not valid above.

[6]There are many methods which come as a consequence of the choice of S (Newton method).

at $x = x^*$ is contained in the unit circle $\mathbb{S} = \{z \in \mathbb{C} : |z| < 1\}$, then the algorithm $x_{i+1} = f(x_i)$ converges to x^* from any point in a suitable neighborhood of x^* (even if the convergence is slow).

To guarantee convergence of $x_{i+1} = g(x_i)$ towards x^*, we need to ensure that

i) $g(x^*) = x^*$;
ii) $\sigma \left[\frac{\partial g}{\partial x}\right]_{|x=x^*} \subset \mathbb{S}$ (spectral norm);
iii) $g(\cdot)$ can be formed from $f(\cdot)$ without *a priori* information about x^*.

Given a constant matrix K with eigenvalues in $\mathbb{S}$, a simple function g satisfying *i)* and *ii)* is the affine function shown in Figure P.6.

The affine function $\hat{g}$ can be written as

$$\hat{g}(x; K) = x^* + K(x - x^*) = Kx + (I - K)x^* \tag{P.11}$$

This choice of function g requires previous knowledge of x^*, which can be obtained, for instance, from f by applying a Taylor Series Expansion (TSE) around x^*, to give

$$\hat{f}(x) = f(x^*) + J(x^*)(x - x^*) = J(x^*)x + \left[f(x^*) - J(x^*)x^*\right] = J(x^*)x + \left[I - J(x^*)\right] x^*$$

and finally

$$x^* = \left[I - J(x^*)\right]^{-1} \left[\hat{f}(x) - J(x^*)x\right] \tag{P.12}$$

Replacing x^* into Equation (P.11) and using approximations $x^* \to x$ and $\hat{f}(x) \to f(x)$, we arrive at Ferrante's solution

$$g(x; K) = Kx + (I - K)\left[I - J(x)\right]^{-1} \left[f(x) - J(x)x\right] \tag{P.13}$$

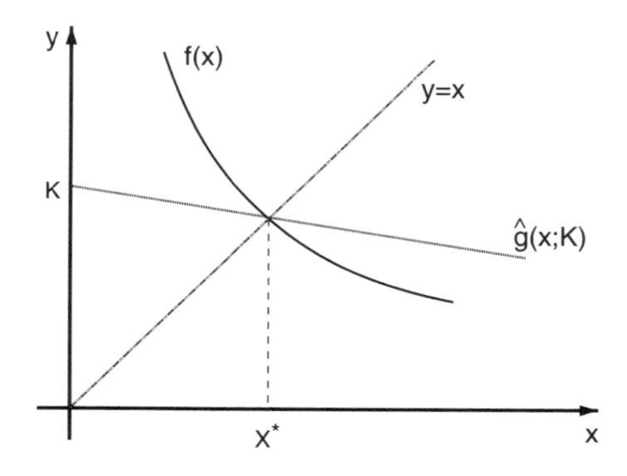

Figure P.6 Illustration of Ferrante's modified FPI method

Observe that $g(x)$ is independent of x^* and therefore $x_{i+1} = g(x_i; K)$ converges to every fixed point x^* of $f(x)$ starting from a suitable neighborhood of x^*. The next example illustrates the power of the so modified fixed point iteration.[7]

Example P.3 *Ensure convergence of the fixed point iteration for the scalar function*

$$f(x) = \frac{1}{2}x^3 - x + 1$$

for which the fixed points are $\mathbf{x}^* = [-2.2143, 0.5392, 1.6751]^{\mathrm{T}}$.

Solution. Since the derivative $f'(x) = \frac{3}{2}x^2 - 1$ is not bounded, the convergence of standard FPI not guaranteed a priori. To ensure convergence, apply (P.13), to yield ($K \rightarrow k > 1$) and

$$g(x; K) = \frac{-(2+k)\,x^3 + 4kx + 2\,(1-k)}{4 - 3x^2}$$

The size of the interval of attraction (basin of attraction, region of convergence), depends on k. For instance, for $k = 0$

$$g(x; 0) = 2\frac{1 - x^3}{4 - 3x^2}$$

Also, depending on the initial value x_0, FPI can converge to any of the three fixed points. Figure P.7 illustrates that indeed $g(x^*) = f(x^*) \Leftrightarrow f(x) = x$. For $k = 2$ we have quadratic convergence and for x_0 close to x^*, the fixed point is reached in only one step.

A comprehensive account of FPT background material can be found in [59, 157, 174, 321].

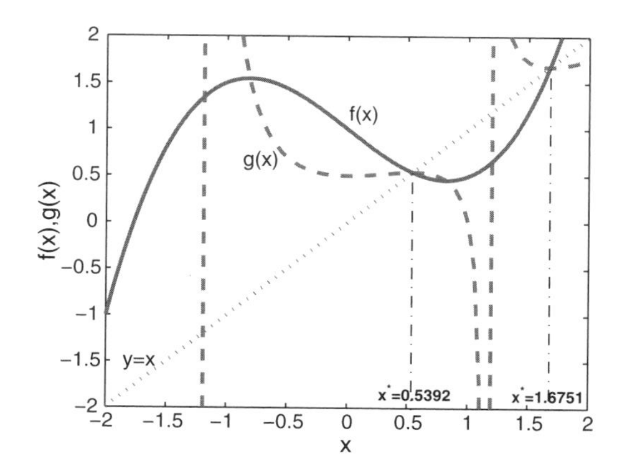

Figure P.7 The modified fixed point iteration. Solid line: $f(x)$, broken line: $g(x)$

[7]This example is adopted from [71].

P.3 Fractals and Mandelbrot Set

Fractals are mathematical objects that obey self-similarity, that is, the parts of the object are somehow self-similar to the whole. This self-similarity implies that fractals are scale invariant, and we cannot distinguish a small part from the larger structure. Classical examples include the Sierpinski triangle and the Koch snowflake, however, many phenomena in nature are self-similar and scale invariant. Trees and ferns are fractal in nature and can be modelled using a recursive algorithm [176].

Perhaps the most famous fractal is the classical Mandelbrot set M, a subset of the complex plane, which despite its complicated structure is defined by a mathematical rule of sheer simplicity, that is, as a fixed point iteration

$$z \rightarrow z^2 + c \tag{P.14}$$

where c is a complex number. The iteration starts with $z = 0$ and continues as [233]

$$0, \quad c, \quad c^2 + c, \quad c^4 + 2c^3 + c^2 + c, \quad \ldots$$

until

- z goes to infinity, and point c is coloured white;
- z wanders around in some restricted region, and point c is coloured black.

The black regions give us the Mandelbrot set. The Earth also exhibits fractal nature because it shows self-similarity at many different scales (lightning, forest, ice shapes, fern, river and mountain systems), for more detail see [37]. Examples of fractals and iterated maps are given in Figure P.8.

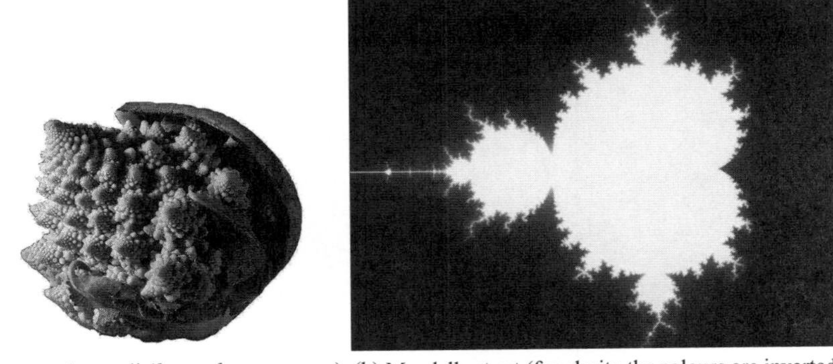

(a) Romanesco broccoli (*broccolo romanesco*) (b) Mandelbrot set (for clarity the colours are inverted)

Figure P.8 Examples of iterated maps. Top: Natural fractal; bottom: Mandelbrot set

References

[1] http://www-history.mcs.st-andrews.ac.uk/biographies/cardan.html.

[2] T. Aboulnasr and K. Mayyas. A robust variable step-size LMS-type algorithm: Analysis and simulations. *IEEE Transactions on Signal Processing*, **45**(3):631–639, 1997.

[3] K. Aihara. Chaos engineering and its application to parallel distributed processing with chaotic neural networks. *Proceedings of the IEEE*, **90**(5):919–930, 2002.

[4] I. Aizenberg, E. Myasnikova, M. Samsonova, and J. Reinitz. Temporal classification of D*rosophila* segmentation gene expression patterns by the multi-valued neural recognition method. *Mathematical Biosciences*, **176**:145–159, 2002.

[5] I. N. Aizenberg, N. N. Aizenberg, and J. Vandewalle. *Multi-valued and Universal Binary Neurons: Theory, learning, applications*. Kluwer Academic Publishers, 2000.

[6] N. N. Aizenberg and I. N. Aizenberg. NN based on multi-valued neuron as a model of associative memory for gray-scale images. In *Proceedings of the the Second IEEE International Workshop on Cellular Neural Networks and Their Application*, pp 36–41, 1992.

[7] N. N. Aizenberg and Y. L. Ivaskiv. *Multiple-Valued Threshold Logic*. Naukova Dumka, Kiev (in Russian), 1977.

[8] N. N. Aizenberg, Y. L. Ivaskiv, and D. A. Pospelov. About one generalization of the threshold function. *Doklady Akademii Nauk SSSR (in Russian)*, :**196**(6):1287–1290, 1971.

[9] M. Umair Bin Altaf, T. Gautama, T. Tanaka, and D. P. Mandic. Rotation invariant complex Empirical Mode Decomposition. In *Proceedings of the International Conference on Acoustics, Speech and Signal Processing (ICASSP'07)*, vol. III, pp 1009–1012, 2007.

[10] S.-I. Amari. Natural gradient works efficiently in learning. *Neural Computation*, **10**(2):251–176, 1998.

[11] P. O. Amblard, M. Gaeta, and J. L. Lacoume. Statistics for Complex Variables and Signals – Part I: Variables. *Signal Processing*, **53**:1–13, 1996.

[12] P. O. Amblard, M. Gaeta, and J. L. Lacoume. Statistics for Complex Variables and Signals – Part II: Signals. *Signal Processing*, **53**:15–25, 1996.

[13] W.-P. Ang and B. Farhang-Boroujeny. A new class of gradient adaptive step-size LMS algorithms. *IEEE Transactions on Signal Processing*, **49**(4):805–810, 2001.

[14] H. Aoki. Applications of complex valued neural networks to image processing. In A. Hirose, (ed.), *Complex Valued Neural Networks*, pp 181–204. World Scientific Publishing, Singapore, 2004.

[15] H. Aoki, M. R. Azimi-Sadjadi, and Y. Kosugi. Image association using a complex-valued associative memory model. *Transactions of the IEICE*, **83**A:1824–1832, 2000.

[16] T. M. Apostol. *Modular Functions and Dirichlet Series in Number Theory*. Springer-Verlag, 2nd edn, 1997.

[17] P. Arena, L. Fortuna, G. Muscato, and M. G. Xibilia. *Neural Networks in Multidimensional Domains*. Springer-Verlag, 1998.

[18] P. Arena, L. Fortuna, R. Re, and M. G. Xibilia. Multilayer perceptrons to approximate complex valued functions. *International Journal of Neural Systems*, **6**:435–446, 1995.

Complex Valued Nonlinear Adaptive Filters: Noncircularity, Widely Linear and Neural Models
Danilo P. Mandic and Vanessa Su Lee Goh

[19] J. Arenas-Garcia, A. R. Figueiras-Vidal, and A. H. Sayed. Mean-square performance of a convex combination of two adaptive filters. *IEEE Transactions on Signal Processing*, **54**(3):1078–1090, 1996.

[20] Aristotle. *Posterior Analytics* (Translated in to English by J. Barnes). Clarendon Press, Oxford, 1975.

[21] S. Barnett and C. Storey. *Matrix Methods in Stability Theory*. Nelson, 1970.

[22] A. Barrlund, H. Wallin, and J. Karlsson. Iteration of Möbius transformations and attractors on the real line. *Computers and Mathematics with Applications*, **33**(1–2):1–12, 1997.

[23] J. Benesty, T. Gansler, Y. Huang, and M. Rupp. Adaptive algorithms for MIMO acoustic echo cancellation. In Y. Huang and J. Benesty, (eds), *Audio Signal Processing for Next-Generation Multimedia Communication Systems*, pp 119–147. Springer, 2007.

[24] A. Benveniste, M. Metivier, and P. Priouret. *Adaptive Algorithms and Stochastic Approximation*. Springer-Verlag: New York, 1990.

[25] N. Benvenuto, M. Marchesi, F. Piazza, and A. Uncini. A comparison between real and complex valued neural networks in communication applications. In *Proceedings of the International Conference on Artificial Neural Networks (ICANN'92)*, pp 1177–1180, 1991.

[26] N. Benvenuto, M. Marchesi, F. Piazza, and A. Uncini. Nonlinear satellite radio links equalised using blind neural networks. In *Proceedings of the International Conference on Acoustics, Speech and Signal Processing, (ICASSP'91)*, pp 1521–1524, 1991.

[27] N. Benvenuto and F. Piazza. On the complex backpropagation algorithm. *IEEE Transactions on Signal Processing*, **40**(4):967–969, 1992.

[28] S. A. Billings. Identification of nonlinear systems–a survey. *IEE Proceedings–D*, **127**(6):272–285, 1980.

[29] V. D. Blondel and J. N. Tsitsiklis. A survey of computational complexity results in systems and control. *Automatica*, **36**:1249–1274, 2000.

[30] A. Van Den Bos. Complex gradient and Hessian. *IEE Proceedings – Vision, Image and Signal Processing*, **141**(6):380–382, 1994.

[31] A. Van Den Bos. The multivariate complex normal distribution: A generalization. *IEEE Transactions on Information Theory*, **41**(2):537–539, 1995.

[32] C. Boukis, D. P. Mandic, and A. G. Constantinides. Towards bias minimisation in acoustic feedback cancellation systems. *Journal of the Acoustic Society of America*, **121**(3):1529–1537, 2007.

[33] G. E. Box and G. M. Jenkins. *Time series analysis: forecasting and control*. Holden-Day, 2nd edn, 1976.

[34] C. Boyer and C. Merzbach. *A History of Mathematics*. Wiley, 1968.

[35] D. Brandwood. A complex gradient operator and its application in adaptive array theory. *IEE Proceedings F: Communications, Radar and Signal Processing*, **130**(1):11–16, 1983.

[36] G. A. D. Briggs, J. M. Rowe, A. M. Sinton, and D. S. Spencer. Quantitative methods in acoustic microscopy. In *Proceedings of the 1988 Ultrasonic Symposium*, pp 743–749, 1988.

[37] J. Briggs. *Fractals: The Patterns of Chaos*. Simon & Schuster, 1992.

[38] W. Brown and R. Crane. Conjugate linear filtering. *IEEE Transactions on Information Theory*, **15**(4):462–465, 1969.

[39] F. Cajori. *A History of Mathematical Notations*. Open Court, 1929.

[40] V. Calhoun, T. Adali, and D. Van De Ville. *The Signal Processing Handbook for Functional Brain Image Analysis*. Springer, 2009.

[41] V. D. Calhoun and T. Adali. Unmixing fMRI with independent component analysis. *IEEE Engineering in Medicine and Biology Magazine*, **25**(2):79–90, 2006.

[42] V. D. Calhoun, T. Adali, G. D. Pearlson, P. C. Van Zijl, and J. J. Pekar. Independent component analysis of fMRI data in the complex domain. *Magnetic Resonance in Medicine*, **48**:180–192, 2002.

[43] P. Caligiuri, M. L. Giger, and M. Favus. Multifractal radiographic analysis of osteoporosis. *Medical Physics*, **21**(4):503–508, 1994.

[44] M. C. Casdagli and A. S. Weigend. Exploring the continuum between deterministic and stochastic modeling. In A. S. Weigend and N. A. Gershenfield, editors, *Time Series Prediction: Forecasting the Future and Understanding the Past*. Addison-Wesley, 1994.

[45] E. Catmull and R. Rom. A class of local interpolating splines. *Computer Aided Geometric Design, R. E. Barnhill and R. F. Riesenfeld, Eds. New York: Academic*, pp 317–326, 1974.

[46] W. Chang and M. Kam. Systems of random iterative continuous mappings with a common fixed point. In *Proceedings of the 28th Conference on Decision and Control*, pp 872–877, 1989.

[47] Y. S. Choi, H. C. Shin, and W. J. Song. Robust regularisation of normalized NLMS algorithms. *IEEE Transactions on Circuits and Systems II*, **53**(8):627–631, 2006.

[48] P. G. Ciarlet. *Introduction to numerical linear algebra and optimization*. Cambridge University Press, 1989.

[49] A. Cichocki and S.-I. Amari. *Adaptive Blind Signal and Image Processing: Learning Algorithms and Applications*. Wiley, 2002.

[50] T. Clarke. Generalization of neural networks to the complex plane. In *Proceedings of the International Joint Conference on Neural Networks*, No. 2, pp 435–440, 1990.

[51] J. Cockle. A new imaginary in algebra. *London–Edinburg–Dublin Philosophical Magazine*, **33**(3):345–349, 1848.

[52] L. Cohen. Instantaneous anything. *Proceedings of the International Conference on Acoustics, Speech, and Signal Processing (ICASSP'92)*, **5**:105–108, 1993.

[53] P. Comon. Course on Higher-Order Statistics. Technical report, School of Theoretical Physics at Les Houches, France, September 1993.

[54] A. G. Constantinides. Personal communication.

[55] N. E. Cotter. The Stone–Weierstrass theorem and its applications to neural networks. *IEEE Transactions on Neural Networks*, **1**(4):290–295, 1990.

[56] G. Cybenko. Approximation by superpositions of a sigmoidal function. *Mathematics of Control, Signals, and Systems*, **2**:303–314, 1989.

[57] T. Dantzig. *Number: The Language of Science*. Kessinger Publishing, 1930.

[58] A. Van den Bos. Price's theorem for complex variables. *IEEE Transactions on Information Theory*, **42**(1):286–287, 1996.

[59] J. E. Dennis, Jr. and R. B. Schnabel. *Numerical Methods for Unconstrained Optimization and Nonlinear Equations*. Prentice–Hall Series in Computational Mathematics, 1983.

[60] J. Derbyshire. *Unknown Quantity: A Real and Imaginary History of Algebra*. Joseph Henry Press, 2006.

[61] S. C. Douglas. A family of normalized LMS algorithms. *IEEE Signal Processing Letters*, **1**(3):49–51, 1994.

[62] S. C. Douglas and D. P. Mandic. Mean and mean–square analysis of the complex LMS algorithm for non–circular Gaussian signals. In *Proceedings of the IEEE DSP Symposium*, pp, 2008.

[63] S. C. Douglas and M. Rupp. A posteriori updates for adaptive filters. In *Conference Record of the Thirty–First Asilomar Conference on Signals, Systems and Computers*, vol. 2, pp 1641–1645, 1997.

[64] G. Dreyfus and Y. Idan. The canonical form of nonlinear discrete–time models. *Neural Computation*, **10**:133–136, 1998.

[65] A. Dumitras and V. Lazarescu. On viewing the transform performed by a hidden layer in a feedforward ANN as a complex Möbius mapping. In *Proceedings of International Conference on Neural Networks*, vol. 2, pp 1148–1151, 1997.

[66] R. C. Eberhart. Standardization of neural network terminology. *IEEE Transactions on Neural Networks*, **1**(2):244–245, 1990.

[67] J. Eriksson and V. Koivunen. Complex random vectors and ICA models: Identifiability, uniqueness, and separability. *IEEE Transactions on Information Theory*, **52**(3):1017–1029, 2006.

[68] J. B. Evans, P. Xue, and B. Liu. Analysis and implementation of variable step size adaptive algorithms. *IEEE Transactions on Signal Processing*, **41**(8):2517–2535, 1993.

[69] B. Farhang-Boroujeny. *Adaptive Filters: Theory and Applications*. Wiley, 1999.

[70] L. A. Feldkamp and G. V. Puskorius. A signal processing framework based on dynamic neural networks with applications to problems in adaptation, filtering, and classification. *Proceedings of the IEEE*, **86**(11):2259–2277, 1998.

[71] A. Ferrante, A. Lepschy, and U. Viaro. Convergence analysis of a fixed point algorithm. *Italian Journal of Pure and Applied Mathematics*, **9**:179–186, 2001.

[72] E. Fiesler. Neural network formalization. *Computer Standards & Interfaces*, **16**(3):231–239, 1994.

[73] E. Fiesler and R. Beale, editors. *Handbook of Neural Computation*. Institute of Physics Publishing and Oxford University Press, 1997.

[74] A. R. Figueras-Vidal, J. Arenas-Garcia, and A. H. Sayed. Steady state performance of convex combinations of adaptive filters. In *Proceedings of the International Conference on Acoustis, Speech and Signal Processing (ICASSP'05)*, pp 33–36, 2005.

[75] Y. Fisher. *Fractal Image Compression - Theory and Application*. Springer-Verlag, New York, 1994.

[76] J. B. Foley and F. M. Boland. A note on convergence analysis of LMS adaptive filters with Gaussian data. *IEEE Transactions on Acoustics, Speech, and Signal Processing*, **36**(7):1087–1089, 1988.

[77] D. H. Fowler and E. R. Robson. Square root approximation in Babylonian mathematics: YBC 7289 in this context. *Historia Mathematica*, **25**:366–378, 1998.

[78] D. Franken. Complex digital networks: a sensitivity analysis based on the Wirtinger calculus. *IEEE Transactions on Circuits and Systems I: Fundamental Theory and Applications*, **44**(9):839–843, 1997.

[79] K. Funahashi. On the approximate realisation of continuous mappings by neural networks. *Neural Networks*, **2**:183–192, 1992.

[80] W. A. Gardner. Exploitation of spectral redundancy in cyclostationary signals. *IEEE Signal Processing Magazine*, **8**(2):14–36, 1991.

[81] N. S. Garis and I. Pazsit. Control rod localisation with neural networks using a complex transfer function. *Progress in Nuclear Energy*, **34**(1):87–98, 1999.

[82] T. Gautama, D. P. Mandic, and M. M. Van Hulle. On the analysis of nonlinearity in fMRI signals: Comparing BOLD to MION. *IEEE Transactions on Medical Imaging*, **22**(5):636–644, 2003.

[83] T. Gautama, D. P. Mandic, and M. M. Van Hulle. The delay vector variance method for detecting determinism and nonlinearity in time series. *Physica D*, **190**(3–4):167–176, 2004.

[84] T. Gautama, D. P. Mandic, and M. Van Hulle. On the indications of nonlinear structures in brain electrical activity. *Physical Review E*, **67**(4):046204–1 - 046204–5, 2003.

[85] T. Gautama, D. P. Mandic, and M. Van Hulle. A non–parametric test for detecting the complex–valued nature of time series. *International Journal of Knowledge–Based Intelligent Engineering Systems*, **8**(2):99–106, 2004.

[86] T. Gautama, D. P. Mandic, and M. Van Hulle. A novel method for determining the nature of time series. *IEEE Transactions on Biomedical Engineering*, **51**(5):728–736, 2004.

[87] S. Gemon, E. Bienenstock, and R. Doursat. Neural networks and the bias/variance dilemma. *Neural Computation*, **4**(1):1–58, 1992.

[88] G. M. Georgiou and C. Koutsougeras. Complex Domain Backpropagation. *IEEE Transactions on Circuits and Systems - II: Analog and Digital Signal Processing*, **39**(5):330–334, 1992.

[89] D. N. Godard. Self-recovering equalization and carrier tracking in twodimensional data communication systems. *IEEE Transactions on Communications*, **28**(11):1867–1875, 1980.

[90] S. L. Goh, Z. V. Babic, and D. P. Mandic. An adaptive amplitude learning algorithm for nonlinear adaptive IIR filter. *Proceedings of the 6th International Conference on Telecommunications in Modern Satellite, Cable and Broadcasting Services, TELSIKS*, pp 313–316, 2003.

[91] S. L. Goh, M. Chen, D. H. Popovic, K. Aihara, D. Obradovic, and D. P. Mandic. Complex-valued forecasting of wind profile. *Renewable Energy*, **31**(11):1733–1750, 2006.

[92] S. L. Goh and D. P. Mandic. Recurrent neural networks with trainable amplitude of activation functions. *Neural Networks*, **16**:1095–1100, 2003.

[93] S. L. Goh and D. P. Mandic. A general complex valued RTRL algorithm for nonlinear adaptive filters. *Neural Computation*, **16**(12):2699–2731, 2004.

[94] S. L. Goh and D. P. Mandic. Nonlinear adaptive prediction of complex valued non–stationary signals. *IEEE Transactions on Signal Processing*, **53**(5):1827–1836, 2005.

[95] S. L. Goh and D. P. Mandic. An Augmented Extended Kalman Filter Algorithm for Complex-Valued Recurrent Neural Networks. *Neural Computation*, **19**:1039–1055, 2007.

[96] S. L. Goh and D. P. Mandic. An augmented CRTRL for complex–valued recurrent neural networks. *Neural Networks*, **20**:1061–1066, 2007.

[97] S. L. Goh and D. P. Mandic. Stochastic gradient adaptive complex–valued nonlinear neural adaptive filters with a gradient adaptive step size. *IEEE Transactions on Neural Networks*, **18**:1511–1516, 2007.

[98] S. L. Goh, D. Popovic, and D. P. Mandic. Complex valued estimation of wind profile and wind power. In *Proceedings of the IEEE MELECON'04 Conference*, vol. 2, pp 1037–1040, 2004.

[99] G. H. Golub and C. F. Van Loan. *Matrix Computation*. The Johns Hopkins University Press, 3rd edn, 1996.

[100] F. J. Gonzales-Vasquez. The differentiation of functions of conjugate complex variables: Application to power network analysis. *IEEE Transactions on Education*, **31**(4):286–291, 1988.

[101] N. R. Goodman. Statistical Analysis Based on Certain Multivariate Complex Gaussian Distribution. *Annals of Mathematics and Statistics*, **34**:152–176, 1963.

[102] M. S. Grewal and A. P. Andrews. *Kalman Filtering: Theory and Practice*. John Wiley & Sons, 2001.

[103] W. R. Hamilton. *Lectures on Quaternions: Containing a Systematic Statement of a New Mathematical Method*. Hodges and Smith, 1853.

[104] S. M. Hammel, C. K. R. T. Jones, and J. V. Moloney. Global Dynamical Behavior of the Optical Field in a Ring Cavity. *Journal of the Optical Society of America, Optical Physics*, **2**(4):552–564, 1985.

[105] A. Hanna and D. P. Mandic. Complex–valued nonlinear neural adaptive filters with trainable amplitude of activation functions. *Neural Networks*, **16**(2):155–159, 2003.

[106] A. Hanna and D. P. Mandic. A general fully adaptive normalised gradient descent learning algorithm for complex-valued nonlinear adaptive filters. *IEEE Transactions on Signal Processing*, **51**(10):2540–2549, 2003.
[107] A. I. Hanna, I. R. Krcmar, and D. P. Mandic. Perlustration of error surfaces for linear and nonlinear stochastic gradient descent algorithms. In *Proceedings of NEUREL 2002*, pp 11–16, 2002.
[108] A. I. Hanna and D. P. Mandic. Nonlinear FIR adaptive filters with a gradient adaptive amplitude in the nonlinearity. *IEEE Signal Processing Letters*, **9**(8):253–255, 2002.
[109] A. I. Hanna and D. P. Mandic. A Data-Reusing Nonlinear Gradient Descent Algorithm for a Class of Complex-Valued Neural Adaptive Filters. *Neural Processing Letters*, **17**:85–91, 2003.
[110] M. Hayes. *Statistical Digital Signal Processing and Modelling*. John Wiley & Sons, 1996.
[111] S. Haykin. *Neural Networks: A Comprehensive Foundation*. Prentice Hall, 2nd edn, 1999.
[112] S. Haykin, editor. *Kalman Filtering and Neural Networks*. John Wiley & Sons, 2001.
[113] S. Haykin. *Adaptive Filter Theory*. Prentice-Hall, 4th edn, 2002.
[114] S. Haykin and L. Li. Nonlinear adaptive prediction of nonstationary signals. *IEEE Transactions on Signal Processing*, **43**(2):526–535, 1995.
[115] O. Heaviside. *Electromagnetic Theory*. Van Nostrand Co., 2nd edn, 1922.
[116] R. Hecht-Nielsen. Neurocomputing: Picking the human brain. *IEEE Spectrum*, **7**(6):36–41, 1988.
[117] Y. Hirata, D. P. Mandic, H. Suzuki, and K. Aihara. Wind direction modelling using multiple observation points. *Philosophical Transactions of the Royal Society A*, **366**:591–607, 2008.
[118] A. Hirose. Continous complex-valued backpropagation learning. *IEE Electronics Letters*, **28**(20):1854–1855, 1990.
[119] A. Hirose, editor. *Complex-Valued Neural Networks: Theories and Applications*. World Scientific, 2004.
[120] A. Hirose. *Complex-Valued Neural Networks*. Springer, 2006.
[121] A. Hirose and M. Minami. Complex-valued region-based-coupling image clustering neural networks for interferometric radar image processing. *IEICE Transactions on Electronics*, **E84–C**(12):1932–1938, 2001.
[122] A. Hjorungnes and D. Gesbert. Complex–valued matrix differentiation: Techniques and key results. *IEEE Transactions on Signal Processing*, **55**(6):2740–2746, 2007.
[123] M. Hoehfeld and S. E. Fahlman. Learning with limited numerical precision using the cascade-correlation algorithm. *IEEE Transactions on Neural Networks*, **3**(4):602–611, 1992.
[124] R. A. Horn and C. A. Johnson. *Matrix Analysis*. Cambridge University Press, 1985.
[125] K. Hornik. Approximation Capabilities of Multilayer Feedforward Networks. *Neural Networks*, **4**:251–257, 1990.
[126] K. Hornik, M. Stinchcombe, and H. White. Multilayer feedforward networks are universal approximators. *Neural Networks*, **2**:359–366, 1989.
[127] L. L. Horowitz and K. D. Senne. Performance advantage of complex LMS for controlling narrow–band adaptive arrays. *IEEE Transactions on Circuits and Systems*, **CAS–28**(6):562–576, 1981.
[128] N. E. Huang, Z. Shen, S. R. Long, M. C. Wu, H. H. Shih, Q. Zheng, N.-C. Yen, C. C. Tung, and H. H. Liu. The Empirical Mode Decomposition and the Hilbert spectrum for nonlinear and non–stationary time series analysis. *Proceedings of the Royal Society London A*, **454**:903–995, 1998.
[129] R. C. Huang and M. S. Chen. Adaptive Equalization Using Complex-Valued Multilayered Neural Network Based on the Extended Kalman Filter. *Proceeding of the International Conference on Signal Processing, WCCC-ICSP '00*, **1**:519–524, 2000.
[130] Eugene M. Izhikevich. Simple model of spiking neurons. *IEEE Transactions on Neural Networks*, **14**(6):1569–1572, 2003.
[131] M. Jafari, J. Chambers, and D. P. Mandic. Natural gradient algorithm for cyclostationary sources. *Electronics Letters*, **38**(14):758–759, 2002.
[132] S. Javidi, M. Pedzisz, S. P. Goh, and D. P. Mandic. The augmented complex least mean square algorithm with application to adaptive prediction problems. In *Proceedingss of the I Cognitive Information Processing Systems Conference*, CD, 2008.
[133] B. Jelfs, P. Vayanos, M. Chen, S. L. Goh, C. Boukis, T. Gautama, T. Rutkowski, T. Kuh, and D. P. Mandic. An online method for detecting nonlinearity within a signal. In *Proceedings of KES 2006*, pp 1216–1223, 2006.
[134] B. Jelfs, P. Vayanos, S. Javidi, V. S. L. Goh, and D. P. Mandic. Collaborative adaptive filters for online knowledge extraction and information fusion. In D. P. Mandic, M. Golz, A. Kuh, D. Obradovic, and T. Tanaka, editors, *Signal Processing Techniques for Knowledge Extraction and Information Fusion*, pp 3–22. Springer, 2008.
[135] W. K. Jenkins, A. W. Hull, J. C. Strait, B. A. Schnaufer, and X. Li. *Advanced Concepts in Adaptive Signal Processing*. Kluwer Academic Publishers, 1996.

[136] C. R. Johnson, Jr. Adaptive IIR filtering: Current issues and open issues. *IEEE Transactions on Information Theory*, **IT–30**(2):237–250, 1984.
[137] D. H. Johnson. The application of spectral estimation methods to bearing estimation problems. *Proceedings of the IEEE*, **70**(9):1018–1028, 1982.
[138] D. H. Johnson and D. E. Dudgeon. *Array signal processing: Concepts and techniques*. Prentice–Hall, 1993.
[139] S. J. Julier and J. K. Uhlmann. Unscented filtering and nonlinear estimation. *Proceedings of the IEEE*, **92**(3):401–422, 2004.
[140] T. Kailath. *Linear Systems*. Prentice–Hall, 1980.
[141] R. E. Kalman. A new approach to linear filtering and prediction problems. *Transaction of the ASME–Journal of Basic Engineering*, pp 35–45, 1960.
[142] I. Kant. *Critique of Pure Reason* (Translated into English by N. Kemp Smith). St Martin's Press and Macmillan, 1929.
[143] I. L. Kantor and A. S. Solodovnikov. *Hypercomplex numbers: An elementary introduction to algebras*. Springer, English edn, 1989.
[144] H. Kantz and T. Schreiber. *Nonlinear Time Series Analysis*. Cambridge University Press, 2nd edn, 2004.
[145] D. T. Kaplan. Exceptional events as evidence for determinism. *Physica D*, **73**:38–48, 1994.
[146] W. H. Kautz. Transient Synthesis in the Time Domain. *IRE Transactions on Circuit Theory*, **1**(3):29–39, 1954.
[147] G. Kechriotis and E. S. Manolakos. Training fully recurrent neural networks with complex weights. *IEEE Transactions on Circuits and Systems–II: Analogue and Digital Signal Processing*, pp 235–238, 1994.
[148] D. G. Kelly. Stability in contractive nonlinear neural networks. *IEEE Transactions on Biomedical Engineering*, **37**(3):231–242, 1990.
[149] D. I. Kim and P. De Wilde. Performance analysis of signed self-orthogonalizing adaptive lattice filter. *IEEE Transactions on Circuits and Systems II: Analog and Digital Filters*, **47**(11):1227–1237, 2000.
[150] T. Kim and T. Adali. Complex Backpropagation Neural Network Using Elementary Transcendental Activation Functions. *Proceedings of the International Conference on Acoustics, Speech, and Signal Processing, ICASSP' 01*, **2**:1281–1284, 2001.
[151] T. Kim and T. Adali. Fully-Complex Multilayer Perceptron for Nonlinear Signal Processing. *Journal of VLSI Signal Processing Systems for Signal, Image, and Video Technology*, **32**:29–43, 2002.
[152] T. Kim and T. Adali. Approximation by fully complex multilayer perceptrons. *Neural Computation*, **15**(7):1641–1666, 2003.
[153] D. E. Knuth. *Digital Typography*. CSLI Lecture Notes, No. 78, 1999.
[154] A. N. Kolmogorov. Interpolation and extrapolation von stationären zufäfolgen. *Bull. Acad. Sci. (Nauk)*, **5**:3–14, 1941.
[155] A. N. Kolmogorov. On the representation of continuous functions of several variables by superposition of continuous functions of one variable and addition. *Dokladi Akademii Nauk SSSR*, **114**:953–956, 1957.
[156] T. Kono and I. Nemoto. Complex-Valued Neural Networks. *IEICE Tech. Report*, **NC90-69**:7–12, 1990.
[157] K. Kreith and D. Chakerian. *Iterative Algebra and Dynamic Modeling*. Springer-Verlag, 1999.
[158] K. Kreutz-Delgado. The complex gradient operator and the CR calculus, Lecture Supplement ECE275A, pp 1–74, 2006.
[159] C.-M. Kuan and K. Hornik. Convergence of learning algorithms with constant learning rates. *IEEE Transactions on Neural Networks*, **2**(5):484–489, 1991.
[160] D. Kugiumtzis. Test your surrogate data before you test for nonlinearity. *Physical Review E*, **60**:2808–2816, 1999.
[161] V. Kurkova. Kolmogorov's theorem and multilayer neural networks. *Neural Networks*, **5**(3):501–506, 1992.
[162] T. I. Laakso, V. Valimaki, and M. Karjalainen. Splitting the unit delay for FIR and all pass filter design. *IEEE Signal Processing Magazine*, **13**(1):30–60, 1996.
[163] J. P. LaSalle. *The Stability and Control of Discrete Processes*. Springer-Verlag, 1986.
[164] K. N. Leibovic. Contraction mapping with application to control processes. *Journal of Electronics and Control*, pp 81–95, July 1963.
[165] M. Leshno, V. Y. Lin, A. Pinkus, and S. Schocken. Multliayer feedforward networks with a nonpolynomial activation function can approximate any function. *Neural Networks*, **6**:861–867, 1993.
[166] H. Leung and S. Haykin. The complex backpropagation algorithm. *IEEE Transactions on Signal Processing*, **3**(9):2101–2104, 1991.
[167] H. Li and T. Adali. Complex-valued adaptive signal processing using nonlinear functions. *EURASIP Journal on Advances in Signal Processing*, **8**(2):1–9, 2008.

[168] R. P. Lippmann. An introduction of computing with neural nets. *IEEE Acoustics, Speech and Signal Processing Magazine*, **4**(2):4–22, 1987.

[169] S. Lipschutz. *Theory and Problems of Probability*. Schaum's Outline Series. McGraw–Hill, 1965.

[170] J. T. Lo and L. Yu. Recursive neural filters as dynamical range transformers. *Proceedings of the IEEE*, **92**(3):514–535, 2004.

[171] D. Looney and D. P. Mandic. A machine learning enhanced empirical mode decomposition. In *Proceedings of the International Conference on Acoustics, Speech and Signal Processing (ICASSP)*, 2008.

[172] G. G. Lorentz. The 13th problem of Hilbert. In F. E. Browder, editor, *Mathematical Developments Arising from Hilbert Problems*. American Mathematical Society, 1976.

[173] E. N. Lorenz. Deterministic Nonperiodic Flow. *Journal of the Atmospheric Sciences*, **20**(2):130–141, March 1963.

[174] D. G. Luenberger. *Optimization by Vector Space Methods*. Wiley, 1969.

[175] J. Makhoul. Linear prediction: A tutorial overview. *Proceedings of the IEEE*, **63**(4):561–580, 1975.

[176] B. Mandelbrot. *The Fractal Geometry of Nature*. W. H. Freeman and Co., 1982.

[177] D. P. Mandic. NNGD algorithm for neural adaptive filters. *Electronics Letters*, **36**(6):845–846, 2000.

[178] D. P. Mandic. The use of Möbius transformations in neural networks and signal processing. In *Proceedings of the International Workshop on Neural Networks for Signal Processing (NNSP'00)*, pp 185–194, 2000.

[179] D. P. Mandic. Data-reusing recurrent neural adaptive filters. *Neural Computation*, **14**(11):2693–2707, 2002.

[180] D. P. Mandic. A general normalised gradient descent algorithm. *IEEE Signal Processing Letters*, **11**(2):115–118, 2004.

[181] D. P. Mandic, J. Baltersee, and J. A. Chambers. Nonlinear prediction of speech with a pipelined recurrent neural network and advanced learning algorithms. In A. Prochazka, J. Uhlir, P. J. W. Rayner, and N. G. Kingsbury, (eds), *Signal Analysis and Prediction*, pp 291–309. Birkhauser, Boston, 1998.

[182] D. P. Mandic and J. A. Chambers. A posteriori real time recurrent learning schemes for a recurrent neural network based non-linear predictor. *IEE Proceedings–Vision, Image and Signal Processing*, **145**(6):365–370, 1998.

[183] D. P. Mandic and J. A. Chambers. On stability of relaxive systems described by polynomials with time–variant coefficients. Accepted for IEEE Transactions on Circuits and Systems–I: Fundamental Theory and Applications, 1999.

[184] D. P. Mandic and J. A. Chambers. A posteriori error learning in nonlinear adaptive filters. *IEE Proceedings–Vision, Image and Signal Processing*, **146**(6):293–296, 1999.

[185] D. P. Mandic and J. A. Chambers. Relations between the a priori and a posteriori errors in nonlinear adaptive neural filters. Accepted for Neural Computation, **12**(6):1285–1292, 2000.

[186] D. P. Mandic and J. A. Chambers. Toward an optimal PRNN based nolinear predictor. *IEEE Transactions on Neural Networks*, 10(6):1435–1442, 1999.

[187] D. P. Mandic and J. A. Chambers. A normalised real time recurrent learning algorithm. *Signal Processing*, **80**(11):1909–1916, 2000.

[188] D. P. Mandic and J. A. Chambers. On robust stability of time–variant discrete–time nonlinear systems with bounded parameter perturbations. *IEEE Transactions on Circuits and Systems–I: Fundamental Theory and Applications*, **47**(2):185–188, 2000.

[189] D. P. Mandic and J. A. Chambers. Relationship Between the *A Priori* and *A Posteriori* Errors in Nonlinear Adaptive Neural Filters. *Neural Computation*, **12**:1285–1292, 2000.

[190] D. P. Mandic and J. A. Chambers. *Recurrent Neural Networks for Prediction: Learning Algorithms, Architectures and Stability*. John Wiley & Sons, 2001.

[191] D. P. Mandic, M. Chen, T. Gautama, M. M. Van Hulle, and A. Constantinides. On the characterisation of the deterministic/stochastic and linear/nonlinear nature of time series. *Proceedings of the Royal Society A*, **464**(2093):1141–1160, 2008.

[192] D. P. Mandic, S. L. Goh, and K. Aihara. Sequential data fusion via vector spaces: Fusion of heterogeneous data in the complex domain. *International Journal of VLSI Signal Processing Systems*, **48**(1–2):98–108, 2007.

[193] D. P. Mandic, M. Golz, A. Kuh, D. Obradovic, and T. Tanaka. *Signal processing Techniques for Knowledge Extraction and Information Fusion*. Springer, 2008.

[194] D. P. Mandic, S. Javidi, S. L. Goh, A. Kuh, and K. Aihara. Complex–valued prediction of wind profile using augmented complex statistics. *Renewable Energy*, **34**(1):196–210, 2009.

[195] D. P. Mandic, G. Souretis, S. Javidi, S. L. Goh, and T. Tanaka. Why a complex valued solution for a real domain problem. In *Proceedings of the International Conference on Machine Learning for Signal Processing (MLSP'07)*, pp 384–389, 2007.

[196] D. P. Mandic, G. Souretis, W. Y. Leong, D. Looney, and T. Tanaka. Complex empirical mode decomposition for multichannel information fusion. In D. P. Mandic, M. Golz, A. Kuh, D. Obradovic, and T. Tanaka, editors, *Signal Processing Techniques for Knowledge Extraction and Information Fusion*, pp 131–153. Springer, 2008.

[197] D. P. Mandic, P. Vayanos, C. Boukis, B. Jelfs, S. L. Goh, T. Gautama, and T. Rutkowski. Collaborative adaptive learning using hybrid filters. In *Proceedings of the IEEE International Conference on Acoustics, Speech, and Signal Processing, ICASSP'07*, vol. III, pp 921–924, 2007.

[198] D. P. Mandic, P. Vayanos, M. Chen, and S. L. Goh. A collaborative approach for the modelling of wind field. *International Journal of Neural Systems*, **18**(2):67–74, 2008.

[199] D. P. Mandic and I. Yamada. Machine learning and signal processing applications of fixed point theory. *Tutorial in IEEE ICASSP*, 2007.

[200] A. Manikas. *Differential Geometry in Array Processing*. Imperial College Press, 2004.

[201] G. Mantica. On computing Jacobi matrices associated with recurrent and Möbius iterated function systems. *Journal of Computational and Applied Mathematics*, **115**:419–431, 2000.

[202] J. F. Markham and T. D. Kieu. Simulations with complex measures. *Nuclear Physics B*, **516**:729–743, 1998.

[203] J. E. Marsden and M. J. Hoffman. *Basic Complex Analysis*. W.H. Freeman, New York, 3rd edn, 1999.

[204] J. R. R. A. Martins. The complex–step derivative approximation. *ACM Transactions on Mathematical Software*, **29**(3):245–262, 2003.

[205] M. A. Masnadi-Shirazi and N. Ahmed. Optimum Laguerre Networks for a Class of Discrete-Time Systems. *IEEE Transactions on Signal Processing*, **39**(9):2104–2107, 1991.

[206] J. H. Mathews and R. W. Howell. *Complex Analysis for Mathematics and Engineering*. Jones and Bartlett Pub. Inc., 3rd edn, 1997.

[207] V. J. Mathews and Z. Xie. A stochastic gradient adaptive filter with gradient adaptive step size. *IEEE Transactions on Signal Processing*, **41**(6):2075–2087, 1993.

[208] T. Mestl, R. J. Bagley, and L. Glass. Common chaos in arbitrary complex feedback networks. *Physical Review Letters*, **79**(4):653–656, 1997.

[209] H. N. Mhaskar and C. Micchelli. Approximation by superposition of sigmoidal and radial basis functions. *Advances in Applied Mathematics*, **13**:350–373, 1992.

[210] M. Milisavljevic. Multiple environment optimal update profiling for steepest descent algorithms. In *Proceedings of the International Conference on Acoustics, Speech and Signal Processing, ICASSP2001*, vol. VI, pp 3853 –3856, 2001.

[211] K. S. Miller. *Complex Stochastic Processes: An Introduction to Theory and Application*. Addison-Wesley, 1974.

[212] H. Mizuta, M. Jibu, and K. Yana. Adaptive estimation of the degree of system nonlinearity. In *Proceedings IEEE Adaptive Systems for Signal Processing and Control Symposium (AS-SPCC)*, pp 352–356, 2000.

[213] M. Morita, S. Yoshizawa, and K. Nakano. Analysis and Improvement of the Dynamics of Autocorrelation Associative Memory. *IEICE Trans. D-II*, **J73-D-II**:232–242, 1990.

[214] M. C. Mozer. Neural net architectures for temporal sequence processing. In A. S. Weigend and N. A. Gershenfeld, editors, *Time Series Prediction: Forecasting the Future and Understanding the Past*. Addison-Wesley, 1993.

[215] P. J. Nahin. *An Imaginary Tale: The Story of the Square Root of Minus One*. Princeton University Press, 1998.

[216] K. S. Narendra and K. Parthasarathy. Identification and control of dynamical systems using neural networks. *IEEE Transactions on Neural Networks*, **1**(1):4–27, 1990.

[217] J. Navarro-Moreno. ARMA prediction of widely linear systems by using the innovations algorithm. *IEEE Transactions on Signal Processing*, **56**(7):3061–3068, 2008.

[218] T. Needham. *Visual Complex Analysis*. Oxford University Press, 1997.

[219] F. D. Neeser and J. L. Massey. Proper Complex Random Processes with Application to Information Theory. *IEEE Transactions on Information Theory*, **39**(4):1293–1302, 1993.

[220] O. Nerrand, P. Roussel-Ragot, L. Personnaz, and G. Dreyfus. Neural networks and nonlinear adaptive filtering: Unifying concepts and new algorithms. *Neural Computation*, **5**:165–199, 1993.

[221] T. Nitta. Orthogonality of decision boundaries in complex–valued neural networks. *Neural Computation*, **16**:73–97, 2004.

[222] D. Obradovic. On-line training of recurrent neural networks with continous topology adaptation. *IEEE Transactions on Neural Networks*, **7**(1):222–228, 1996.

[223] D. Obradovic, H. Lenz, and M. Schupfner. Fusion of sensor data in Siemens car navigation system. *IEEE Transactions on Vehicular Technology*, **56**(1):43–50, 2007.

[224] S. C. Olhede. On probability density functions for complex variables. *IEEE Transactions on Information Theory*, **52**(3):1212–1217, 2006.

[225] A. V. Oppenheim, J. R. Buck, and R. W. Schafer. *Discrete–time signal processing*. Prentice Hall, 1999.

[226] A. V. Oppenheim, G. E. Kopec, and J. M. Tribolet. Signal Analysis by Homomorphic Prediction. *IEEE Transactions on Acoustic, Speech, and Signal Processing*, **ASSP-24**(4):327–332, August 1976.

[227] A. V. Oppenheim and J. S. Lim. The importance of phase in signals. *Proceedings of the IEEE*, **69**(5):529–541, 1981.

[228] T. Paatero, M. Karjalainen, and A. Harma. Modeling and Equalization of Audio Systems Using Kautz Filters. In *Proceedings of the IEEE International Conference on Acoustics, Speech, and Signal Processing*, pp 3313–3316, 2001.

[229] M. Pavon. A new formulation of stochastic mechanics. *Physics Letters A*, **209**:143–149, 1995.

[230] J. K. Pearson. *Clifford Networks*. PhD Thesis, University of Kent, 1994.

[231] M. S. Pedersen, J. Larsen, U. Kjems, and L. C. Parra. A survey of convolutive blind source separation methods. In J. Benesty, Huang, and M. Sondhi (eds), *Springer Handbook of Speech*, pp tba. Springer Press, 2007.

[232] M. Pedzisz and D. P. Mandic. A homomorphic neural network for filtering and prediction. *Neural Computation*, **20**(4):1042–1064, 2008.

[233] R. Penrose. *The Road to Reality: A Complete Guide to the Laws of the Universe*. Vintage Books London, 2004.

[234] H. Perez and S. Tsujii. A System Identification Algorithm Using Orthogonal Functions. *IEEE Transactions on Signal Processing*, **39**(3):752–755, 1991.

[235] L. Personnaz and G. Dreyfus. Comment on "Discrete–time recurrent neural network architectures: A unifying review". *Neurocomputing*, **20**:325–331, 1998.

[236] B. Picinbono. *Random Signals and Systems*. Englewood Cliffs, Prentice Hall, NJ, USA, 1993.

[237] B. Picinbono. On Circularity. *IEEE Transactions on Signal Processing*, **42**(12):3473–3482, 1994.

[238] B. Picinbono. Second-Order Complex Random Vectors and Normal Distributions. *IEEE Transactions on Signal Processing*, **44**(10):2637–2640, 1996.

[239] B. Picinbono and P. Bondon. Second-order statistics of complex signals. *IEEE Transactions on Signal Processing*, **45**(2):411–420, 1997.

[240] B. Picinbono and P. Chevalier. Widely Linear Estimation with Complex Data. *IEEE Transactions on Signal Processing*, **43**(8):2030–2033, 1995.

[241] T. Poggio and F. Girosi. Networks for approximation and learning. *Proceedings of the IEEE*, **78**(9):1481–1497, 1990.

[242] I. A. Priestley. *Introduction to Complex Analysis*. Oxford University Press, 2nd edn, 2003.

[243] M. B. Priestley. *Spectral Analysis and Time Series*. Academic Press, 1981.

[244] J. C. Principe, N. R. Euliano, and W. C. Lefebvre. *Neural and Adaptive Systems*. Wiley, 2000.

[245] G. V. Puskorius and L. A. Feldkamp. Practical consideration for Kalman filter training of recurrent neural networks. In *Proceedings of the IEEE International Conference on Neural Networks*, pp 1189–1195, 1993.

[246] G. V. Puskorius, L. A. Feldkamp, F. M. Co., and M. I. Dearborn. Decoupled extended Kalman filter training of feedforward layered networks. *IJCNN91 Settle Intertional Joint Conference on Neural Networks*, **1**:771–777, 1991.

[247] R. Reed. Pruning algorithms – a survey. *IEEE Transactions on Neural Networks*, **4**(5):740–747, 1993.

[248] P. A. Regalia. *Adaptive IIR Filtering in Signal Processing and Control*. Marcel Dekker, 1994.

[249] P. A. Regalia, S. K. Mitra, and P. P. Vaidyanathan. The digital all–pass filter: A versatile signal processing building block. *Proceedings of the IEEE*, **76**(1):19–37, 1988.

[250] R. Remmert. *Theory of Complex Functions*. Springer-Verlag, 1991.

[251] G. Rilling, P. Flandrin, P. Goncalves, and J.M. Lilly. Bivariate empirical mode decomposition. *IEEE Signal Processing Letters*, **14**:936–939, 2007.

[252] T. Roman, S. Visuri, and V. Koivunen. Blind frequency synchronisation in OFDM via diagonality criterion. *IEEE Transactions on Signal Processing*, **54**(8):3125–3135, 2006.

[253] D. B. Rowe and B. R. Logan. A complex way to compute fMRI activation. *NeuroImage*, **23**(3):1078–1092, 2004.

[254] S. Roy and J. J. Shynk. Analysis of the data–reusing LMS algorithm. In *Proceedings of the 32nd Midwest Symposium on Circuits & Systems*, vol. 2, pp 1127–1130, 1989.

[255] W. Rudin. *Real and Complex Analysis*. McGraw Hill, New York, USA, 1974.

[256] M. Rupp. Contraction mapping: An important property in adaptive filters. In *Proceedings of the Sixth IEEE DSP Workshop*, pp 273–276, 1994.

[257] H. Saleur, C. G. Sammis, and D. Sornette. Discrete scale invariance, complex fractal dimensions, and log-periodic fluctuations in seismicity. *Journal of Geophysical Research*, 101(B8):17661–17678, 1994.

[258] S. Sastry and M. Bodson. *Adaptive Control: Stability, Convergence, and Robustness*. Prentice–Hall International, 1989.

[259] M. Scarpiniti, D. Vigliano, R. Parisi, and A. Uncini. Generalized Splitting Functions for Blind Separation of Complex Signals. *Neurocomputing*, page in print, 2008.

[260] R. W. Schafer and L. R. Rabiner. System for Automatic Analysis of Voiced Speech. *The Journal of the Acoustical Society of America*, **47**(2):634–648, 1970.

[261] B. A. Schnaufer and W. K. Jenkins. New data-reusing LMS algorithms for improved convergence. In *Conference Record of the Twenty–Seventh Asilomar Conference on Signals and Systems*, vol. 2, pp 1584–1588, 1993.

[262] R. Schober, W. H. Gerstacker, and L. H. J. Lampe. Data–aided and blind stochastic gradient algorithms for widely linear MMSE MAI supression for DS–CDMA. *IEEE Transactions on Signal Processing*, **52**(3):746–756, 2004.

[263] T. Schreiber. Interdisciplinary application of nonlinear time series methods. *Physics Reports*, **308**:23–35, 1999.

[264] T. Schreiber and A. Schmitz. Improved surrogate data for nonlinearity tests. *Physical Review Letters*, **77**:635–638, 1996.

[265] T. Schreiber and A. Schmitz. Surrogate time series. *Physica D*, **142**:346–382, 2000.

[266] P. J. Schreier and L. L. Scharf. Low-Rank Approximation of Improper Complex Random Vectors. In *Conference Record of the Thirty-Fifth Asilomar Conference on Signals, Systems and Computers*, vol. 1, pp 597–601, 2001.

[267] P. J. Schreier and L. L. Scharf. Canonical coordinates for reduced-rank estimation of improper complex random vectors. In *Proceedings of the International Conference on Acoustics, Speech, and Signal Processing, ICASSP '02*, vol. 2, pp 1153–1156, 2002.

[268] P. J. Schreier and L. L. Scharf. Second-order analysis of improper complex random vectors and processes. *IEEE Transactions on Signal Processing*, **51**:714–725, 2003.

[269] P. J. Schreier and L. L. Scharf. *Statistical Signal Processing of Complex-Valued Data*. Cambridge University Press, 2009.

[270] P. J. Schreier, L. L. Scharf, and A. Hanssen. A generalized likelihood ratio test for impropriety of complex signals. *IEEE Signal Processing Letters*, **13**(7):433–436, 2006.

[271] P. J. Schreier, L. L. Scharf, and C. T. Mullis. Detection and estimation of improper complex random signals. *IEEE Transactions on Information Theory*, **51**(1):306–312, 2005.

[272] B. A. Sexton and J. C. Jones. Means of complex numbers. *Texas College Mathematics Journal*, **1**(1):1–4, 2005.

[273] K. Shenoi. *Digital Signal Processing in Telecommunications*. Prentice Hall, 1995.

[274] J. Shynk. Adaptive IIR filtering. *IEEE Acoustics, Speech and Signal Processing (ASSP) Magazine*, **6**(2):4–21, 1989.

[275] J. J. Shynk. A complex adaptive algorithm for IIR filtering. *IEEE Transactions on Acoustics, Speech, and Signal Processing*, **ASSP–34**(5):1342–1344, 1986.

[276] J. Z. Simon and Y. Wang. Fully complex magnetoencephalography. *Journal of Neuroscience Methods*, **149**(1):64–73, 2005.

[277] E. J. Singley, R. Kawakami, D. D. Awschalom, and D. N. Basov. Infrared probe of itinerant ferromagnetism in $Ga_{1-x}Mn_xAs$. *Physical Review Letters*, **89**(9):097203–1 – 097203–4, 2002.

[278] M. Solazzia, A. Uncini, E. Di Claudio, and R. Parisi. Complex discriminative learning Bayesian neural equalizer. *Signal Processing*, **81**:2493?2502, 2001.

[279] E. Soria-Olivas, J. Calpe-Maravilla, J. F. Guerrero-Martinez, M. Martinez-Sober, and J. Espi-Lopez. An easy demonstration of the optimum value of the adaptation constant in the LMS algorithm. *IEEE Transactions on Education*, **41**(1):81, 1998.

[280] W. Squire and G. Trapp. Using complex variables to estimate derivatives of real functions. *SIAM Review*, **40**(1):110–112, 1998.

[281] H. Stark and J. W. Woods. *Probability, random processes and estimation theory for engineers*. Englewood Cliffs, N. J. London: Prentice-Hall, 1986.

[282] A. B. Suksmono and A. Hirose. Adaptive complex–amplitude texture classifier that deals with both height and reflectance for interferometric SAR images. *IEICE Transactions on Electronics*, **E83–C**(12):1912–1916, 2000.

[283] J. Sum, C.-S. Leung, G. H. Young, and W.-K. Kan. On the Kalman filtering method in neural–network training and pruning. *IEEE Transactions on Neural Networks*, **10**(1):161–166, 1999.

[284] T. Tanaka and D. P. Mandic. Complex empirical mode decomposition. *IEEE Signal Processing Letters*, **14**(2):101–104, Feb 2007.

[285] A. Tarighat and A. H. Sayed. Least mean-phase adaptive filters with application to communications systems. *IEEE Signal Processing Letters*, **11**(2):220–223, 2004.

[286] P. Tass, M. G. Rosenblum, J. Weule, J. Kurths, A. Pikovsky, J. Volkmann, A. Schnitzler, and H. J. Freund. Detection of *n:m* phase locking from noisy data: Application to magnetoencephalography. *Physical Review Letters*, **81**:3291–3294, 1998.

[287] J. Theiler, S. Eubank, A. Longtin, B. Galdrikian, and J.D. Farmer. Testing for nonlinearity in time series: The method of surrogate data. *Physica D*, **58**:77–94, 1992.

[288] J. Theiler and D. Prichard. Constrained-realization Monte-Carlo method for hypothesis testing. *Physica D*, **94**:221–235, 1996.

[289] A. N. Tikhonov, A. S. Leonov, and A. G. Yagola. *Nonlinear ill–posed problems*. Applied mathematics and mathematical computation. Chapman & Hall, London, 1998.

[290] J. Timmer. What can be inferred from surrogate data testing? *Physical Review Letters*, **85**(12):2647, 2000.

[291] H. L. Van Trees. *Digital Communications: Fundamentals and Applications*. John Wiley & Sons, 2001.

[292] H. L. Van Trees. *Optimum Array Processingy*, volume IV: Detection, Estimation, and Modulation Theory. Springer-Verlag, 2002.

[293] J. R. Treichler, C. R. Johnson, Jr., and M. G. Larimore. *Theory and Design of Adaptive Filters*. John Wiley & Sons, 1987.

[294] E. Trentin. Network with trainable amplitude of activation functions. *Neural Networks*, **14**(5):471–493, 2001.

[295] A. Uncini, L. Vecci, P. Campolucci, and F. Piazza. Complex-valued neural networks with adaptive spline activation function for digital radio links nonlinear equalization. *IEEE Transactions on Signal Processing*, **47**(2):505–514, 1999.

[296] G. Vaucher. A complex valued spiking machine. In *Proceedings of ICANN 2003*, pp 967–976, 2003.

[297] P. Vayanos, S. L. Goh, and D. P. Mandic. Online detection of the nature of complex–valued signals. In *Proceedings of the IEEE Workshop on Machine Learning for Signal Processing*, pp 173–178, 2006.

[298] F. Vitagliano, R. Parisi, and A. Uncini. Generalized splitting 2D flexible activation function. In *Proceedings of the 14th Italian Workshop on Neural Nets, WIRN VIETRI 2003*, pp 85–95, 2003.

[299] B. Wahlberg. System Identification Using Laguerre Models. *IEEE Transactions on Automatic Control*, **36**(5):551–562, 1991.

[300] W. A. Wallace. *Galileo's Logical Treatises (A Translation, with Notes and Commentary, of His Appropriated Latin Questions on Aristotle's Posteriori Analytics)*. Kluwer Academic Publishers, 1992.

[301] E. A. Wan. Time series prediction by using a connectionist network with internal delay lines. In A. S. Weigend and N. A. Gershenfeld, editors, *Time Series Prediction: Forecasting the Future and Understanding the Past*. Addison-Wesley, 1993.

[302] T. J. Wang. Complex–valued ghost cancellation reference signal for TV broadcasting. *IEEE Transactions on Consumer Electronics*, **37**(4):731–736, 1991.

[303] A. S. Weigend and N. A. Gershenfeld (eds). *Time Series Prediction: Forecasting the Future and Understanding the Past*. Santa Fe Institute Studies in the Sciences of Complexity. Addison-Wesley Publishing Company, 1994.

[304] R. C. Weimer. Can the complex numbers be ordered? *The Two–Year College Mathematics Journal*, **7**(4):10–12, 1976.

[305] E. T. Whittaker and G. N. Watson. *A Course in Modern Analysis*. Cambridge University Press, 4th edn, 1927.

[306] B. Widrow and M. A. Lehr. 30 years of adaptive neural networks: Perceptron, madaline, and backpropagation. *Proceedings of the IEEE*, **78**(9):1415–1442, 1990.

[307] B. Widrow, J. McCool, and M. Ball. The complex LMS algorithm. *Proceedings of the IEEE*, **63**(3):712–720, 1975.

[308] B. Widrow and S. D. Stearns. *Adaptive Signal Processing*. Prentice–Hall, 1985.

[309] N. Wiener. *The Extrapolation, Interpolation and Smoothing of Stationary Time Series with Engineering Applications*. Wiley, 1949.

[310] J. H. Wilkinson. *The Algebraic Eigenvalue Problem*. Oxford University Press, 1965.

[311] R. Williams and D. Zipser. A learning algorithm for continually running fully recurrent neural networks. *Neural Computation*, **1**:270–280, 1989.

[312] R. C. Williamson and U. Helmke. Existence and uniqueness results for neural network approximations. *IEEE Transactions on Neural Networks*, **6**:2–13, 1995.

[313] W. Wirtinger. Zur formalen theorie der funktionen von mehr komplexen veränderlichen. *Mathematische Annalen*, **97**(1):357–375, 1927.
[314] H. O. A. Wold. *A Study in the Analysis of Stationary Time Series*. Almquist and Wiksell: Uppsala, 1938.
[315] C. H. Wolters. The finite element method in EEG/MEG source analysis. *SIAM News*, **40**(2):1–2, 2007.
[316] R. A. Wooding. The multivariate distribution of complex normal variables. *Biometrica*, **43**:212–215, 1956.
[317] Z. Wu and N. E. Huang. Ensemble empirical mode decomposition: A noise-assisted data analysis method. Technical Report 193, Center for Ocean-Land-Atmosphere Studies, 2004.
[318] I. Yamada, K. Sakaniwa, and S. Tsujii. A multidimensional isomorphic operator and its properties – a proposal of finite-extent multi-dimensional cepstrum. *IEEE Transactions on Signal Processing*, **42**(7):1766–1785, 1994.
[319] W.-H. Yang, K.-K. Chan, and P.-R. Chang. Complex–valued neural network for direction of arrival estimation. *Electronics Letters*, **30**(7):653–656, 1994.
[320] G. U. Yule. On a method of investigating periodicities in disturbed series,with special reference to Wölfer's sunspot numbers. *Phil. Trans. Royal Soc. (London)*, **A226**:267–298, 1927.
[321] E. Zeidler. *Nonlinear Functional Analysis and its Applications*, vol. 1. Fixed–point theorems. Springer-Verlag, 1986.
[322] D. P. Mandic, S. Still, and S. Douglas. Duality Between Widely Linear and Dual Channel Adaptive Filtering. *Proceedings of ICASSP*, 2009.

Index

Complex Valued Nonlinear Adaptive Filters: Noncircularity, Widely Linear and Neural Models
Danilo P. Mandic and Vanessa Su Lee Goh